Medical Modelling

Related titles

Medical modelling: The application of advanced design and development techniques in medicine
(ISBN 978-1-84569-138-7)

Dental biomaterials, imaging testing and modelling
(ISBN 978-1-84569-296-4)

Rapid prototyping of biomaterials
(ISBN 978-0-85709-599-2)

Woodhead Publishing Series in Biomaterials:
Number 91

Medical Modelling

The Application of Advanced Design and Rapid Prototyping Techniques in Medicine

Second Edition

Richard Bibb, Dominic Eggbeer and Abby Paterson

AMSTERDAM • BOSTON • CAMBRIDGE • HEIDELBERG
LONDON • NEW YORK • OXFORD • PARIS • SAN DIEGO
SAN FRANCISCO • SINGAPORE • SYDNEY • TOKYO
Woodhead Publishing is an imprint of Elsevier

Woodhead Publishing is an imprint of Elsevier
80 High Street, Sawston, Cambridge, CB22 3HJ, UK
225 Wyman Street, Waltham, MA 02451, USA
Langford Lane, Kidlington, OX5 1GB, UK

First edition 2006, Woodhead Publishing Limited
Second edition 2015, Elsevier Ltd
Copyright © 2015 Elsevier Ltd. All rights reserved.

Notice
No responsibility is assumed by the publisher for any injury and/or damage to persons
or property as a matter of products liability, negligence or otherwise, or from any use or
operation of any methods, products, instructions or ideas contained in the material herein.
Because of rapid advances in the medical sciences, in particular, independent verification
of diagnoses and drug dosages should be made.

British Library Cataloguing-in-Publication Data
A catalogue record for this book is available from the British Library

Library of Congress Control Number: 2014955926

ISBN 978-1-78242-300-3 (print)
ISBN 978-1-78242-313-3 (online)

For information on all Woodhead Publishing publications
visit our website at http://store.elsevier.com/

Typeset by TNQ Books and Journals
www.tnq.co.in

Working together
to grow libraries in
developing countries

www.elsevier.com • www.bookaid.org

Contents

Woodhead Publishing Series in Biomaterials   ix
Preface   xiii
Acknowledgements   xv

1   Introduction   **1**
    1.1   Background   1
    1.2   The human form   2
    1.3   Basic anatomical terminology   3
    1.4   Technical terminology   5

2   Medical imaging   **7**
    2.1   Introduction to medical imaging   7
    2.2   Computed tomography (CT)   8
    2.3   Cone beam CT (CBCT)   17
    2.4   Magnetic resonance (MR)   20
    2.5   Noncontact surface scanning   24
    2.6   Medical scan data   30
    2.7   Point cloud data   32
    2.8   Media   32
        References   33
        Recommended reading   33

3   Working with medical scan data   **35**
    3.1   Pixel data operations   35
    3.2   Using CT data: a worked example   39
    3.3   Point cloud data operations   44
    3.4   Two-dimensional formats   48
    3.5   Pseudo 3D formats   48
    3.6   True 3D formats   51
    3.7   File management and exchange   58

4   Physical reproduction   **65**
    4.1   Background to rapid prototyping   65
    4.2   Stereolithography   75
    4.3   Digital light processing   79

4.4   Fused deposition modelling   81
4.5   Laser sintering   84
4.6   Powder bed 3D printing   86
4.7   Material jetting technology   88
4.8   Laminated object manufacture   93
4.9   Computer numerical controlled machining   93
4.10   Cleaning and sterilising medical models   95

5   Case studies   99

Implementation   **101**
5.1   Implementation case study 1: computed tomography guidelines
       for medical modelling using rapid prototyping techniques   101
5.2   Implementation case study 2: the development of a collaborative
       medical modelling service – organisational and technical
       considerations   110
5.3   Implementation case study 3: medical rapid prototyping
       technologies – state of the art and current limitations for
       application in oral and maxillofacial surgery   120

Surgical applications   **137**
5.4   Surgical applications case study 1: planning osseointegrated
       implants using computer-aided design and rapid prototyping   137
5.5   Surgical applications case study 2: rapid manufacture of
       custom-fit surgical guides   145
5.6   Surgical applications case study 3: use of a reconstructed
       three-dimensional solid model from computed tomography to aid in
       the surgical management of a total knee arthroplasty   155
5.7   Surgical applications case study 4: custom-made titanium orbital
       floor prosthesis in reconstruction for orbital floor fractures   160
5.8   Surgical applications case study 5: use of three-dimensional
       technology in the multidisciplinary management of facial
       disproportion   167
5.9   Surgical applications case study 6: appropriate approach to
       computer-aided design and manufacture of reconstructive
       implants   173
5.10   Surgical applications case study 7: computer-aided planning and
       additive manufacture for complex, mid-face osteotomies   194

Maxillofacial rehabilitation   **201**
5.11   Maxillofacial rehabilitation case study 1: an investigation of the
       three-dimensional scanning of human body surfaces and its use in
       the design and manufacture of prostheses   201
5.12   Maxillofacial rehabilitation case study 2: producing burns therapy
       conformers using noncontact scanning and rapid prototyping   208

5.13   Maxillofacial rehabilitation case study 3: an appropriate approach
       to computer-aided design and manufacture of cranioplasty plates   216
5.14   Maxillofacial rehabilitation case study 4: evaluation of advanced
       technologies in the design and manufacture of an implant retained
       facial prosthesis   228
5.15   Maxillofacial rehabilitation case study 5: rapid prototyping
       technologies in soft-tissue facial prosthetics – current state of
       the art   241
5.16   Maxillofacial rehabilitation case study 6: evaluation of direct and
       indirect additive manufacture of maxillofacial prostheses using
       3D printing technologies   256
5.17   Maxillofacial rehabilitation case study 7: computer-aided methods
       in bespoke breast prosthesis design and fabrication   273

Orthotic rehabilitation applications   **283**
5.18   Orthotic rehabilitation applications case study 1: a review of
       existing anatomical data capture methods to support the mass
       customisation of wrist splints   283
5.19   Orthotic rehabilitation applications case study 2: comparison of
       additive manufacturing systems for the design and fabrication of
       customised wrist splints   294
5.20   Orthotic rehabilitation applications case study 3: evaluation of a
       digitised splinting approach with multiple-material functionality
       using additive manufacturing technologies   319
5.21   Orthotic rehabilitation applications case study 4: digitisation of the
       splinting process – development of a CAD strategy for splint
       design and fabrication   335
5.22   Orthotic rehabilitation applications case study 5: evaluation of a
       refined 3D CAD workflow for upper extremity splint design to
       support AM   344

Dental applications   **353**
5.23   Dental applications case study 1: the computer-aided design and
       rapid prototyping fabrication of removable partial denture
       frameworks   353
5.24   Dental applications case study 2: trial fitting of an RDP
       framework made using CAD and RP techniques   364
5.25   Dental applications case study 3: direct additive manufacture of
       RPD frameworks   371
5.26   Dental applications case study 4: a comparison of plaster, digital
       and reconstructed study model accuracy   380
5.27   Dental applications case study 5: design and fabrication of a
       sleep aponea device using CAD/AM technologies   401
5.28   Dental applications case study 6: computer-aided design,
       CAM and AM applications in the manufacture of dental appliances   410

Research applications   **419**
5.29   Research applications case study 1: bone structure models using
       stereolithography   419
5.30   Research applications case study 2: recreating skin texture relief
       using computer-aided design and rapid prototyping   427
5.31   Research applications case study 3: comparison of additive
       manufacturing materials and human tissues in computed
       tomography scanning   439
5.32   Research applications case study 4: producing physical models from
       computed tomography scans of ancient Egyptian mummies   450
5.33   Research applications case study 5: trauma simulation of massive
       lower limb/pelvic injury   458
5.34   Research applications case study 6: three-dimensional bone
       surrogates for assessing cement injection behaviour in cancellous
       bone   465

6   Future developments   **473**
6.1   Background   473
6.2   Scanning techniques   473
6.3   Data fusion   474
6.4   Rapid prototyping   474
6.5   Tissue engineering   475

Glossary and explanatory notes   **477**
Bibliography   **481**
Index   **487**

Woodhead Publishing Series in Biomaterials

1 **Sterilisation of tissues using ionising radiations**
  Edited by J. F. Kennedy, G. O. Phillips and P. A. Williams
2 **Surfaces and interfaces for biomaterials**
  Edited by P. Vadgama
3 **Molecular interfacial phenomena of polymers and biopolymers**
  Edited by C. Chen
4 **Biomaterials, artificial organs and tissue engineering**
  Edited by L. Hench and J. Jones
5 **Medical modelling**
  R. Bibb
6 **Artificial cells, cell engineering and therapy**
  Edited by S. Prakash
7 **Biomedical polymers**
  Edited by M. Jenkins
8 **Tissue engineering using ceramics and polymers**
  Edited by A. R. Boccaccini and J. Gough
9 **Bioceramics and their clinical applications**
  Edited by T. Kokubo
10 **Dental biomaterials**
  Edited by R. V. Curtis and T. F. Watson
11 **Joint replacement technology**
  Edited by P. A. Revell
12 **Natural-based polymers for biomedical applications**
  Edited by R. L. Reiss et al
13 **Degradation rate of bioresorbable materials**
  Edited by F. J. Buchanan
14 **Orthopaedic bone cements**
  Edited by S. Deb
15 **Shape memory alloys for biomedical applications**
  Edited by T. Yoneyama and S. Miyazaki
16 **Cellular response to biomaterials**
  Edited by L. Di Silvio
17 **Biomaterials for treating skin loss**
  Edited by D. P. Orgill and C. Blanco
18 **Biomaterials and tissue engineering in urology**
  Edited by J. Denstedt and A. Atala
19 **Materials science for dentistry**
  B. W. Darvell
20 **Bone repair biomaterials**
  Edited by J. A. Planell, S. M. Best, D. Lacroix and A. Merolli
21 **Biomedical composites**
  Edited by L. Ambrosio

22 **Drug–device combination products**
    Edited by A. Lewis
23 **Biomaterials and regenerative medicine in ophthalmology**
    Edited by T. V. Chirila
24 **Regenerative medicine and biomaterials for the repair of connective tissues**
    Edited by C. Archer and J. Ralphs
25 **Metals for biomedical devices**
    Edited by M. Ninomi
26 **Biointegration of medical implant materials: Science and design**
    Edited by C. P. Sharma
27 **Biomaterials and devices for the circulatory system**
    Edited by T. Gourlay and R. Black
28 **Surface modification of biomaterials: Methods analysis and applications**
    Edited by R. Williams
29 **Biomaterials for artificial organs**
    Edited by M. Lysaght and T. Webster
30 **Injectable biomaterials: Science and applications**
    Edited by B. Vernon
31 **Biomedical hydrogels: Biochemistry, manufacture and medical applications**
    Edited by S. Rimmer
32 **Preprosthetic and maxillofacial surgery: Biomaterials, bone grafting and tissue engineering**
    Edited by J. Ferri and E. Hunziker
33 **Bioactive materials in medicine: Design and applications**
    Edited by X. Zhao, J. M. Courtney and H. Qian
34 **Advanced wound repair therapies**
    Edited by D. Farrar
35 **Electrospinning for tissue regeneration**
    Edited by L. Bosworth and S. Downes
36 **Bioactive glasses: Materials, properties and applications**
    Edited by H. O. Ylänen
37 **Coatings for biomedical applications**
    Edited by M. Driver
38 **Progenitor and stem cell technologies and therapies**
    Edited by A. Atala
39 **Biomaterials for spinal surgery**
    Edited by L. Ambrosio and E. Tanner
40 **Minimized cardiopulmonary bypass techniques and technologies**
    Edited by T. Gourlay and S. Gunaydin
41 **Wear of orthopaedic implants and artificial joints**
    Edited by S. Affatato
42 **Biomaterials in plastic surgery: Breast implants**
    Edited by W. Peters, H. Brandon, K. L. Jerina, C. Wolf and V. L. Young
43 **MEMS for biomedical applications**
    Edited by S. Bhansali and A. Vasudev
44 **Durability and reliability of medical polymers**
    Edited by M. Jenkins and A. Stamboulis
45 **Biosensors for medical applications**
    Edited by S. Higson
46 **Sterilisation of biomaterials and medical devices**
    Edited by S. Lerouge and A. Simmons
47 **The hip resurfacing handbook: A practical guide to the use and management of modern hip resurfacings**
    Edited by K. De Smet, P. Campbell and C. Van Der Straeten
48 **Developments in tissue engineered and regenerative medicine products**
    J. Basu and J. W. Ludlow

49   Nanomedicine: Technologies and applications
     Edited by T. J. Webster
50   Biocompatibility and performance of medical devices
     Edited by J-P. Boutrand
51   Medical robotics: Minimally invasive surgery
     Edited by P. Gomes
52   Implantable sensor systems for medical applications
     Edited by A. Inmann and D. Hodgins
53   Non-metallic biomaterials for tooth repair and replacement
     Edited by P. Vallittu
54   Joining and assembly of medical materials and devices
     Edited by Y. (Norman) Zhou and M. D. Breyen
55   Diamond-based materials for biomedical applications
     Edited by R. Narayan
56   Nanomaterials in tissue engineering: Fabrication and applications
     Edited by A. K. Gaharwar, S. Sant, M. J. Hancock and S. A. Hacking
57   Biomimetic biomaterials: Structure and applications
     Edited by A. J. Ruys
58   Standardisation in cell and tissue engineering: Methods and protocols
     Edited by V. Salih
59   Inhaler devices: Fundamentals, design and drug delivery
     Edited by P. Prokopovich
60   Bio-tribocorrosion in biomaterials and medical implants
     Edited by Y. Yan
61   Microfluidic devices for biomedical applications
     Edited by X-J. James Li and Y. Zhou
62   Decontamination in hospitals and healthcare
     Edited by J. T. Walker
63   Biomedical imaging: Applications and advances
     Edited by P. Morris
64   Characterization of biomaterials
     Edited by M. Jaffe, W. Hammond, P. Tolias and T. Arinzeh
65   Biomaterials and medical tribology
     Edited by J. Paolo Davim
66   Biomaterials for cancer therapeutics: Diagnosis, prevention and therapy
     Edited by K. Park
67   New functional biomaterials for medicine and healthcare
     E. P. Ivanova, K. Bazaka and R. J. Crawford
68   Porous silicon for biomedical applications
     Edited by H. A. Santos
69   A practical approach to spinal trauma
     Edited by H. N. Bajaj and S. Katoch
70   Rapid prototyping of biomaterials: Principles and applications
     Edited by R. Narayan
71   Cardiac regeneration and repair Volume 1: Pathology and therapies
     Edited by R-K. Li and R. D. Weisel
72   Cardiac regeneration and repair Volume 2: Biomaterials and tissue engineering
     Edited by R-K. Li and R. D. Weisel
73   Semiconducting silicon nanowires for biomedical applications
     Edited by J. L. Coffer
74   Silk biomaterials for tissue engineering and regenerative medicine
     Edited by S. Kundu
75   Biomaterials for bone regeneration: Novel techniques and applications
     Edited by P. Dubruel and S. Van Vlierberghe

76   **Biomedical foams for tissue engineering applications**
     Edited by P. Netti
77   **Precious metals for biomedical applications**
     Edited by N. Baltzer and T. Copponnex
78   **Bone substitute biomaterials**
     Edited by K. Mallick
79   **Regulatory affairs for biomaterials and medical devices**
     Edited by S. F. Amato and R. Ezzell
80   **Joint replacement technology Second edition**
     Edited by P. A. Revell
81   **Computational modelling of biomechanics and biotribology in the musculoskeletal system: Biomaterials and tissues**
     Edited by Z. Jin
82   **Biophotonics for medical applications**
     Edited by I. Meglinski
83   **Modelling degradation of bioresorbable polymeric medical devices**
     Edited by J. Pan
84   **Perspectives in total hip arthroplasty: Advances in biomaterials and their tribological interactions**
     S. Affatato
85   **Tissue engineering using ceramics and polymers Second edition**
     Edited by A. R. Boccaccini and P. X. Ma
86   **Biomaterials and medical device-associated infections**
     Edited by L. Barnes and I. R. Cooper
87   **Surgical techniques in total knee arthroplasty (TKA) and alternative procedures**
     Edited by S. Affatato
88   **Lanthanide oxide nanoparticles for molecular imaging and therapeutics**
     G. H. Lee
89   **Surface modification of magnesium and its alloys for biomedical applications Volume 1: Biological interactions, mechanical properties and testing**
     Edited by T. S. N. Sankara Narayanan, I. S. Park and M. H. Lee
90   **Surface modification of magnesium and its alloys for biomedical applications Volume 2: Modification and coating techniques**
     Edited by T. S. N. Sankara Narayanan, I. S. Park and M. H. Lee
91   **Medical modelling: The application of advanced design and rapid prototyping techniques in medicine Second edition**
     Edited by R. Bibb, D. Eggbeer and A. Paterson
92   **Switchable and responsive surfaces for biomedical applications**
     Edited by Z. Zhang
93   **Biomedical textitles for orthopaedic and surgical applications: fundamentals, applications and tissue engineering**
     Edited by T. Blair
94   **Surface coating and modification of metallic biomaterials**
     Edited by C. Wen

Preface

The principal aim of this book is to provide a genuinely useful text that can help professionals from a broad range of disciplines to understand how advanced product design and development technologies, techniques and methods can be employed in a variety of medical applications. The book describes the technologies; methods and potential complexities of these activities as well as suggesting solutions to some of the commonly encountered problems and highlighting potential benefits. This book is based on the collective experience of the authors spanning 20 years of research and practice in medical applications. The majority of the research has been conducted through the activities of the Surgical & Prosthetic Design team of PDR and Loughborough University's Design School through collaboration with clinical, academic and industrial partners.

The book is presented in two main sections. In the first section, the technical chapters provide an introduction to the various technologies involved ranging from medical scanning to physical model manufacture. The second section provides a number of interesting and varied case studies that collectively cover the application of most, if not all, of the technologies introduced in the previous chapters. To ensure that these case studies are relevant and appropriate they have been drawn from work previously published in internationally peer-reviewed journals or conference proceedings with full acknowledgement, proper citation and permission where appropriate. Where appropriate these papers have been updated to reflect recent technological advances.

This text also aims to encourage what is, by its very nature, a multidisciplinary and collaborative field. The case studies selected reflect this by describing a broad range of techniques and applications. Although much work has been done in this area, there is a tendency for people to publish in the journals, language and context of their own professional practice. Whilst this text does not purport to be the most comprehensive review of the work done to date, it is a conscious effort to overcome these professional interfaces and encourage multidisciplinary collaboration by providing a single source of useful reference material accessible to readers from any relevant background.

Therefore it is hoped that this book will appeal equally to medical and technical specialties, including for example: designers, biomedical engineers, clinical engineers, rehabilitation engineers, medical physicists, radiologists, radiographers, surgeons, prosthetists, orthotists, orthodontists, anatomists, medical artists and anthropologists, and perhaps even veterinarians, archaeologists and palaeontologists.

The text will also provide an excellent resource for postgraduate students, researchers and doctoral candidates working in this rapidly developing, important and exciting area.

Richard Bibb
Loughborough

Acknowledgements

This book would not have been possible without the help and support of many people. It is appropriate therefore to offer our thanks to our colleagues at PDR (Cardiff Metropolitan University) and Loughborough University for their assistance and support. We would like to thank Prof. Robert Brown and Prof. Alan Lewis for providing the Surgical and Prosthetic Design Group at PDR with the support needed to establish and grow the group.

As the central theme underpinning this book is multidisciplinary collaboration, it is important to recognise the input of all who have contributed to it. We thank all of our collaborators and co-authors without whom none of the work reported in this book would have been possible. Each case study is fully acknowledged and we would also like to thank the various publishers for their kind permission to reproduce our previous papers and articles.

We have been fortunate enough to establish a number of significant and long-term partnerships and we would like to offer our particular thanks to the following. Professor Robert Williams for his enthusiasm and collaboration in dental technology; Professor Stephen Richmond and his colleagues at Cardiff Dental School, for their collaboration in imaging and orthodontic applications; Lucia Ramsey and Ella Donnison for their contribution to orthotics research; Dr John Winder for his collaboration on medical modelling and imaging and Prof. Julian Minns for his collaboration on orthopaedic applications. We would also like to thank all those involved in CARTIS, especially Adrian Sugar, Alan Bocca and Peter Evans from Morriston Hospital who did so much to help establish research in advanced technologies in head and neck reconstruction.

We would like to thank Sean Peel for the computer rendering used for the cover image.

Finally, we would also like to thank everyone at Woodhead/Elsevier for all their help and professional expertise in turning our manuscript into the book you see here.

Richard Bibb, Dominic Eggbeer and Abby Paterson

Introduction

1

1.1 Background

The purpose of this book is to describe some of the many possibilities, techniques, and challenges involved when using advanced design and product development technologies in medicine. This ranges from the creation of models of anatomy to the design and manufacture of bespoke medical devices and prostheses.

The origin of this field lies in medical modelling, sometimes called biomodelling. "Medical modelling" is the term used to describe the creation of highly accurate physical models of human anatomy directly from medical scan data. The process involves capturing human anatomy data, processing the data to isolate individual tissue or organs, optimising the data for the manufacturing technology to be used, and finally building the model using rapid prototyping (RP) techniques. RP is the general name coined to describe computer-controlled machines that are able to manufacture physical items directly from three-dimensional (3D) computer data. Originally, these machines were developed to enable designers and engineers to build prototype models of objects which they had designed using computer-aided design software (CAD). These so-called RPs allowed them to ensure that what they had designed on-screen fit together with all the other components of the product being developed. Therefore, the machines were quickly developed to produce models of high accuracy as rapidly as possible.

In the 1990s, it was realised that RP machines could use other types of 3D computer data, such as those obtained from medical scanners. Software was developed to enable medical scan data to interface with RP machines, and medical modelling began. Since then, the field has developed to cover all kinds of applications ranging from forensic science to reconstructive surgery. Early success and clear demonstration of benefits have led to widespread interest in the technologies from many medical specialties. However, with each development more and more clinicians, surgeons, engineers, and researchers are realising the potential benefits of RP techniques, which in turn places new challenges on people whose job it is to build these models.

This book aims to describe the stages required to produce high-quality medical models and offers an insight into the techniques and technologies that are commonly used. Chapters 2–4 follow the logical sequence of stages in the medical modelling process as shown in Table 1.1. Each chapter describes the technologies and processes used in each stage in general terms for those who are not familiar with them or are new to the field, whereas the case studies illustrate a number of diverse applications carried out in recent years. Where appropriate, case studies include cross-references to particular sections of Chapters 2–4 as a reminder, to eliminate repetition or enable the reader to begin by reading case studies and then find the relevant technical information easily without necessarily having to read the whole book in chapter order.

Medical Modelling. http://dx.doi.org/10.1016/B978-1-78242-300-3.00001-9

Table 1.1 **Stages of the medical modelling process**

Step 1
Medical imaging for rapid prototyping (RP)
Select the optimal modality
Set appropriate protocols
Scan patient
Step 2
Export data media and format
Export the data from the scanner in an appropriate format
Transfer data to the RP laboratory
Step 3
Working with medical scan data
Isolate data relating to the tissues or organs to be modelled
Save and transfer data in the correct format the RP process
Step 4
Physical reproduction—RP
Build the model
Clean, finish, or sterilise as required
Deliver the model to the clinician

By its very nature, medical modelling has brought together the fields of engineering and medicine. Consequently, this book aims to satisfy the needs of both fields as they work together on medical modelling. Therefore, although it is not possible to cover every medical or technical definition here, this chapter offers a brief introduction to some anatomical terminology for the benefit of engineers new to the field, as well as an introduction to some specialist technical terms. Where a longer or more detailed description is required, an explanatory note may be found in "Glossary and explanatory notes", which also contains glossaries of technical and medical terms and abbreviations. There are also recommendations for further reading at the end of some chapters, to enable those with particular interests to develop their knowledge further.

This book is essentially technical in nature but it is important to consider that it also addresses genuine human needs, and consequently there is due consideration for patients and ethics. Therefore, throughout this book, illustrations and case studies have been made anonymous and where necessary, permission has been granted.

1.2   The human form

The human body is the most significant physical form that we possess or encounter. Our physical form is inextricably bound up with our minds and behaviour. It influences but also responds to our lifestyle choices and, combined with our character, defines us as individuals. Our physical form defines how we appear to others and it affects our perception of ourselves. It displays our health and fitness and even our attractiveness to our loved ones. It enables us to recognise any one individual among the seven billion fellow humans with which we share the planet.

In addition to its undeniable importance, our physical form is perhaps one of the most complex shapes we encounter in life. Its importance to us makes us sensitive to the tiniest of details and the subtlest of contours. This complexity, combined with our sensitivity to the human form, has provided perhaps the preeminent challenge to artists in our history. Through drawing, painting, and sculpture, artists have strived to capture what it is that makes us human and how that is expressed through our physical form and appearance.

In terms of medicine, the human body is both subject and object. The study of the human form is the basis for all medicine as it strives to correct our malfunctions and degradation. It is from these noble aims that we constantly try to apply the latest in technology to improve our treatment of all kinds of illness. When such unfortunate occurrences as disease or trauma damage our physical form, they not only physically debilitate us but also affect our psychological health. Therefore, the ability to capture and reproduce human anatomy to the infinite subtlety that we desire is a pursuit that is as important as it is challenging.

The age of computer technology has not necessarily eased this process. The reconstruction and rehabilitation of people can consist of any combination of dressing, rehabilitation, prosthesis, and surgery. Skills employed range from the artistry of the prosthetist to the engineering of artificial implants. Until recent times, reconstruction and rehabilitation have relied almost solely on the dexterity and artistry of a small but highly dedicated range of health professionals. However, in the current age, the pressures on these people grow as survival rates increase and surgical interventions become ever more sophisticated. It is therefore not surprising that medicine looks toward advanced technologies to provide the effort, time, and cost savings that have been so successfully achieved in product design and engineering.

This book aims to describe some of the product design technologies that have been successfully used in the field of human reconstruction and rehabilitation and to illustrate their application through case studies. As we will discover, there are many benefits to be found from applying modern technologies, yet they are not without obstacles. The nature of the human form makes the transfer of techniques that are well suited to product design and engineering a particularly challenging yet ultimately rewarding field of work.

1.3 Basic anatomical terminology

Although this book is not intended to be used as a guide to human anatomy, the descriptions of techniques, medical conditions, and treatments require the use of accepted anatomical nomenclature. For readers with medical training, this nomenclature will be well known. However, for those from a technical, design, or engineering background, some basic terminology will prove useful. This section will introduce some basic

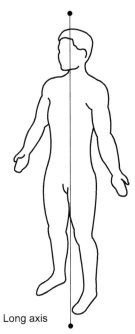

Long axis

Figure 1.1 The anatomical position and the long axis.

terms that will enable the reader to proceed with the rest of the text, but further read-ing on anatomy and physiology is strongly recommended. There are many excellent texts on anatomy and a selection of titles is provided in the bibliography. Attending a short course in human anatomy and physiology would be highly recommended to any engineer or designer wishing to specialise in clinical or medical applications; many universities offer such courses.

When referring to human anatomy, the relative positions of organs, limbs, and features are only useful if the body is in a known pose. Therefore, it is standard practice to assume that the body is in the "anatomical position" when describing relative positions of anatomy. The anatomical position is with the body and limbs straight, feet together, head looking forward, and arms at the sides of the torso with the palms facing forward and fingers straight. The principal axis of the human is through the centre of the body running from head to feet; this is referred to as the long axis, shown in Figure 1.1.

Once the anatomical position is known, perpendicular planes can divide the body. The plane through the body perpendicular to the long axis is known as the axial plane. The planes perpendicular to this are known as the coronal and sagittal. Directions and dis-tances are described as they relate to the centre of the body. Parts nearer to the body are known as proximal and those farther from the body centre are distal. Parts that are nearer the midline of the body are known as medial and those farther from it are described as lateral. These terms are summarised in Table 1.2 and illustrated in Figures 1.2 and 1.3.

Table 1.2 **Anatomical directional terms**

Term	Definition
Superior	Toward the head (upper)
Inferior	Away from the head (lower)
Anterior	Toward the front of the body
Posterior	Toward the rear of the body
Medial	Nearer the midline of the body
Lateral	Farther from the midline of the body
Contralateral	On the opposite side of the body
Ipsilateral	On the same side of the body
Proximal	Relating to limbs—nearer to the body
Distal	Relating to limbs—farther from the body
Superficial	Toward the surface of the body
Deep	Into the body away from the surface

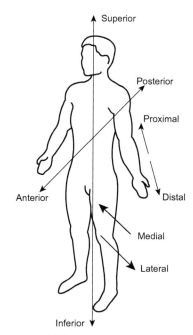

Figure 1.2 The direction terms used in human anatomy.

1.4 Technical terminology

A number of technical terms used throughout this book are introduced here so that they can subsequently be abbreviated in later chapters. Much of the book is devoted to explaining RP technologies. "Rapid prototyping" is a term coined in the 1980s to refer to a number of automated manufacturing technologies that produce physical items in

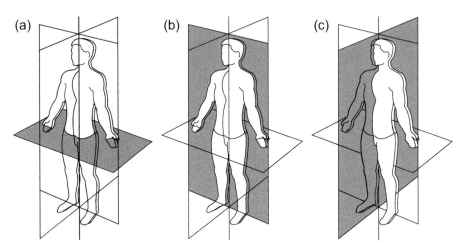

Figure 1.3 The major reference planes used in human anatomy: axial (a), coronal (b), and sagittal (c).

an additive layer-by-layer manner from 3D computer data. In the 1980s and 1990s, these technologies revolutionised the production of prototype parts during new product development that was increasingly being done using CAD. These technologies enabled the direct manufacture of prototype parts that previously were handmade by skilled technicians in a slow and laborious process. Hence, the technologies became known as RP. In the early 2000s, the layer additive manufacturing principle was applied to a variety of materials and applications including the manufacture of end-use products, and consequently the phrase "rapid manufacturing" was frequently used.

To reflect an increasing shift toward manufacturing as opposed to prototyping, more recently the umbrella term "additive manufacturing" has become widely accepted. RP can therefore be considered as a subset of additive manufacturing. The mainstream media have adopted the term "three-dimensional" (3D) printing, which in the past meant a specific type of process but is perhaps a more descriptive term for the lay reader. The term "3D printing" is now in common use and is frequently associated with the cheapest and most basic machines that may be purchased for a few hundred dollars and are increasingly popular in schools and colleges, and as a hobby. However, to maintain consistency with previous editions, this book will continue to use the term "RP" to refer to layer-additive technologies.

The term "CAD" is in common use and refers to a vast array of software applications that enable the design and definition of objects. In the simplest terms, CAD could refer to simply drawing on a computer screen. The technologies and processes described in this book rely on 3D data, and consequently the term "3D CAD" will be used; where the term "CAD is used," 3D is usually inferred. 3D CAD applications work by providing a manner of defining a 3D surface using a computer programme. In some versions, the surfaces are generated using mathematical geometry; others allow a more arbitrary construction or manipulation of the data. Where these differences are significant, they will be explained in the following chapters.

Medical imaging

2

2.1 Introduction to medical imaging

To manufacture a virtual or physical model of any human anatomy, it must first be captured in three dimensions in a manner that can be utilised by the computer processes to be used. A number of scanning modalities can be used, ranging from substantial hospital facilities normally found in radiology departments to small hand-held scanners that can be used in the laboratory or clinic.

There are essentially two main categories of scanning modality for human bodies, those that capture data from the whole body both internally and externally and those that capture only external data. Most hospital-based scanners capture data from the whole body both internally and externally. These are normally large, sophisticated medical imaging machines capable of scanning the complete human body. Examples include computed tomography (CT), magnetic resonance imaging (MRI) and positron emission tomography. Each modality uses a different physical effect to generate cross-sectional images through the human body. Typically, the patient is placed lying down on a table that is fed through the scanner whilst the images are taken. The cross-sectional images are arranged in an order so that the computer can construct a three-dimensional (3D) data set of the patient. Software can then be used to isolate particular organs or tissues. This data can then be used to reconstruct an exact replica of the organ using computer-aided techniques. The different physical effects used by each type of scanner result in different types of tissue being imaged. These machines require highly specialised staff to operate and require large capital investment. The use of the two most common modalities will be described in more detail later in this chapter.

The other type of scanning is used to capture only the external surface of a patient. A wide range of technologies can be used for capturing 3D surface data. Three-dimensional surface scanning, sometimes referred to as digitising, has been used in engineering and product design for many years as a method of integrating the surfaces of existing physical objects with computer-aided design models. Consequently, this process is often referred to as 'reverse engineering'. There are many types of surface scanner or digitiser available to the engineer or designer. They can be separated into two main categories; 'contact' or 'touch probe digitisers' and noncontact scanners. Touch probe digitisers use a pressure sensitive probe tip and calibrated motion to map out the surface of an object point-by-point. Depending on the quality of its manufacture, they can be extremely accurate. However, it is also a very slow and laborious process, sometimes taking hours to capture the surface of an object. Although this is acceptable when scanning inanimate objects, it is clearly not appropriate to capture the surface of human anatomy. Therefore, noncontact scanners are typically used when

Medical Modelling. http://dx.doi.org/10.1016/B978-1-78242-300-3.00002-0

capturing surface data from people. Noncontact scanners use light and digital camera technologies to capture many thousands of data points on the surface of an object in a matter of seconds. The fast capture of data and the harmless light used make these types of scanner ideal for capturing human anatomy. These scanners are typically like very large cameras and may be tripod mounted or in some instances even hand held. Despite the variety of surface scanners available on the market, the general principles of their operation and application are the same and these principles are described later in this chapter.

It is not intended to provide a definitive description of the technology and practice of each scanning modality here but to establish some criteria and guidelines that may be employed to optimise their use in the production of virtual or physical medical models. Many texts are available that describe each modality fully and some are listed in the recommended reading list at the end of this chapter.

2.2 Computed tomography (CT)

2.2.1 Background

CT works by passing focussed X-rays through the body and measuring the amount of the X-ray energy absorbed. The amount of X-ray energy absorbed by a known slice thickness is proportional to the density of the body tissue. By taking many such measurements from many angles, the tissue densities can be composed as a cross-sectional image using a computer. The computer generates a grey scale image where the tissue density is indicated by shades of grey. The Hounsfield scale is a quantitative scale for describing radiodensity in medical CT and provides an accurate density for the type of tissue. On the Hounsfield scale, air is represented by a value of -1000 (black on the grey scale) and bone between $+700$ (cancellous bone) to $+3000$ (dense bone) (white on the grey scale). As bones are much denser than surrounding soft tissues, they show up very clearly in CT images, as can be seen in Figure 2.1. This makes CT an important imaging modality when investigating skeletal anatomy. Similarly, the density difference between soft tissues and air is great allowing, for example, the nasal airways to be clearly seen. Soft tissues and organs represent narrow Hounsfield value ranges and are therefore more difficult to differentiate between adjacent structures, such as between fat and muscle when viewing and segmenting CT data. Artificial contrast agents that absorb X-ray energy may be introduced into the body, which makes some structures stand out more strongly in CT images.

As CT uses ionising radiation in the form of X-rays, exposure should be minimised, particularly to sensitive organs such as eyes, thyroid and gonads. The X-rays are generated and detected by a rotating circular array through which a moving table can travel. Typically, the patient lies on his or her back and is passed through the circular aperture in the scanner. The detector array acquires cross-sections perpendicular to the long axis of the patient. The images acquired are therefore usually termed the axial or transverse images.

Most modern scanners perform a continuous spiral around the long axis of the patient. This innovation enables 3D CT scanning to be performed much more rapidly

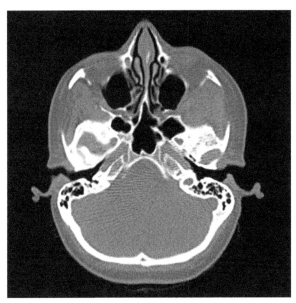

Figure 2.1 A CT image of the head.

and consequently 3D CT scans are frequently referred to as helical CT. In addition, modern CT scanners employ multiple arrays to enhance the rate of data capture and improve 3D volume acquisition.

CT images are generated as a grey scale pixel image, just like a bitmap computer image. If the distance between a series of axial images, called the slice thickness, is known, they can be interpolated from one image to the next to form cuboids, known as voxels. Therefore, a 3D CT scan generates a voxel representation of the human body. Software can be used to reslice these voxel data sets in axes perpendicular to the long axis enabling different cross-sectional images to be generated from the original axial data. This is typically done in the sagittal and coronal planes; however, images may be generated in any plane.

The radiographers who conduct CT scans have specific parameters and settings for different types of scan. These are standardised and referred to as protocols. When embarking on using CT data for medical modelling, it is helpful to discuss it first with the radiographers and they may well develop a protocol specifically for medical modelling.

CT scans are time-consuming, expensive and potentially harmful, so every care must be taken to ensure that the scan is conducted correctly the first and only time.

2.2.2   Partial pixel effect

When CT data is captured, the resulting images are divided up into a large number of pixels (typically a 512×512 matrix, but more modern scanners have a higher 1024×1024 matrix). Each pixel is a shade of grey that relates to the density of the

Figure 2.2 Original object shape.

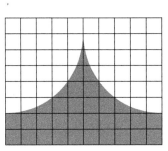

Figure 2.3 The effect of pixel size.

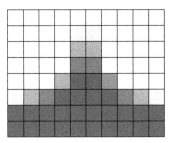

Figure 2.4 The resulting tomographic image (blurring of edges).

tissue at that location. The resulting images are therefore an approximation of the original tissue shapes according to their density. The quality of that approximation is a function of the number and relative size of the pixels as well as other aspects of the CT scanner. The discrete size of the pixels means that edges between different anatomical structures are to some degree affected by this image quality.

Boundaries between different tissues can be effectively 'blurred' due to what is referred to as the partial pixel effect. If the boundary between two different structures crosses a given pixel, that single pixel cannot represent both densities. Instead that pixel displays an intermediate density which is somewhere between the two. The effect can be illustrated by considering the shape in Figure 2.2, which consists of two densities low being grey and high being white. When the CT scan is performed, the cross-section is broken down into pixels as shown in Figure 2.3. In this view, it can be seen that some squares contain both high- and low-density areas. These pixels will therefore be shown as an intermediate grey depending on their relative proportions. This leads to the partial pixel effect which can be seen in the resulting tomographic image as shown in Figure 2.4.

2.2.3 Anatomical coverage

The coverage of a CT volume is defined in two ways, by the number of axial images taken and the field of view used for those images. In the long axis, it is defined by the table position where the first and last images are acquired. It is clear that the series of axial images must begin and end either side of the anatomy of interest but it is often important to begin and end the scan some distance either side of the anatomy of interest. When conducting 3D series scans, the data should be continuous. Noncontinuous sets of data may be satisfactorily combined in software later as the patient has usually shifted position slightly and the separate series may not align perfectly.

The area covered by each axial image is referred to as the field of view and is typically square. The axial image consists of a fixed number of pixels, typically 512×512. The field of view is the physical distance over which the image is taken. Therefore, altering the field of view will alter the pixel size in the axial image. Usually, a field of view large enough to capture the whole cross-section of the anatomy is used. However, where specific small areas of anatomy are required, the field of view may be reduced to capture only that area. This results in a smaller pixel size, which increases the physical accuracy of the scan. For example, a typical field of view used to CT the head would be 25 cm resulting in a pixel size of 0.49 mm (assuming a 512×512 array) whilst a field of view of 13 cm may be used to capture data relating only to the mouth, which would result in a pixel size of 0.25 mm. Whilst small pixel size is a desirable factor, it is more important that the images cover all of the anatomy of interest plus some margin.

Although exposure to X-rays should be minimised wherever possible, it is more important that the scan covers all of the required anatomy plus some additional margin. It is better to perform one extensive CT scan than a minimal one that is later found to be inadequate and necessitates subsequent scans. Basic mistakes in capturing the required coverage can be avoided through clear communication between the surgeon and radiographer.

2.2.4 Slice thickness

This is the distance between the axial scans taken to form the 3D scan series. In the case of helical CT scanning, the parameter applies to the distance between the images calculated during the scan (distance between cuts). To maximise the data acquired, this distance should be minimised. Some scanners can go as low as 0.5 mm, which gives excellent results but this must be balanced against increased X-ray dose. Typically, distances of 1–1.5 mm produce acceptable results. A scan distance of 2 mm may be adequate for larger structures such as the long bones or pelvis. A scan distance greater than 2 mm will give poorer results as the scan distance increases.

Collimation is the term used to describe the thickness of the X-ray beam used to take the cross-sectional image. In combination with scan distance, consideration may be given to collimation and overlap. In most circumstances, the scan distance and collimation should be the same. However, using a slice distance that is smaller than the collimation gives an overlap. When scanning for very thin sections of bone that lie in the axial plane, such as the orbital floor or palate, an overlap may give improved results.

Even with a very small scan distance, some detail may be lost where thin sections of bone exist between the scan planes (although this is not true for volumetric acquisition). Typically, these are areas in the skull such as the palette and orbital floors. In addition, parts that are very small, or connected by thin sections, may not survive subsequent data or physical manufacturing processes.

2.2.5 Gantry tilt

Typically for the purposes of virtual or physical modelling, gantry tilt should be avoided as it does not significantly improve the quality of the acquired data and provides an opportunity for error when reading the images. Large gantry tilt angles are clearly apparent on visual inspection and can be corrected. However, small angles may not be easy to check visually and may be compensated for incorrectly. Even the use of automatic import of the medical image standard digital imaging and communications in medicine (DICOM) is no guarantee as although the size of gantry tilt angle is included in the format the *direction* of tilt is not. Failure to compensate for the direction of the tilt correctly will lead to an inaccurate model, wasting time and money and potentially leading to errors in surgery or prosthesis manufacture. This effect is illustrated by the example shown in Figure 2.5.

2.2.6 Orientation

Anterior-posterior and inferior-superior orientation is usually obvious but lateral (left–right) orientation may be ambiguous. This is not a problem with automatically imported data but when manually importing data it is important that the correct lateral orientation can be ascertained. If there is an obvious lateral defect, then a note from the clinician describing it is usually sufficient. Where the lateral orientation cannot be easily determined from the anatomy extra care should be taken to verify the orientation before building a potentially expensive medical model.

2.2.7 Artefacts

This is the general term for signals within an image that do not correspond to anatomy. These may result from patient movement or X-ray scatter. Examples of medical

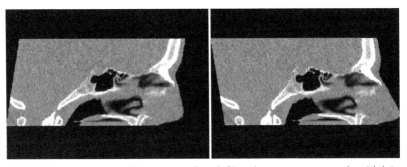

Figure 2.5 Incorrect 5° gantry tilt compensation (left) and correct compensation (right).

modelling problems that have been encountered because of artefacts are discussed in the case studies in Chapter 5, in the Implementation section. Scattering is typically caused by very dense bones or metal objects such as dental fillings, plates, screws or even shrapnel.

1. Movement

A good-quality CT scan depends on the patient remaining perfectly still throughout the acquisition. Movement during the acquisition will lead to distortions in the data (analogous to a blurred photograph). This has become less of a problem as acquisition times have decreased with the advent of helical multislice CT. However, it can still present a problem in some cases. For example, involuntary movement of the chest, neck, head or mouth can occur through breathing or swallowing. Movement can be particularly difficult to control when scanning babies, small children and claustrophobic patients in which case a sedative or even general anaesthetic may be required.

2. X-ray image scatter by metal implants

Dense objects such as amalgam or gold fillings, braces, bridges, screws, plates and implants scatter X-rays resulting in a streaked appearance in the scan image. The scatter results in significant image errors where false data appears with corresponding false missing data or shadows. Because of the nature of X-rays, little can be done to eliminate these effects although manufacturers are now offering image-processing algorithms that can reduce its effect. Figure 2.6 shows an axial CT image with significant artefact from scatter. Figure 2.7 (left) shows how the scatter will be demonstrated on a 3D reconstruction of the data, apparent as spikes radiating from the source of the scatter. These effects can be manually edited in software to produce a normal looking model (right). However, this does depend to some degree of the expertise of

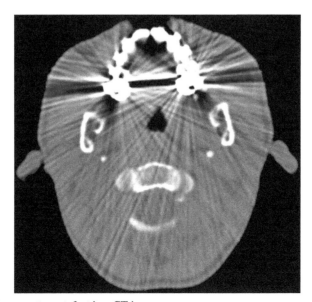

Figure 2.6 X-ray scatter artefact in a CT image.

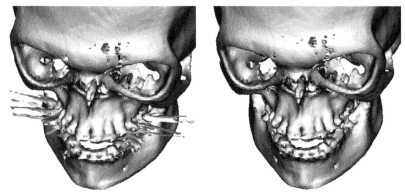

Figure 2.7 3D reconstruction (left) and edited 3D reconstruction (right).

the operator and consequently the accuracy of the model in the affected areas cannot be guaranteed. In most cases, this does not usually affect the usefulness of the whole model. As cases showing artefacts usually occur in and around the teeth, a dental cast is typically used in conjunction with the medical model and indeed may be combined with a physical model.

3. Noise

Noise is a fundamental component of a CT image and is especially prevalent in dense tissues. Although these images may be visually acceptable, they are impractical for modelling. Good modelling from CT data depends on identifying a smooth boundary between bone and soft tissue. Noise reduces the boundary, which results in poor 3D reconstructions and consequently poor models. This commonly affects areas through the shoulders and hence vertebrae in the lower neck and upper back (C6 to T4). A typical example is shown in Figure 2.8. The effect becomes much more apparent when zooming into image data as can be seen in the close-up view of the same data shown in Figure 2.9. Three-dimensional reconstructions from this data will lead to a poor result as shown in Figure 2.10. Typically, such reconstructions will appear rough surfaced or porous.

If the effect only occurs in a few images, it may be possible to edit them out to produce a normal looking model. However, this does mean that the accuracy of the model in these areas cannot be guaranteed. If the whole data set is affected, this editing is unfeasible and the resulting model may be too poor to be useful. Image processing may be used to 'filter out' these effects but this may affect the accuracy of the original data.

2.2.8   Kernels

Modern CT scanning software allows different kernels (digital filters) to be used. These modify the data to give better 3D reconstructions and can help to reduce noise. Typically, the options will range from 'sharp' to 'smooth'. Sharpening filters increase edge sharpness but at a cost of increasing image noise. Smoothing filters reduce noise content in images but also decrease edge sharpness.

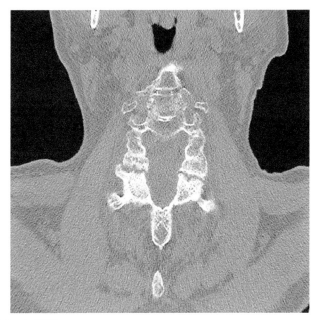

Figure 2.8 Noise in CT image.

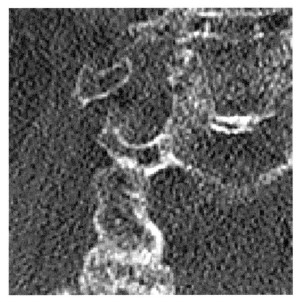

Figure 2.9 Close-up of noise.

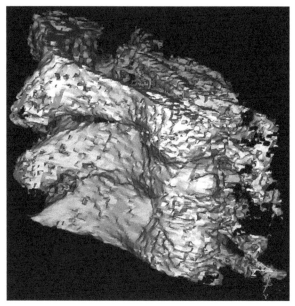

Figure 2.10 3D reconstruction of a noisy data set.

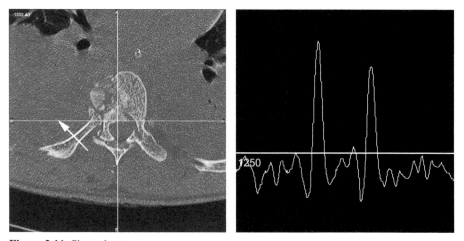

Figure 2.11 Sharp data.

In general, when building medical models, smooth filters tend to give better results and are easier to work with. The effect of sharp versus smooth filtering is illustrated in Figures 2.11 and 2.12 where the arrow represents a density profile, which is shown as the graph on the right.

Although the smooth image contrast appears poor on screen, taking a density profile shows that the actual contrast is good and that the smooth data allows a much lower threshold to be used.

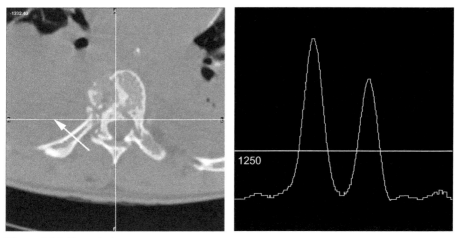

Figure 2.12 Smooth data.

2.3   Cone beam CT (CBCT)

2.3.1   Background

Cone beam CT (CBCT) is a more recent development in medical imaging that has become popular in orthodontic, dental and maxillofacial applications, but also has foundation in interventional radiology, image-guided surgery and radiation therapy. It uses the same basic principle as conventional CT scanning, but in a different way that results in far lower patient radiation doses. As well as fundamental technical differences in the way X-rays are delivered, detected and data is processed there are more obvious differences in the appearance of dental/craniofacial CBCT scanners. The first obvious difference is that the patient is usually scanned in the seated position as opposed to lying down (Figure 2.13). Whereas conventional CT uses a narrow fan of X-rays and a narrow detector, the X-rays in CBCT are emitted in a cone shape and detected by a flat panel detector. Multiple images (usually hundreds) from different angles are captured through a single rotation of the emitter and detector around the head. Software algorithms are used to calculate volumetric data based on these multiple angles of capture around the fixed central point (termed isocentre).

2.3.2   Advantages

The primary advantages of CBCT over conventional CT are: lower radiation doses, higher pixel/voxel resolution, ability to focus in great detail on a small area without irradiating surrounding tissue, lower operational cost and the smaller footprint.

1. Reduced radiation dose

CBCT offers significantly lower radiation dose than conventional CT when scanning the same area; however, the actual difference will depend on the machine and the parameters used, therefore care should be taken. Studies that have used dosimetry

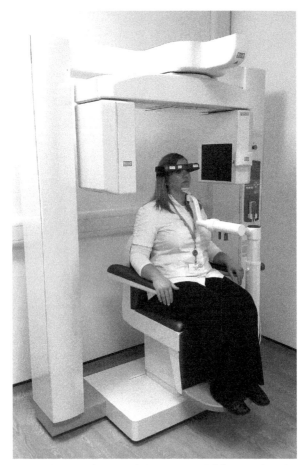

Figure 2.13 The Morita 3D Accuitomo 170 Cone Beam CT Scanner based at Morriston Hospital, Swansea.

phantoms based on a craniofacial (upper and lower jaw) model cite reductions in the order of 15 times less than conventional CT for the equivalent scanned area.

2. Voxel size – resolution

Whereas voxel size for conventional CT can be as small as 0.5 mm (at the compromise of increased radiation dose), CBCT produces isotropic (equal in the x–y–z dimensions) voxels as low as 0.076 mm (Kodak 9000 3D manufacturer specifications). This is a distinct advantage when evaluating bone quality for procedures such as implant planning.

3. Ability to focus on small areas

The field of view is dependent on the detector size and shape, beam projection geometry, and the ability to collimate the beam. Small fields of view generally correlate to increased resolution in a very local area. A small or focussed field of view would be in the order of 5 cm or less, whereas 15 cm or more may be required for craniofacial applications. As with conventional CT, the reduction in field of view and associated radiation dose must be balanced against acquiring a sufficient region of interest for the intended application.

2.3.3 Limitations

The advantages of CBCT must be balanced against a number of limitations, including: artefact, the relatively limited field of view (limited to the anterior portions of the head, usually focussed on upper and lower jaws), grey scale values that change throughout the data volume (thereby causing an inconsistency in being able to measure Hounsfield units) and lower soft tissue contrast than conventional CT (further compounding difficulties in differentiating between tissue types across the data set). These limitations are also highly dependent on the machine type, scanning and reconstruction algorithms used, making it even more difficult to compare and measure data sets from different CBCT scans.

1. Artefact

Like conventional CT, CBCT also suffers from artefact such as those discussed in Sections 2.2.2 and 2.2.7. The way the data affected however slightly different, particularly in the presence of dense metallic objects (such as metal teeth fillings). Because of the cone shape of the X-ray beam and large detector, scatter in CBCT can cover the whole data set rather than being limited to the single slice of a conventional CT image. Like conventional CT, various proprietary software algorithms and scan strategies have been developed and adopted to minimise and overcome the problems associated with artefact.

2. Field of view

Whereas conventional, helical CT scanners can continuously scan large areas in one session, CBCT scanners are relatively restricted in their field of view to less than 20 cm height depending on the emitter and detector size, and scan protocols.

3. Inconsistency of grey scale values across the volume

A major limitation is the inconsistency of grey scale values across the data volume. This is a particular problem at the extremities of the field of view, where noise increases and contrast decreases. An example reconstructed axial image slice is shown in Figure 2.14. Contrast between soft tissues is also relatively poor, making CBCT less useful for the evaluation of some diseases and pathologies. Both of these issues can make subsequent data segmentation in postprocessing software using automatic algorithms challenging.

2.3.4 Applications

Despite the limitations of CBCT scanning, it has found widespread application in implant dentistry, orthodontics, temporomandibular joint dysfunction and increasingly,

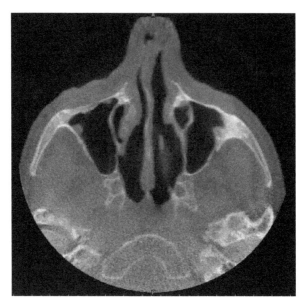

Figure 2.14 Axial slice from cone beam showing varied grey scale values.

areas of maxillofacial and craniofacial surgery, where the limited field of view and inconsistent grey scale values are less of a problem. Like conventional CT, very dense tissues (such as cortical bone or enamel) contrasts well against air, making it suitable for acquiring volumes around the anterior midface. This makes CBCT a valuable tool in the process of planning intraoral implant placement, which is now one of the most common applications. From a practical perspective, the small machine footprint and relatively low purchase and operating costs also make it feasible for even small laboratories and clinics to operate scanners. CBCT scan data can also be complemented well with colour surface 3D topography scans, which can be visualised in addition to the underlying bony data (Jayaratne, McGrath, & Zwahlen, 2012). This development is in response to the desire to better visualise and predict soft-tissue movements in relation to surgical procedures, where there has been significant research in recent years.

There have also been attempts at using CBCT mounted on a C-arm arrangement for use in operating theatres (Erovic et al., 2014). This has led to applications in interventional radiology and image-guided surgery, where it can be necessary to rapidly evaluate spatial locations of anatomical structures or surgical instruments, without having to transport the patient.

2.4   Magnetic resonance (MR)

2.4.1   Background

MRI exploits the phenomenon that all atoms have a magnetic field that can be affected by radio waves. Atoms have a natural alignment and MR works by using powerful

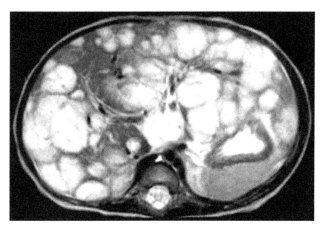

Figure 2.15 A typical magnetic resonance imaging through the abdomen.

radio waves to alter this alignment temporarily. When the radio waves are turned off, the atoms return to their natural alignment and release the energy they absorbed as radio waves. To construct an MRI, the strength of the radio waves emitted by the atoms is measured at precise locations. By collecting signals from many locations, a cross-sectional image can be created. As in CT scanning the resulting cross-sectional image is a grey scale pixel image, the shade of grey being proportional the strength of the signal.

As the human body is composed mostly of water, MRI scanning targets the hydrogen nuclei present in water molecules. Therefore, locations that have a high water content show up in lighter shades of grey and areas containing little or no water show up darker. For example, air shows up black as will the densest bone, whilst tissues highest in water content, such as fat, will show up white.

As the water content of different soft tissues differs, MRI is an excellent modality for investigating the anatomy of soft tissue organs, as can be seen in a typical MRI of the abdomen shown in Figure 2.15. However, unlike CT, MRI is not good for visualising bone. The boundary between air and soft tissue is also good, allowing models to be made of the skin surfaces of patients. Unlike CT, by altering the direction of the radio waves used, MRI can be acquired in cross-sectional slices at any angle. MRI also differs from CT scanning in that there are more parameters that can be altered to improve the results for specific tissues. It is therefore important that the radiographer knows precisely which tissue type is being targeted before conducting the scans.

Due to the strong magnetic fields encountered during MRI, the presence of metal may cause problems. Therefore, jewellery and watches must be removed and patients' notes must be checked to ensure that they do not have attached or implanted devices that may be adversely affected. As with CT scanning, movement will lead to distorted images and babies, small children and claustrophobic patients may require sedation or anaesthesia. MRI scanners also generate an enormous amount of noise, which even with ear protection is not pleasant for the patient.

Although the MRI does not use ionising radiation, it may present a risk for certain patients. They are time-consuming and expensive so every care must be taken

to ensure that the scan is conducted correctly the first and only time. It is also important to consider the dangers that any magnetic metal implants may have before conducting MRI scans.

2.4.2   Anatomical coverage

As with any radiographic procedure, basic mistakes can be made through poor communication between the clinician and radiographer. Detail can be lost when the scans
do not cover the whole anatomy of interest or do not include sufficient margins surrounding the anatomy of interest. Detail may also be lost by using a field of view that is
too small. When conducting 3D series scans, the data should be continuous. Noncontinuous sets of data may not be satisfactorily combined in software later as the patient
has usually shifted position slightly and the separate series may not align perfectly.

2.4.3   Missing data

Even with a very small scan distance, some detail may be lost where thin sections
of tissue exist between the scan planes. In addition, parts that are very small, or connected only by thin sections may not survive subsequent data processing or physical
manufacture. Parts that are not connected will not be present unless they are artificially
attached to surrounding anatomy.

Due to the time taken to acquire each image, flowing fluids will have moved
between the excitation and emission stages of the scan. With multiple images being
taken, this may result in the signal being reduced or reinforced. Therefore, blood vessels, for example, may appear too dark or too bright.

2.4.4   Scan distance

This is the distance between the scans taken to form the 3D scan series (unlike CT
data capture is not limited to the axial plane). This may also be referred to as 'pitch' or
'distance between cuts'. To maximise the data available to produce a smooth model,
this distance should be minimised. Typically, distances of 1–1.5 mm produce good
results. A scan distance greater than 2 mm will give increasingly poor results as the
scan distance increases. However, taking thinner slices results in less signal strength
per pixel being detected by the scanner. Therefore, more echoes are required to boost
the signal strength that results in significantly longer scan times.

Unlike most CT scanners, the number of pixels used in a cross-section is a variable
parameter. Typically, the cross-section will be broken down in to a relatively small
number of larger pixels compared to CT. For example, a typical CT image may be
512×512 pixels at a pixel size of 0.5 mm, whereas an MRI scan may be 256×256
pixels at a pixel size of 1 mm. The main reason for this is to maintain signal strength. A
larger number of smaller pixels result in less signal strength per pixel. Again therefore,
more echoes are required to boost the signal strength, increasing scan times.

For 3D modelling, it may be necessary to alter the compromise between scan time
and signal strength compared with the protocols normally used for diagnostic imaging.

MRI is often a preferred imaging methodology due to its inherent safety compared to CT. However, the application of MRI for 3D modelling should be carefully considered due to the increased scan times. Although MRI is safe, the procedure may be uncomfortable and perhaps distressing for the patient and the added costs and delays incurred by the radiography department should be considered.

2.4.5 Orientation

As with all medical imaging, anterior-posterior and inferior-superior orientation is usually visually obvious but lateral orientation is ambiguous. Usually this is not a problem with automatically imported data but when manually importing data it is important that the correct orientation can be ascertained.

2.4.6 Image quality and protocol

MRI data is typically taken for diagnostic reasons; to investigate areas of specific illness or locate pathology, such as a tumour. Usually, the minimum number of images required to identify the problem are acquired, consequently it is not common practice to undertake 3D MRI scans. Conducting a 3D MRI scans may take significantly longer than a normal session and the compromise between scan time and the necessity of the 3D data has to be considered. As conducting MRI scans is expensive and a critical resource in most hospitals, increasing the scanning time may cause problems and increases the inconvenience for the patient. Close collaboration with the radiographer is recommended to ensure that the data is of sufficient quality without creating problems.

The configuration of the MRI machine may be enhanced by the addition of more coils, which has the effect of increasing the signal strength. Such configurations may be used by specific specialities such as neurosurgery.

Unlike CT images, altering the protocol of an MRI scan can dramatically alter the nature of the image. By varying the sequence and timing of excitation and emission of the radio waves, different effects can be achieved. These may serve to improve the image quality for specific tissues or improve contrast between similar adjacent tissues.

2.4.7 Artefacts

This is the general term for corrupted or poor data in MRI scans. These may result from patient movement or magnetic effects.

1. Movement

A good quality MRI scan depends on the patient remaining perfectly still throughout the acquisition. Movement during the acquisition will lead to distortions in the data (analogous to a blurred photograph). For example, involuntary movement of the chest, neck, head or mouth can occur through breathing or swallowing. When taking multiple scans, it may be possible to synchronise the timing of the sequences used with breathing. Movement can be particularly difficult to control when scanning babies, small children and claustrophobic patients, in which case a sedative or even general anaesthetic may be required.

2. Shadowing by metal implants

Dense metal objects such as amalgam or gold fillings, braces, bridges, screws, plates and implants affect the magnetic field around them leading to the appearance of arte-facts in the scan image. The effect is normally apparent as a lack of data shown as dark patches or shadows surrounding the location of the metal object. The extent of this effect depends on the type of metal present.

3. Noise

Noise occurs in all MRI scans and may reduce image quality but taking multiple acquisitions can reduce this. Noise blurs the boundaries between different adjacent soft tissues. When observing the image, the human eye can account for this (we are very good at recognising shapes we are familiar with) and the boundaries appear visi-ble. However, for successful modelling from the data the boundary has to be clear and distinct to a much higher degree than when the images are only used visually.

Taking multiple acquisitions reinforces the signal improving the quality of the image data; however, this may double or quadruple the time taken to complete a scan session. In practice, a compromise between noise reduction and scan time must be reached with your radiographer.

2.5 Noncontact surface scanning

2.5.1 Background

When attempting to capture human topography, that is the external shape or skin surface, it is frequently more practical and comfortable to use noncontact scanning systems, which typically rely on light-based data acquisition. Noncontact surface scanning uses light-based techniques to calculate the exact position in space of points on the surface of an object. Computer software is then used to create surfaces from these points. These surfaces can then be analysed in their own right or integrated with computer-aided design models.

Unlike CT and MRI, these techniques capture only the exterior topography of the patient. This allows models to be made of the skin surfaces of patients. Although increasingly common, these techniques are not yet considered routine medical imaging modalities and it is therefore possible that a non-medical scanning facility will have to be used to capture the data. Many product development facilities have access to this kind of equipment although the nature of the equipment can vary significantly.

Noncontact surface scanning can be time-consuming and potentially expensive but is completely safe. The noncontact nature of the scanning means there is less discom-fort for the patient and no distortion of soft tissues caused by the pressure applied when taking casts or impressions (sometimes called a moulage). This advantage in combination with the ability to manipulate data makes the approach particularly well suited to applications in prosthetic reconstruction and rehabilitation. It is difficult, for example, to take a satisfactory impression of a breast, therefore noncontact scanning may be used in the creation of symmetrical prostheses for mastectomy patients. However, care should be taken when scanning the body surface to ensure it is in the position that relates to the intended use. For example, body parts that are weight bearing will distort according to the position and posture of the patient.

Data manipulation and high accuracy can be a significant aid in prosthesis manufacture, especially for large or complex cases. These techniques may be a valuable aid to shaping and positioning the prosthesis but the skill and knowledge of the prosthetist will always be required to determine the best method of creating, colour matching and attaching the prosthesis to the patient.

Case studies illustrating some applications of non-contact scanning can be found in Chapter 5, in particular the sections on Maxillofacial rehabilitation, Orthotic rehabilitation applications, and Dental applications.

2.5.2 Anatomical coverage

Basic mistakes can be made through poor communication between the clinician and technician. Detail can be lost when the scans do not cover the whole region of interest or do not include sufficient margins around the anatomy of interest. Most scanners have a limited field of view and several scans may be required.

In addition, these techniques rely on 'line of sight'. This means that areas that are obscured or at too great an angle to the line of sight will not appear in the scan data. Therefore, several scans may have to be taken from different viewpoints to ensure all of the required details are captured. This can be achieved by moving either the patient or the scanner and repeating the process. Depending on the shape of the object, many scans may be necessary. When scanning faces for example, a single scan will not acquire data where the nose casts a shadow. Figure 2.16 shows the three overlapping

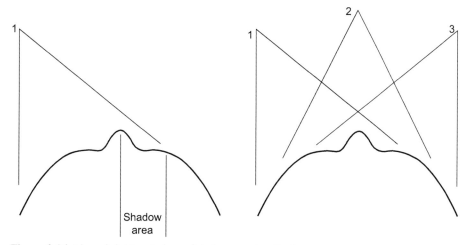

Figure 2.16 Line of sight and viewpoints for scanning the face.

scans: 1 from the left, 2 from straight ahead and 3 from the right. However, this is overcome by taking several overlapping scans.

The data from each of these scans can then be aligned using software to give a single coherent data set. When conducting a series of scans it is therefore important that they overlap so that the individual scans can be put together accurately.

Despite using multiple scans, some areas may remain difficult to capture. For example, it is very difficult to capture data from behind the ear, at the nostrils and between the digits of the hands and feet.

2.5.3 Missing data

Because these techniques use light to calculate the points, transparent, dark or highly reflective surfaces can cause problems. Usually human skin performs very well in this respect but steps may be required to dry particularly greasy or moist skin. These effects may also be overcome somewhat by applying a fine powder, such as talcum powder, that will give the object an opaque matt finish. The eyes may cause problems due to their shiny surface and watering. All optical scanners should be inherently safe, however, care should be taken when scanning the eyes to ensure that bright light or laser light does not directly enter the pupil.

Another inherent problem is caused by hair. Long, dense hair does not form a coherent surface and absorbs or randomly scatters the light from the scanner. However, the fine hair present on most body surfaces does not normally affect the captured data, although excessively thick body hair may reduce the quality of the data captured.

In most cases, the presence of hair will lead to gaps in the data. Fine or downy body hair often does not cause significant problems but there may be little that can be done about a full head of hair and it would not be normal to consider shaving the head for such a procedure unless the clinical benefits were overwhelming. In situations where the entire head must be scanned, a latex cover such as a swimming cap may be used to conceal hair, but one must bear in mind the likely effects of altering the topography as a result of constrained hair, particularly individuals with long, thick hair encased on the cap. Another consequence is the induced ripples which typically arise along the sagittal plane, but such effects may be corrected in editing software such as Geomagic Studio (Geomagic, 3D Systems, USA). When considering scanning the face, gaps are likely to arise from areas of significant facial hair, such as beards and moustaches but may also be encountered around the eyebrows and lashes. These gaps can be clearly seen as black areas in the surface scan data as shown in Figure 2.17.

2.5.4 Movement

As with other scanning modalities, any patient movement during the scan will lead to the capture of poor data. The length of time required varies depending on the exact type and specification of the scanner being used but may range from a fraction of a second to as much as 10 min. Consequently, it is important that the subject be scanned in a comfortable and steady posture. A consequence of movement may be noise, false representations of anatomic topography and subsequent difficulty if aligning multiple

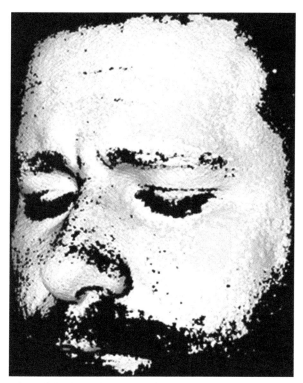

Figure 2.17 Scan data of the face of a male subject.

scans. This does not normally pose a problem when dealing with cooperative adults and older children but small children, babies, the elderly or patients with conditions such as Parkinson's disease may be difficult to keep still during the scanning. Unlike MRI or CT scanning, it is highly unlikely that sedation would be justified for this type of scan. Capturing a specific posture may benefit from rigging or supporting structures, but requires careful consideration to avoid soft-tissue deformation, as shown in Figure 2.18 which shows flattening of the soft tissues on the hand and forearm.

With most scanning software, if there is sufficient overlap between adjacent scans they can be aligned by suitable computer software. This means the patient does not have to remain perfectly still between successive scans. However, the same body posture or facial expression should be maintained throughout. For the best results, the patient may have to be braced in a comfortable position with suitable rigging apparatus during each scan. Depending on the scanner being used, it may be simpler to keep the subject still and move the scanner around them. Alternatively, a series of multiple scanners can be positioned around the subject to capture different views simultaneously or in rapid succession.

However, despite these steps some movement is likely to be encountered such as breathing, swallowing, blinking or involuntary tremors, so care should be taken when scanning the face, neck, chest and extremities. If the patient is able, a breath-hold may help if the scan time is only a matter of seconds.

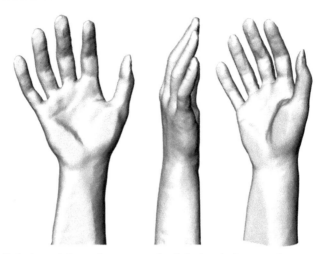

Figure 2.18 Soft-tissue deformation as a result of rigging during scanning.

2.5.5 Noise

Noncontact scanners capture many thousands of points at a time. The vast majority of these points will fall accurately on the surface of the object being scanned. However, due to tolerances and optical affects some of these points will deviate from the object surface. If enough points deviate from the surface by a sufficient amount, it will affect the quality of the data. These errant points are usually referred to as 'noise' in the data. The amount of noise present in captured data will depend on the type of scanner and the optical properties on the surface being scanned. Smooth, matt surfaces usually produce less noise than reflective or textured surfaces. Although it is usually necessary to take multiple overlapping scans to cover the whole surface of an object, large overlapping areas are likely to result in increased levels of noise.

Noise can be reduced by data processing usually called 'filtering'. Filtering selectively removes data points that deviate greatly from the vast majority of neighbouring points. This is illustrated schematically in steps 1–5 in Figure 2.19: (1) the object surface to be scanned, (2) the scan data points, (3) using all of the data points creates a poor surface, (4) points are selectively filtered according to their deviation from the majority of neighbouring points, and (5) deleting filtered points leaves a closer fitting surface the actual object surface. The magnitude of the deviation can be defined by the user to vary the effect of the filtering. Filtering functions are typically included in the software that is used to operate the scanner.

The effect of noise can be seen in Figure 2.20. The image on the left shows a 3D polygon model created from optical scan data of a dental cast (in this example the stereolithography (STL) file format is used). Noise in the original scan data has resulted in a polygon surface that appears rough and pitted. Postprocessing software can be used to improve the quality of the surface model. These functions work by averaging out the angular differences across neighbouring polygons within

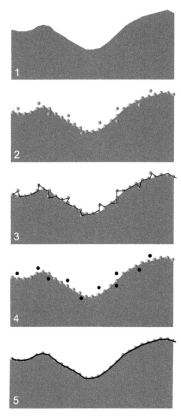

Figure 2.19 The effect of filtering to remove noise from optical scan data.

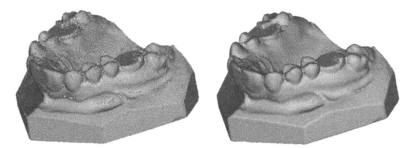

Figure 2.20 Scan data showing noise (left) and the same data with noise reduced (right).

user-defined tolerances. The effect is to smooth out the surface, but without losing critical characteristics of the artefact's topography. When dealing with human anatomy the smooth, curved surfaces typically encountered mean that this approach often leads to a more accurate model despite the additional data operations. The image on the right of Figure 2.20 shows the same 3D model that has been 'smoothed' to mitigate the effects of noise. Point cloud data is in itself of little use and it is normally used as the basis for a 3D surface or solid model. This is described in more detail in Chapter 3.

2.5.6   Low-cost and open-source methods for surface capture

In the past, organisations paid a premium for adequate resolution optical surface capture hardware and software. Now in the wake of the open sourcing era for hardware and software algorithms, the use of low cost hardware such as the Microsoft Xbox Kinect™ are being increasingly used for digitising artefacts and environments, providing a suitable supporting program/algorithm is in place. The low cost of hardware makes the approach more accessible to individuals or organisations on a restricted budget, and depending on the application, the resolution may be suitable for end use of the intended application or part if used to support additive manufacture. However, one must bear in mind that problems with optical data capture still exist, such as line-of-sight.

2.6   Medical scan data

Medical scanners such as CT and MRI produce pixel-based images in a series of slices whilst noncontact surface scanners produce 3D point clouds. Therefore, the formats used to describe them are completely different and separate software technologies are required to use them.

   CT and MRI scanners are highly complex pieces of equipment made in low volumes by a small number of high technology firms. The hardware and software interface is usually part of the whole system. Therefore, radiographers are often limited in the output formats that they can deliver. In many cases, the only option available is the archiving system. In some cases, these output formats are also compressed and cannot be read by third party software.

   However, this should be less of a problem over the coming years. Many radiology departments are embracing teleradiology and therefore can make the image data available over hospital networks. Good third party radiography software should be able to import most of the commonly used proprietary formats. The industry standard is called DICOM and it should be used whenever possible, as almost all software should support it. An example of the practical implications of transferring data from a hospital to a service provider can be found in case study in Section 5.2.

2.6.1   Digital imaging and communications in medicine

DICOM is an internationally agreed standard for all medical imaging modalities. The standard was initiated in response to the development of computer-aided imaging in the 1970s by a joint committee from the American College of Radiology (ACR) and the National Electrical Manufacturers Association (NEMA). They first published an ACR-NEMA standard in 1985 and updated it in 1988. Version 3 saw the name changed to DICOM and was published in 1993. The standard now covers all kinds of medical images but also includes other data such as patient name, reference number, study number, dates and reports. Most manufacturers adhere to the standard and data transfer problems are much less likely to occur than was previously the case.

The DICOM standard (ISO 12052) enables the transfer of medical images to and from software and scanners from different manufacturers and aided the development of picture archiving and communication systems, which can be incorporated with larger medical information or records systems.

More information on DICOM can be found at http://medical.nema.org/.

2.6.2 Automatic import

When using medical data manipulations software, such as Mimics for example, there is usually an automatic import facility. The facility will automatically recognise many manufacturers' formats and will almost certainly recognise data written in the international image standard DICOM. The software will automatically read in the data and convert it into its own format ready for manipulation.

However, depending on the specification of the automatic import software being used, there may be some instances where the user will require some knowledge of the scan parameters in order to complete the import. Usually these factors are not fully described in the DICOM standard. Gantry tilt is one such example. Although DICOM supports the magnitude of gantry tilt, it does not provide the direction. The software will automatically apply a correction but the user will be required to either confirm or reverse the direction to ensure that it the data is correct. The second example is the anatomical orientation of the data set. DICOM will provide left, right, and anterior-posterior orientation but may not provide inferior-superior information. The user may be required to select this orientation during import. The inferior-superior orientation is usually anatomically obvious so this rarely creates a problem.

2.6.3 Compression

Image data can be very large and compression is sometimes used, especially for archiving. Data compression may incur a loss of information, called 'lossy' compression or retain all data but write it in a more efficient manner, called 'lossless' compression. For modelling purposes, compression should be lossless or preferably avoided completely. Any loss of information may reduce the accuracy of models made from the data set. In addition, some compressed formats are unreadable by third-party software, rendering the data useless. Compression for medical modelling often is not necessary. In normal circumstances, although the image data may be several hundred megabytes, it can usually be accommodated on a single CD-ROM.

2.6.4 Manual import

When the data format is not in a DICOM-compatible format or from a manufacturer that is not directly supported by the software application being used, manual import must be used. Error-free manual import of medical scan data this will require access

Table 2.1 **Information required for manual data import**

Number of images	Integer
File header size	In bytes
Inter image file header size	In bytes
Image size	In pixels e.g., 512×512
Slice distance	In millimetres
Field of view or pixel size	In millimetres
Gantry tilt	In degrees
Scan orientation	Left-right or right-left

to *all* of the parameters of the data. Table 2.1 lists the parameters that must be known to import data successfully.

2.7 Point cloud data

The data captured by a noncontact surface scanner is merely a 'point cloud'. This is a collection of thousands of point coordinates in 3D space, and may range from hundreds to many thousands of points. It is typical to convert this into a more useful format before applying the data to analysis, manipulation or visualisation. There are a number of formats used for point cloud data and the format will depend on the type of scanner being used.

The simplest forms of export format are polygon meshes, such as the commonly used STL file format (the STL file format is fully described in Chapter 3). Many modern scanners will output data directly into the STL file format rather than point clouds and typically a number of export options will be available. However, the export data format used will depend very much on the anticipated use of the data; this area is explored in greater depth in the next chapter. Some techniques can be applied directly to the point cloud data. These are typically done to remove erroneous points, decrease noise or simply reduce the number of them. This has the effect of cleaning up the data set and reducing the file size.

2.8 Media

In combination with the data format, the output media type is also a part of the whole system. Advances in computer networks and telecommunications have enabled the phasing out of storage media such as magnetic optical disks previously used for archiving large data sets. These were expensive and required specialist hardware and software to translate data. In addition, there has been widespread adoption of Windows operating systems, PC hardware, CDs and DVD as data storage media. With teleradiology increasing, many hospitals can now put medical image data onto the hospital local area network.

The Picture Archiving and Communication System, is commonly used to store and manage medical image data within hospitals. This enables clinicians to remotely access medical images data in their own departments rather than relying on a limited number of films printed in the radiography department. Once the data is available over the hospital network it can be burned on convenient and cheap storage media such as CDs or transferred over secure, Internet-based systems to external parties. Given that DICOM data typically contains sensitive patient information, there are tight regulations on the security and transfer of it using Internet-based or postal systems. For example, information on patients within the European Union should not be stored outside the European Union and there should be compliance with standards such as ISO/IEC 27001:2005, which specifies a management system that is intended to bring information security under explicit management control. Data that is archived to a CD/DVD can be password protected if being sent by postal systems, but this is becoming less desirable in favour of Internet-based transfer.

When dealing with point cloud data, media formats are normally those typically used in the design and engineering community and translation does not pose the problems associated with radiological data; it is far simpler to strip data of sensitive information that could relate it to an individual patient.

References

Erovic, B. M., Chan, H. H., Daly, M. J., Pothier, D. D., Yu, E., Coulson, C., et al. (2014, January). Intraoperative cone-beam computed tomography and multi-slice computed tomography in temporal bone imaging for surgical treatment. *Otolaryngol Head and Neck Surgery, 150*(1), 107–114. http://dx.doi.org/10.1177/0194599813510862.

Jayaratne, Y. S. N., McGrath, C. P. J., & Zwahlen, R. A. (2012). How accurate are the fusion of cone-beam CT and 3-D stereophotographic images? *PLoS One, 7*(11), e49585. www.plosone.org. DOI: 10.1371/journal.pone.0049585.

Recommended reading

The case studies in Chapter 5, and the following works:

Bianchi, S., Anglesio, S., Castellano, S., Rizzi, L., & Ragona, R. (2001). Absorbed doses and risk in implant planning: comparison between spiral CT and cone beam CT. *Dentomaxillofacial Radiology, 30*, S28.

Drage, N. A., & Sivarajasingam, V. (2009). The use of cone beam computed tomography in the management of isolated orbital floor fractures. *British Journal of Oral and Maxillofacial Surgery, 47*, 65–66. http://dx.doi.org/10.1016/j.bjoms.2008.05.005.

Gibbons, A. J., Duncan, C., Nishikawa, H., Hockley, A. D., & Dover, M. S. (2003). Stereolithographic modelling and radiation dosage. *British Journal of Oral and Maxillofacial Surgery, 41*, 416.

Henwood, S. (1999). *Clinical CT: Techniques and practice London, UK:* Greenwich Medical Media Ltd, ISBN: 1900151561.

Hofer, M. (2000). *CT teaching manual*. New York: Thieme-Stratton Corp, ISBN: 0865778973.

Kalander, W. (2000). *Computed tomography*. Weinheim: Wiley-VCH, ISBN: 3895780812.

Li, G. (2013). Patient radiation dose and protection from cone-beam computed tomography. *Imaging Science in Dentistry*, *43*, 63–69. http://dx.doi.org/10.5624/isd.2013.43.2.63.

Loubele, M., Bogaerts, R., Van Dijck, E., Pauwels, R., Vanheusden, S., & Suetens, P., et al. (2009, September). Comparison between effective radiation dose of CBCT and MSCT scanners for dentomaxillofacial applications. *European Journal of Radiology*, *71*(3), 461–468.

Scarfe, W. C., & Farman, A. G. (2008). What is cone-beam CT and how does it work? *The Dental Clinics of North America*, *52*, 707–730. Available at http://endoexperience.com/user-files/file/unnamed/New_PDFs/cbCT/CBCT_how_does_it_work_Scarfe_et_al_2008.pdf.

Swann, S. (1996). Integration of MRI and stereolithography to build medical models: a case study. *Rapid Prototyping Journal*, *2*, 41–46.

Working with medical scan data

3.1 Pixel data operations

As described in earlier chapters, both computed tomography (CT) and magnetic resonance (MR) images are made up of grey scale pixels. In CT, the grey scale is proportional to the X-ray density. In MR, the grey scale will be proportional to the magnetic resonance of the soft tissues. In many cases, it is advisable to work with the original data rather than any three-dimensional (3D) reconstruction derived from it. Therefore, CT and MR image data is often manipulated in the pixel format. Much of this data manipulation is similar in concept to popular photo-editing software such as Adobe Photoshop. Many software packages are available that use such pixel manipulation to allow specific individual anatomical structures to be isolated from a CT or MR data set and exported in an appropriate format. Many of these packages operate in a similar manner to the software that radiographers routinely use to generate images in radiology departments.

3.1.1 Thresholding

Thresholding is the term used for selecting anatomical structures depending on their density, or grey scale value. By specifying upper and lower density thresholds, tissues of a certain density range can be isolated from surrounding tissues. Due to the partial pixel effect described in Chapter 2, small variations in the thresholds may affect the quality of the anatomical structures isolated. The effect may be to make them slightly larger or smaller as illustrated in Figures 3.1–3.3. However, thresholding will select all pixels within the specified density range regardless of their relationship to individual anatomical structures. This may be overcome using region growing.

The effect can be clearly seen in the real example shown in Figure 3.4. In this example, bone is selected by setting a high upper threshold and an appropriate lower threshold, with the resulting region shown in Figure 3.5. However, the effects of varying the lower threshold can be seen in Figures 3.6 and 3.7. Notice that in Figure 3.6, it is clear that bone is present beyond the boundary of the selected region. However, in Figure 3.7, we can see that other areas, unconnected to our region of interest, have been selected. It is therefore essential that thresholds be accurately selected where accuracy is of high importance. This becomes particularly critical when very thin or narrow objects are of interest as small changes in threshold can result in these areas not appearing in the selected region.

Medical Modelling. http://dx.doi.org/10.1016/B978-1-78242-300-3.00003-2

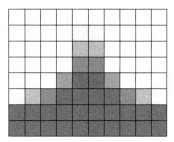

Figure 3.1 Original CT image.

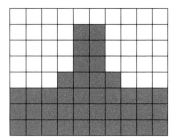

Figure 3.2 Effect of a low threshold.

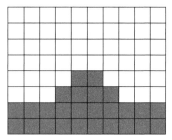

Figure 3.3 The effect of a high threshold.

3.1.2 Region growing

To select single anatomical structures from all of those present within the specified thresholds, a technique called region growing is typically used. This works by allowing the user to select a single pixel within a region already specified by thresholding. The software then automatically selects every pixel within the specified thresholds that is connected to the one selected by the user. This results in single anatomical structures being isolated from neighbouring, but unconnected, structures.

However, this is not as simple as it might first appear. It only requires one single pixel (representing perhaps 0.25 mm) to connect two regions for the software to assume that they are the same structure. Therefore, structures that are separate but in close proximity or contact may need to be separated manually before region growing will be successful. This may occur, for example, in joints.

Figure 3.4 Original CT image.

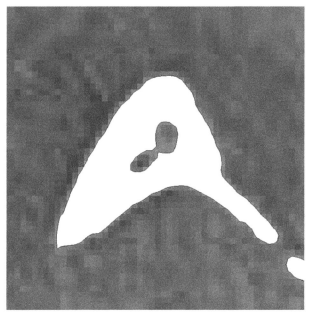

Figure 3.5 Region selected using appropriate threshold values.

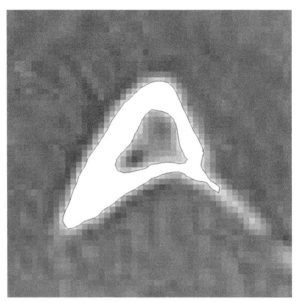

Figure 3.6 Region selected using an increased lower threshold value.

Figure 3.7 Region selected using a decreased lower threshold value.

3.1.3   Other techniques

Many other techniques may be available depending on the software being used. Manual techniques can be used and are similar to those found in photo editing software, such as draw, delete, cavity fill, etc. These allow the user to edit data to remove artefacts or connect neighbouring structures.

Other techniques are sophisticated variations on the thresholding and region growing functions. These may incorporate local variations to thresholds, or add or remove pixels from selected regions to alter the boundaries. The exact nature of these functions will depend on the software being used; therefore, it is not appropriate to attempt to describe them all here.

3.2   Using CT data: a worked example

To illustrate how such software can be used, let us work through a simple example using the popular software package called Mimics (Materialise NV, Technologielaan 15, 3001 Leuven, Belgium).

The first step is to import the data. In most cases, the data will be recorded in a format that is compliant with the internationally recognised digital imaging and communications in medicine format (DICOM). If this is the case, the software has an automatic import function. During importing, the software converts the images into a format it recognises as its own and displays the resulting axial images on the screen. The axial images are the original scan images from the CT data. Mimics software also uses the axial pixel data and the slice distance to calculate images in the sagittal and coronal planes.

As described earlier, in CT images the grey scale is proportional to the density of the tissue. Therefore, the denser the tissue is, the lighter the shade of grey that corresponds to it will be (see Figure 3.8). Mimics, and software like it, use these grey scale values to differentiate between different tissue types. By selecting upper and lower grey scale values, specific tissue types can be selected. These levels are typically referred to as thresholds. On importing a new data set, Mimics displays the images using a default threshold for bone, which is shown in green in Figure 3.9. Selecting the desired tissue type is accomplished by varying the upper and lower thresholds until the required tissue type is isolated. This process is usually referred to as segmentation.

Once the desired tissue type, in this case bone, has been segmented, it may be necessary to limit the selected data to one particular structure. Region growing allows the user to select a certain pixel within the desired structure and the software then automatically selects all other pixels that are connected to it. As this function operates in all three dimensions, single structures can quickly be segmented from the whole data set. This can be seen in the yellow structure highlighted in Figure 3.10.

The software can also be used to view 3D-shaded images of the selected data. The segmented data is then exported in the format used to create the computer files necessary to build the medical model (see Figure 3.11).

Scans are usually performed in slices or a spiral form in a plane perpendicular to the long axis of the patient. The interval between the slices may be in the order of a millimetre

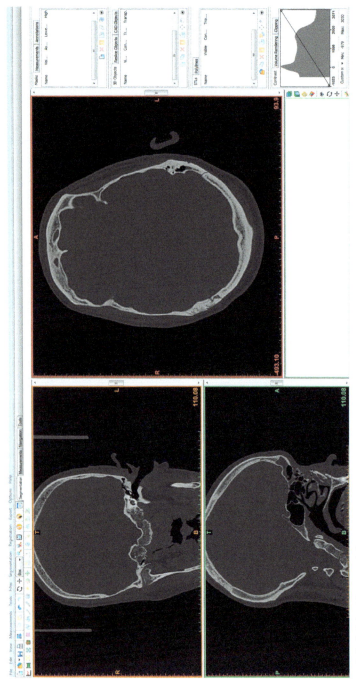

Figure 3.8 Original imported data.

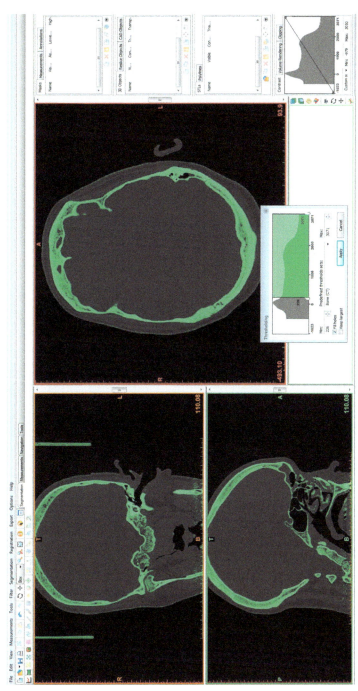

Figure 3.9 Default threshold applied.

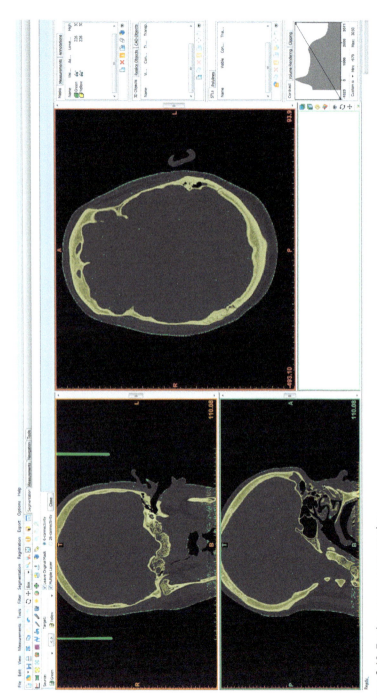

Figure 3.10 Region grown segmentation.

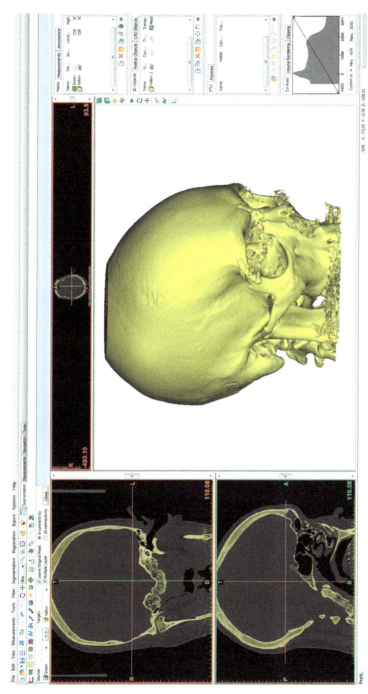

Figure 3.11 A 3D-shaded image of the segmented data.

or two, whereas the models are built in layers between 0.1 and 0.2 mm in thickness. Therefore, additional software is used to interpolate intermediate slices between the scan data slices. Another interpolation is carried out within the plane of the scan to improve resolution. Together, these operations result in the natural and accurate appearance of the model. This interpolated data can then be exported in a number of formats that may be transferred to computer-aided design (CAD) packages or rapid prototyping (RP) preparation software.

Some of the different output data formats are described later in this chapter.

3.3   Point cloud data operations

Noncontact scanning typically produces a large number of points that correspond to 3D coordinate points on the surface of the target object. The collection of points taken of an object is usually referred to as a 'point cloud'. The nature of point cloud data is completely different to the pixel image data we obtain from CT and MR scanning and therefore requires quite different software.

Point clouds in themselves are of little use. They are therefore usually converted into 3D surfaces. The simplest method is polygonisation. This involves taking points and using them as vertices to construct polygon facets. The collection of facets is usually known as a polygon mesh. The simplest form of polygon is the triangle that is frequently used. The steps from point cloud data to polygon surface are shown in Figure 3.12. Triangular faceted meshes can be easily stored in the stereolithography (STL) file format. Applying other shapes of polygon mesh, such as square or hexagonal may be more appropriate, particularly for use in finite element analysis (FEA) techniques.

Often scan data of an artefact's topography will extend beyond the principal area of interest and unnecessary points should be removed immediately. This reduces the memory requirement for any subsequent operations. Once a mesh has been generated

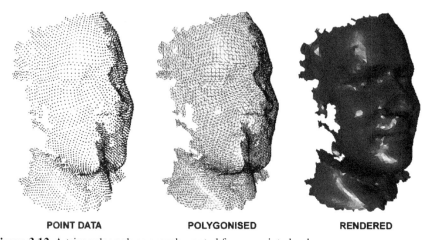

POINT DATA   POLYGONISED   RENDERED

Figure 3.12 A triangular polygon mesh created from a point cloud.

from the available points, any gaps will become apparent. At this point, holes and gaps can be filled or corrected to form a single coherent or 'manifold' mesh.

As an alternative or extension to mesh modelling, the user may wish to use the scan data to act as a template to construct a series of mathematically defined curves and surfaces. This is typically done in reverse engineering, enabling the artefact's topography to be taken into traditional engineering CAD systems. This requires a great deal of skill and expensive software and is only necessary when it is desirable to perform precise changes to the surfaces. If the ultimate aim is to produce a physical model by RP techniques then this step often is redundant, as the data will only have to be converted into an appropriate format (e.g., STL) again.

Whichever route is taken, if a physical model is to be made, the surface has to be turned into a single bound volume. This can be achieved by offsetting the surface and filling the gap, resulting in a model with a definite thickness. This can be seen in Figure 3.13, which shows two views of a 3D model created from scan data of a face. The oblique view clearly shows the thickness of the model. To reduce file size, the rear surface may be simplified as it is of no consequence except to give the model bulk.

3.3.1   Data clean-up

Clean-up tools for scan data have improved significantly over recent years, with new algorithms for more complex functions. Often the simplest way to filter and clean scan data is to first ensure the scan has been through the polygonisation process, i.e., advanced from a point cloud into a mesh, as the visualisation of the scan is easier to comprehend to the user in virtual space. There are several tools that the user may decide to use in addition to filtering to enhance the mesh for the intended application. Furthermore, smoothing of meshes may also be necessary through filtering or even relaxation of the overall topography.

Decimation may be required to reduce the number of facets of an overall mesh or a previously selected portion of a mesh. Subsequently, the size of the file reduces due

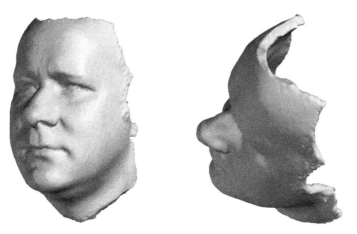

Figure 3.13 A bound volume stereolithography file created from an offset surface.

to fewer entities (points, edges and facets), and future data processing may be accelerated as a result. However, one must consider carefully an acceptable level of accuracy required to balance the file size. Figure 3.14 shows the effects of decimation, by a 50% and 75% reduction in triangular facets, respectively.

Conversely, the user may wish to increase the number of facets around topographies with high curvature and surface detail to increase surface quality. This is achieved by subdividing the existing mesh by a given factor (e.g., 4). The tessellation of meshes may also be adjusted to suit preference; with relatively consistent curvature, a mesh may be retriangulated to make facet size more uniform (relative to edge length of triangles as a guideline), or by varying the spacing between points. Figure 3.15 shows

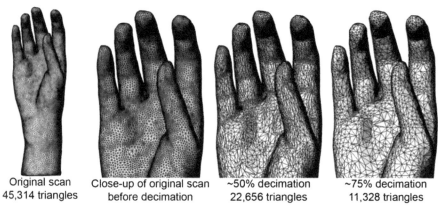

| Original scan 45,314 triangles | Close-up of original scan before decimation | ~50% decimation 22,656 triangles | ~75% decimation 11,328 triangles |

Figure 3.14 Effects of decimation on geometry and facet count.

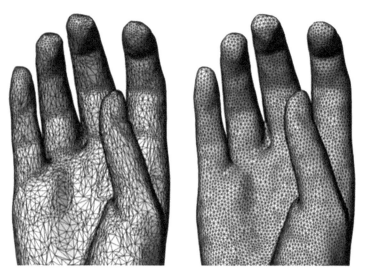

Figure 3.15 Effect of remeshing, left: decimated scan, right: remeshed scan with a maximum 2 mm edge length.

effects of retriangulation or 'remeshing', with the goal of creating equilateral triangular facets.

Depending on the data capture method used or the data manipulation performed, a mesh may present several holes; this is particularly prevalent for laser scanning artefacts which have overlaps and areas which may not be captured due to line-of-sight limitations. Therefore, several tools are available to fill holes within suitable clean-up software. The user may specify varying levels of continuity when filling holes, from flat fill, to tangency and curvature. Parameters may also be set to automatically fill holes below a specified diameter, to increase efficiency and to allow the user to filter out larger more problematic holes, which may require more attention to fill.

3.3.2   CAD data generation

There are several options to generating alternate CAD data from point clouds, but in most cases, require the initial conversion to mesh data to ensure that the desired topography is achieved. Visualisation of a point cloud can be misinterpreted unless a highly dense point cloud can be achieved. Misinterpretation is often a result of misunderstood point placement and subsequently may create an undesirable output once converted to mathematical surfaces known as nonuniform rational B-spline (NURBS) surfaces. Characteristics of NURBS surfaces can be found later in this chapter. Most data editing software packages now feature auto-surfacing tools, which create a series of four-sided NURBS patches over the topography of the artefact. Figure 3.16 shows the automated creation of NURBS surface patches from a polygon mesh; the close-up shows each patch is four-sided. Various parameters are available to the user, such as patch count and number of control points. The benefit of this approach is that NURBS patches present a smoother appearance due to their mathematical construction and therefore may offer a more realistic appearance compared to mesh modelling, which may offer an approximate resolution to the original artefact.

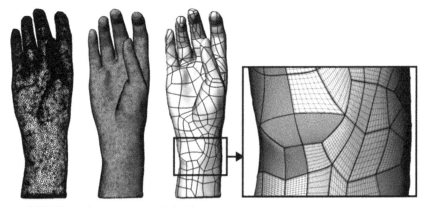

Figure 3.16 Cloud data to auto-surfacing image.

3.4   Two-dimensional formats

Typically in radiography, the output of medical scanning modalities are two-dimensional (2D) images. These are usually prepared from the scan data by the radiographer according to instructions from doctors and surgeons. These images may be from the slices taken through the body or 3D reconstructions. Often these images are printed on film and treated in much the same way as X-ray films. As medical scan data images are made up of pixels, the images can be exported in familiar computer graphics formats such as bit maps or JPEGs.

3.5   Pseudo 3D formats

Data can be exported in formats that allow 3D operations to be undertaken without being true 3D forms. The objects are defined by a series of 2D contours arranged in increments in the third dimension. These types of file are often referred to as two-and-a-half dimensional data or 'slice' formats. More typically, however, these formats are used as an intermediate step in creating true 3D CAD representations.

The formats typically are in the form of lines delineating the inner and out boundaries of structures isolated by thresholding and region growing techniques. The lines are usually smooth curves, or polylines that are derived from the pixel data. This technique results in smooth contours that more closely approximate the original anatomical shape than the pixelated data. For example, if we consider the original CT data that is shown in Figure 3.17, we can see there is a high-density bone structure surrounded by lower density soft tissue. The effect of specifying upper and lower threshold and

Figure 3.17 Original pixelated CT image.

region growing is shown in Figure 3.18. The inner and outer boundaries of the selected region are smooth polylines.

The pseudo-3D effect arises when the 2D polylines are stacked in correct orientation and spacing to provide a layered model, similar in effect to a contour map. Figure 3.19 shows a 3D rendering derived from CT data of the proximal tibia alongside the

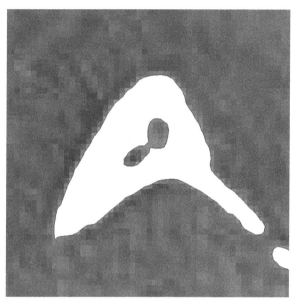

Figure 3.18 Segmented region showing smooth polyline boundaries.

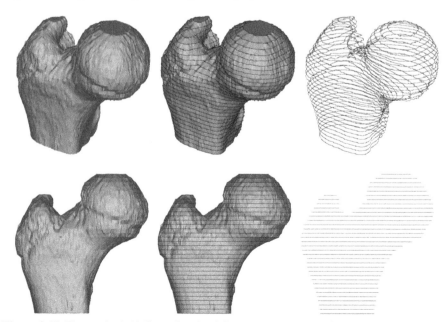

Figure 3.19 Two-and-a-half dimensional polyline data compared to 3D rendered model.

same data exported in a two-and-a-half dimensional polyline or 'slice' format. When such formats are used in CAD, it is common to create surfaces between the slices to generate true 3D surfaces. However, when these formats are used to interface directly with RP machines, it is common to interpolate intermediate layers between the original slices so that data exists at layer intervals that correspond with the build layer thickness of the machine being used.

The formats that follow are essentially the same and appear similar to that shown in Figure 3.19. The differences between them are concerned with the order and amount of information stored in them.

3.5.1   Initial graphics exchange specification contours

Initial graphics exchange specification (IGES) is an international standard CAD data exchange format that has been used for many years. More information about the standard can be found at http://www.nist.gov/iges/. The standard consists of lists of geometric entities, which can be assigned with size and position properties in three dimensions. When dealing with layer data, 2D contours can be described as simple polylines at increments in the third dimension (Figure 3.19). These contours can be imported into CAD packages that can then form true 3D surfaces on them. IGES files have the three-letter suffix, IGS.

3.5.2   Slice file formats

Derived from the word 'slice' the SLC is a relatively simple file format that describes the perimeter of individual slices through 3D forms. The inner and outer boundaries are expressed as a set of vectors. The format therefore is an approximation but in practice, the vectors can be generated with sufficient resolution, such that curved surfaces on RP models derived from them appear smooth.

The format was developed by 3D Systems as an alternative input file for their STL technology but some other RP machines are able to accept the format as input. If an SLC file can be generated from the original source data, it effectively bypasses the need to create STL files. For a given data set, an SLC file will be much smaller than an STL file of comparable resolution. The SLC therefore provides a highly efficient data transfer format for medical modelling using STL. The format is little used compared to the typical STL file.

Other varieties of slice format are available depending on the software being used but are increasingly obsolete. The common layer interface is a contour file format that describes cross-sectional slices through an object in much the same way as the SLC file. However, the format is commonly available to many software developers. Also derived from the word slice, the SLI file format is similar to the SLC file format, but rather than being an input file, it is an intermediate file format created during the preparation of stereolithography builds. The file, developed by 3D Systems, not only describes the perimeter of the slices but also includes raster lines that make up the cross-sectional area within the boundary. These raster scan lines are usually referred to as hatches. The file format also includes different types of hatch for up-facing and down-facing layers. This is important in stereolithography because up-facing and down-facing layers are built differently to optimise accuracy. As with the SLC file format, the ability to generate this file format directly from medical scan data provides an efficient data transfer when generating STL files. The extra hatch information contained in the file increases its size in comparison to the SLC.

3.6   True 3D formats

3.6.1   Polygon faceted surfaces

Unlike the contour or slice-based formats described previously, true 3D data formats generate computer models that have a surface. If a computer model has a surface, then it can be rendered and visualised on a screen and be manipulated with a higher degree of sophistication than two-and-a-half dimensional data. However, file sizes are typically higher.

One of the simplest methods of generating a 3D computer model is to create a polygon faceted mesh. This is achieved by approximating the original data as a large number of tessellating polygon facets. As the triangle is the simplest polygon, it is frequently exploited in polygon faceted representations. However, other polygons are also used, particularly in FEA.

3.6.2   Finite element meshes

FEA packages rely on breaking down 3D objects into a large number of small discrete elements. These elements may make up only the surface of the object; for example, the STL file described previously may be considered a triangular surface mesh. However, some meshes may break the whole object into discrete 3D elements. These may be tetrahedral, cuboids or other 3D tessellating polygons.

Depending on the software package used, the voxel data of a CT or MR scan may be exported or translated into a solid or surface mesh suitable for use in FEA. There are usually variables to set when conducting the translation that affect the quality of the mesh and the resulting file size. This is similar to specifying the quality versus file size compromise for STL files described previously.

Many FEA packages have their own formats for meshing and some medical software packages can produce and export the correct format for a given analysis package. However, this may require the purchase of specific translators or additional modules for the software package.

3.6.3   Mesh optimisation

A collection of manifold facets is referred to as a mesh. There are several methods to optimise a mesh in order to balance surface representation (accuracy and resolution) with file size. These optimisation methods have been described in more detail later in this chapter.

3.6.4 *Mathematical curve-based surfaces*

Unlike faceted polygon surfaces, curve-based surfaces use complex mathematical routines to produce smooth curves in 3D space, such as NURBS curves (Figure 3.20). NURBS curve construction involves control points, knots and knot vectors.

NURBS curves with more than two control points have different geometric continuity (G) states defined by a degree; G0, G1 and G2 states can be seen in Figure 3.21. Assuming the user creates a curve with seven control points, placed in the same position (Figure 3.21(a)); a G0 curve has no degree of continuity and thus appears as six distinct sections (Figure 3.21(b)). G1 displays tangent continuity, where sections are perpendicular but may still display sharp changes in trajectory; this can be seen in the curvature map on Figure 3.21(c), where map direction can have a sudden stepped change. G2, however, has perpendicular continuity but also ensures a smoother trajectory across sections, and this can be seen with the curvature map in Figure 3.21(d) which ensures a smooth transition across sections without stepping.

Some of these routines also operate in three dimensions to create complex curved surfaces called 'patches'. NURBS patches are typically formed from four joined curves with G0 continuity, and principles of curve continuity may be extended to joining surfaces for boundary representation (B-Rep) modelling, for example. An object may require a number of patches to cover the whole object surface. The patches differ according to the complexity of the mathematical curve routine that is used.

This kind of surface modelling produces highly sophisticated surface models that are typically used in the automotive, motor sports and aerospace industries. Usually, objects are designed using surface modelling packages. However, surface patches are also often used in reverse engineering to create useful CAD geometry from digitised physical objects.

The modelling behaviour of NURBS surface modelling is very different to that of polygon modelling. As described previously, NURBS surfaces are typically constructed from four boundary curves. B-Rep models are constructed of a series of trimmed NURBS surfaces, but each surface still retains its underlying data, as shown in Figure 3.22.

Curves and surfaces are mathematically described and as such are typically relatively smooth and simple surfaces. As such, they are typically used to define the smooth but accurate surfaces of objects in product, aerospace and automotive design. This makes them less well-suited to the highly complex surfaces of human anatomy.

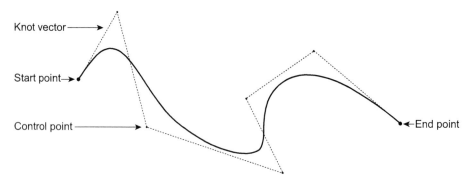

Figure 3.20 Nonuniform rational B-spline curve features.

However, there are many cases where it may prove to be a useful approach, particularly when attempting to integrate human anatomy with the design of products that must accommodate or fit around people.

The curves and surfaces are created by positioning the control points and boundary lines that define the surface patch onto the surface of the source data. The surface patch itself is then mathematically created according to the type of surface the software uses. The most complex type of surface patch is defined by NURBS surfaces.

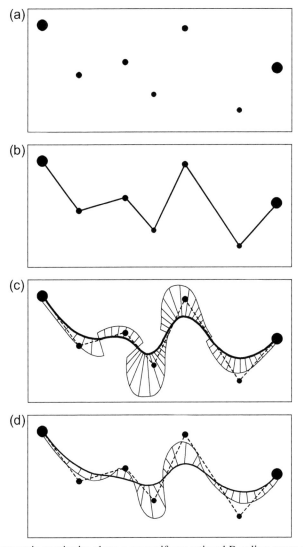

Figure 3.21 Geometric continuity along a nonuniform rational B-spline curve with seven control points. (a) Control point placement. (b) Geometric continuity degree 0 – positional/coincidence (G0). (c) Geometric continuity degree 1 – tangent continuity (G1). (d) Geometric continuity degree 2 – curvature continuity (G2).

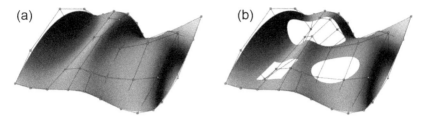

Figure 3.22 Nonuniform rational B-spline geometry. (a) Untrimmed surface with construction features (control points and knot vectors). (b) Trimmed surface showing maintained underlying construction features.

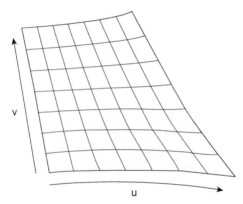

Figure 3.23 Nonuniform rational B-spline surface with isoparms defining the surface topography.

The structure of mathematically defined surfaces can be communicated with isometric parameters or 'isoparms'; integrated curves which assist in defining the topography of the surface. The isoparms typically travel in two opposite directions, termed 'u' and 'v' (Figure 3.23). Whilst the surface patch may have a given number of isoparms in u and v directions, the user can also adjust the visual count to help with communicating the form in wireframe; in this instance, the user views 'isocurves'.

If used to replicate the topography of a scan, for example, then the degree to which this kind of surface patch matches the source data can be controlled by altering the number of control points in the surface. More control points enable the surface patch to be more complex and therefore follow the original data more closely. The number of control points is a variable set by the user when creating the patch.

The effect of altering the number of control points can be seen in Figure 3.24, which shows an IGES surface patch created from noncontact scan data of a human hip with control points varying from 6 to 100. The areas of darker grey show where the IGES surface patch closely approximates the scan surface whilst the lighter areas show that the IGES surface patch is below the scanned surface. The surface patch is also shown as a mesh showing the complexity of the surface as the number of control points increases.

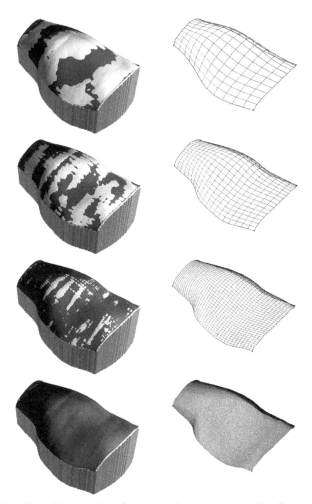

Figure 3.24 The effect of the number of control points on patch quality (from top): six control points, 10 control points, 20 control points and 100 control points.

Depending on the software being used, the quality of the fit between the surface and the original data can be visually inspected on screen as shown in Figure 3.24 or numerically quantified. For the surface with only six control points, the average gap between original surface and the patch is 0.818 mm. In comparison, the surface shown with 100 control points is a much closer fit with the average gap almost negligible at 0.00678 mm. As NURBS surfaces are controlled by mathematical equations, the patches themselves have to obey certain criteria in order for the equations to solve. Failure to obey these constraints will result in no surface patch being generated or a patch that flips, creases or twists. Typically, surfaces that exhibit these faults cannot be physically manufactured or they give rise to other problems.

To create solid models from surface patches, all of the surface area of the object must be covered. In addition, the patches must meet at common edges and not overlap. These surfaces can then be 'stitched' together to form a true 3D solid computer model of the object.

Advances in surface generation tools now allow for automation of surface patches; the user may specify a patch count and topography may then be created. Auto-surfacing is described in more detail earlier in this chapter.

3.6.5   T-splines

Patented by Autodesk (111 McInnis Parkway, San Rafael, USA), T-splines is a novel modelling approach which bridges polygon and NURBS modelling strategies. Significant benefits include simple transfer between meshes to NURBS surfaces, as well as a combination of unique modelling capabilities not previously possible with NURBS or polygon modelling; star points can be included which involve more than four facets or patches aligning at a single point. T-junctions are also possible, as well as localising subdivision for focusing complex features. Finally, surface quality is not compromised unlike polygon modelling. Figure 3.25 shows a close-up view of a T-spline model showing a star point and T-junction. The use of T-splines in the medical sector is still limited but an expanding area in product design.

3.6.6   Voxel modelling

A major impediment to the application of CAD in medical applications, such as prosthetics and implant design is the fact that it requires the integration of existing anatomical forms with the creation of complex, naturally occurring free form shapes. Until recently, CAD has been driven and developed specifically to define geometry for engineering processes using techniques described in the previous sections of this chapter. Consequently, the way they operate makes it extremely difficult to integrate human anatomy and create similar forms. In addition, the methods used to define shapes in CAD are based almost entirely on mathematical geometry (straight lines, angles, arcs,

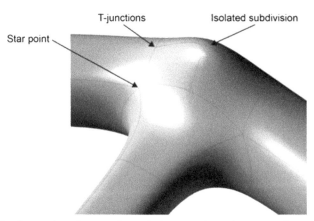

Figure 3.25 T-splines unique features for assisting organic surface modelling.

splines, etc.). In contrast to engineers, prosthetists have highly developed visual and tactile skills that allow them to handle materials to create accurate free form shapes. This creates a significant barrier to the application of current engineering-based CAD software by experienced prosthetists or others designing bespoke-fitting medical products.

Voxel-based modelling software has the potential to overcome many of the geometric constraints of other, more engineering-focussed CAD software. Unlike traditional engineering-based CAD systems, voxel-modelling software has been developed to visualise and manipulate solid, complex, unconstrained 3D shapes and forms by manipulating many thousands of tiny cubes (akin to volumetric pixels). The effect is more like manipulating clay or wax, but in a digital environment. This provides the opportunity to add higher levels of complex details, such as physical textures and enables more precise blending of designed structures into surrounding anatomy.

There is a limited range of voxel-modelling software. Three of the foremost are Z-Brush (Pixologic Inc, USA), 3D-Coat (http://3d-coat.com/) and FreeForm® (Geomagic, 3D Systems, USA). Of these, FreeForm® has the most widely reported use in medical applications. FreeForm® has a unique interface (Phantom® Desktop™ haptic interface); the hardware consists of a 3D stylus developed from research carried out at Massachusetts Institute of Technology. The device gives all six axes of movement via a hand-held stylus, similar to a pen or sculpting tool. Crucially this device not only has six axes of freedom but also incorporates tactile feedback. Thus, when moving the cursor on screen in three dimensions when the cursor comes to the surface of an object on screen the user feels the contact through the stylus. The resistance can be varied to simulate materials of different hardness and consistency.

Unlike traditional engineering-based CAD systems, FreeForm® and the other voxel-modelling software noted has been developed to visualise and manipulate solid, complex, unconstrained 3D shapes and forms. Material can be ground away, drilled, stretched, pushed or added in a manner analogous to sculpting with clay or wax.

3.6.7   STL modelling

In many cases, CAD users may start their initial CAD modelling strategies with alternate forms of 3D data such as a point clouds, quadrilateral mesh modelling or B-Rep modelling, before progressing onto STL. In most cases, STL data is normally exported with the intent to produce valid data for RP. In this case, it is vital that STL data is 'watertight' or closed before transferring to a RP system. Therefore, STL modelling is often necessary to produce adequate files in support of RP. Software such as Geomagic Studio (Geomagic Solutions, USA) and Materialise Magics (Belgium) are specifically tailored for reverse engineering tasks, such as point cloud manipulation and STL modelling. Clean-up tasks including polygonisation and decimation are described in more detail within Section 3.3.1.

Some CAD packages enable modelling directly in mesh formats and can be used for modelling arbitrary forms for artworks, sculpture, jewellery or medical modelling. Examples include 3D Studio Max (Autodesk) and 3-Matic (Materialise).

3.7   File management and exchange

3.7.1   *Stereolithography*

Derived from the word stereolithography, the STL file is a simple file format that describes objects as a series of triangular facets that form its surface. For example, if we view a simple object as an STL file we can see the triangles. It can be seen that large flat areas require few facets, whereas curved surfaces require more facets to approximate the original surface closely (see Figure 3.26). The format was originally developed by 3D Systems to provide a transfer data format from CAD systems to their STL technology, but it has subsequently been adopted as the de facto standard in the RP industry.

The STL file simply lists a description of each of the triangular facets, which make up the surface of a 3D model. Figure 3.27 shows the beginning and end of an STL file in text format. The first line describes the direction of the facet normal. This indicates which surface is the outside of the facet. The next three lines give the coordinates of the three corners, or vertices, of the facet.

The simplicity of the triangle makes mathematical operations such as scaling, rotation, translation, and surface area and volume calculations straightforward. The format also allows the angle of facets to be identified, which is necessary for STL.

STL files can be in binary or text (ASCII) format. Binary format files are much smaller and should be used unless there is a specific reason why the text format is required. STL files can vary in size from around 50 KB to hundreds of megabytes. Highly complex parts may result in excessively large STL files, which may make data

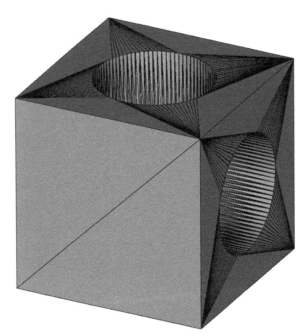

Figure 3.26 Cube with circular holes showing the triangular facets.

operations and file transfer difficult. Figure 3.28 shows a close-up view of an STL file illustrating the vast number of triangles required to provide accurate representation of an anatomical form. However, highly effective compression software is freely available that can reduce an STL file to a small fraction of its normal size to enable easy transfer of files over the Internet or by email.

To be used successfully in RP, the STL file must form a single enclosed volume; meaning that it should have no gaps between facets and all the facets should have their normals facing away from the part (i.e., identifying which is the inside and outside surface). Usually, small problems with an STL file can be corrected with specialist software or the RP machine preparation software. STL files can be generated from practically all 3D CAD systems. Solid modelling CAD systems rarely have problems creating STL files but surface modellers can pose problems if the surfaces are not all properly stitched and trimmed.

When exporting an STL file from a CAD system, the user will normally specify a resolution or quality parameter to the STL file. This is normally achieved by

```
solid  FILENAME
    facet  normal  1. 000000e+00  0. 000000e+00  0. 000000e+00
      outer  loop
         vertex  0. 000000e+00  −1. 204845e+00  −1. 658504e+00
         vertex  0. 000000e+00  −1. 235913e+00  −3. 804270e+00
         vertex  0. 000000e+00  −4. 000000e+00
      endloop
    endfacet

and  so  on...

    facet  normal  1.000000e+00  0.000000e+00  0.000000e+00
      outer  loop
         vertex  1. 000000e+00  4. 000881e+00  1. 221143e−04
         vertex  1. 000000e+00  3. 535500e+00  3. 535500e+00
         vertex  1. 000000e+00  3. 999953e+00  0. 000000e+00
      endloop
    endfacet
endsolid  FILENAME
```

Figure 3.27 The beginning and end of an stereolithography file in text format.

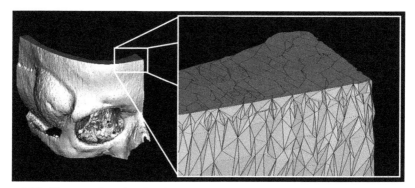

Figure 3.28 Close-up view showing facets.

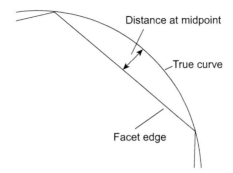

Figure 3.29 Facet deviation.

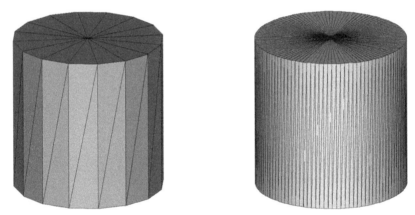

Figure 3.30 The effect of deviation settings (left) large deviation = small file and (right) small deviation = large file.

specifying a maximum deviation. The deviation will be the perpendicular distance between a facet and the original CAD data where the facet forms a chord at a curved surface, as shown in Figure 3.29. In essence, a smaller deviation will give a more accurate representation of the CAD model, but this will result in an STL file with a greater number of smaller facets. The file size depends only upon the number of facets and so a smaller deviation will give rise to a larger file. The effect is illustrated in Figure 3.30.

When creating STL files from medical scan data a similar effect can be achieved. Typically, rather than setting a maximum deviation, the STL file resolution is determined by relating the triangulation to the voxel size of the CT data. If every voxel is triangulated the maximum resolution will be produced; however, the file size will be correspondingly large. By triangulating multiple voxels, a simpler but smaller STL file will be produced. Typically, a compromise is achieved between file size and surface quality. Some software will simplify this by allowing the user to specify low-, medium- or high-quality settings.

For example, for the data previously illustrated in Figure 3.9, the highest possible resolution settings result in an STL file of more than 94 MB. Compare this to the

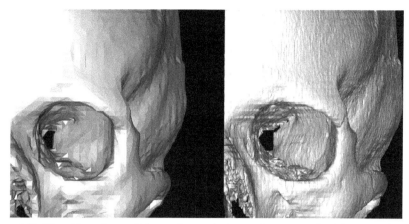

Figure 3.31 The effect of stereolithography resolution settings on medical data showing (left) low resolution and (right) the highest possible resolution.

results of the default software setting that range from 'low' at 4.3 MB, 'medium' at 11.7 MB and 'high' at 34 MB. The two extremes of the effect of these settings can be seen in the close-up views of the resulting STL files shown in Figure 3.31.

STL files can be postprocessed in several ways usually to enable them to be produced by RP methods more efficiently. Many software applications are available that allow STL files to split into separate parts, or reoriented in space. In addition, the quality of the STL file can be manipulated somewhat. File size can be reduced by triangle reduction techniques and smoothing algorithms can be applied.

3.7.2 Object

Capable of transferring NURBS geometry (curves and surfaces) and polygon meshes, the Object (.obj) file format is one of the most powerful and versatile transfer methods. Furthermore, properties relating to a file may also be included such as object names, materials and colour.

3.7.3 Virtual reality modelling language/X3D

Early versions of virtual reality modelling language (VRML) are also triangular faceted surface representations. They are in essence very similar to STL files but are typically created at very small file sizes and are coded in a more efficient manner to make them suitable for transfer over the Internet. VRML also allow the inclusion of colour and texture mapping to models, and may even contain animation/sound data. An extended version of VRML, termed X3D, may include additional data such as geo-locations as well as improved rendering. Applications in the medical sector include interactive teaching aids and procedural training/planning visualisations, which may be beneficial to health care professionals with varying levels of expertise, from medical students to consultants.

3.7.4   Standard for the exchange of product model data

The Standard for the Exchange of Product Model Data format (ISO10303) or STEP format is used to transfer solid models between high-end CAD packages. The advantages of the step format are that it includes not only the geometrical features that make up the model but it also includes the history tree – that is, the list of operations or construction stages that led to the model. This enables a model developed in one CAD package to be edited, revised or redesigned in another CAD package. http://en.wikipedia.org/wiki/ISO_10303-21.

3.7.5   Initial graphics exchange specification

As stated previously, the IGES format is an international standard that describes CAD data as mathematically defined geometries positioned in the 3D space (see http://www.nist.gov/iges/). By converting data from the original source through an intermediary 3D format, such as the STL file some CAD packages may be able to generate geometry such as vertices, curves and surfaces based on the original data. The nature of the surface and the degree to which it accurately reproduces the original anatomy will depend greatly on the data formats and CAD packages used. IGES file exchange also allows for the transfer of model properties such as labelling, notes and colour.

Some CAD packages also enable the user to create curves and surfaces from point cloud data obtained using touch probe or noncontact surface scanners.

3.7.6   Additive manufacturing format/STL 2.0

Fundamental flaws of the STL format for file management are that STL only communicates an approximate form of a mesh relative to its resolution and cannot include additional build properties or metadata. Given the expansion of the market leading to development of new technologies and supporting software, a data management system was needed to communicate characteristics of CAD models such as multimaterial and multicolour constructs as well as complex lattices and scaffolds. Intellectual property has also been a concern since STL formats do not permit limited access, leading to unauthorised sharing and access of the file content.

The additive manufacturing format (AMF) or STL 2.0 (ASTM F2915-11) overcomes these limitations. In tackling lack of detail through approximation, the AMF triangles no longer remain planar like STL facets; AMF triangles effectively become curved patches. This adaptation encompasses tangency in facets by noting the normal direction at mesh vertices (Figure 3.32) and subsequently subdividing facets (Figure 3.33). This reduces the file size, since larger triangular facets may be used to communicate a larger surface area, whilst also increasing the accuracy of the part relative to the original data. Uptake of the AMF format by software developers for 3D CAD software has been limited to date, but shows much promise in future applications due to its versatility.

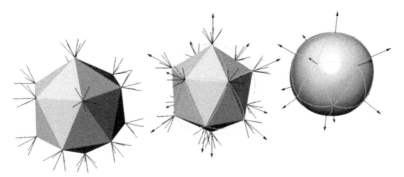

Figure 3.32 Stereolithography mesh to additive manufacturing format conversion. Left: triangular polygon sphere (24 faces) with individual vertex normal. Middle: combined normal at vertices (arrowed) and approximated curves. Right: curved facets through combined vertex normal of surrounding faces.

Figure 3.33 Left to right: mesh face subdivision through additive manufacturing format formatting. Mesh faces are subdivided by a factor of four, and may be performed a maximum of four times to create 256 triangles.

Physical reproduction

4

4.1 Background to rapid prototyping

4.1.1 Introduction

'Rapid prototyping' (RP) is a term that was coined in the 1980s to describe new technologies that produced physical models directly from a three-dimensional (3D) computer-aided design (CAD) of an object. Many other phrases have been used over the years, including solid freeform fabrication, layer-additive manufacturing, 3D printing and advanced digital manufacturing. In the late 1990s, the application of these technologies to tooling was investigated and the term 'rapid tooling' was commonly used to cover these direct and indirect processes. More recently, these technologies have been applied to product manufacture as well as prototyping and 'rapid manufacturing' is increasingly being used to describe this kind of application. Perhaps the most accurate term would be 'additive manufacturing', because this covers all of the processes and distinguishes them from previous technologies such as machining.

However, despite all of the different applications, 'RP' has been adopted by the industry to cover all these technologies and their applications, and therefore it is the one used here.

The objects created by RP processes may be used as models, prototypes, patterns, templates, components or even end-use products. However, for simplicity the objects created by the RP processes described in this chapter will be referred to as models regardless of their eventual use.

4.1.2 A brief history of RP

Processes that build models directly from computer data have become generally referred to as RP techniques. Rapid prototyping systems were originally developed in the 1980s as a method of building exact physical models of products and components designed using CAD. Initial applications concentrated on the automotive and aerospace sectors, because these industries first exploited what was at the time expensive CAD/computer-aided manufacturing (CAM) systems. Since then, the incredible progress that has been made in computer processing power and software sophistication has led to the adoption of CAD/CAM in almost all industries. This rapid growth in the application of CAD/CAM systems fuelled the growth in the demand for RP systems. During the 1980s and 1990s, the number of RP systems increased dramatically and a range of material and process approaches were introduced.

Initially, many of these processes were inaccurate and unreliable, and many promising ideas failed to reach the commercial market. Several processes secured funding

Medical Modelling. http://dx.doi.org/10.1016/B978-1-78242-300-3.00004-4

and developed into commercial manufacturing companies producing RP machines for sale across the world. Since then, the established technologies have been developed to produce effective, accurate and reliable machines using a variety of technologies. As with any other emerging industry, there have been some business failures and some consolidation of the market with mergers, acquisitions and licensing agreements.

One of the first casualties from this period was Helisys, Inc., manufacturers of laminated object manufacturing machines.

DTM Corp. had been developing selective laser sintering based on technology from the University of Austin, Texas, since the 1980s, and by the mid-1990s had a successful product and good sales. This success was partly because the machine used thermoplastic materials and therefore produced strong and functional prototypes. However, the company became focused on developing metal materials for use in tooling applications. This technology was costly and failed to capture the interest of the tooling market, which in the main remained faithful to increasingly efficient high-speed milling machines. This, and protracted legal action against the German RP manufacturer EOS GmbH, adversely affected the company and it was then acquired by 3D Systems. EOS itself has also suffered from the effects of legal action with 3D Systems, which has since been resolved.

This has led to an industry dominated by two large US-based corporations, 3D Systems and Stratasys. 3D Systems was the first major player in the RP market with SL technology. Good development of the technology and, importantly, the materials led to the highly successful SLA-250 model, which provided the company with an installed customer base and subsequent larger and more efficient machines. Today SLA is one of the most commonly used RP process. The company has strongly followed a business development strategy, which involved setting up national subsidiaries in many countries, aggressively protecting intellectual property, developing complementary technologies, and acquiring competitors. This led the US Justice Department to declare that 3D Systems was effectively a monopoly and they forced the licensing of stereolithography patents to Sony, who had been producing SL machines for the domestic Japanese market for many years.

Meanwhile, Stratasys has steadily developed a single technology, fused deposition modelling, to produce a comprehensive and reliable range of machines from the cheapest desktop modelling machines to large-capacity functional prototyping machines. Concentration on a single technology and the ability to sell cheaper machines has sustained the company through the economic downturn of recent years.

Recent newcomers to the market include Z Corp (USA) (now acquired by 3D Systems), EnvisionTEC (Germany) and Objet (Israel) (now merged with Stratasys). Z Corp, now part of 3D Systems, produced a range of three-dimensional (3D) printing machines using technology licensed from the Massachusetts Institute of Technology. These machines have now become part of the 3D Systems ProJet range of technologies and are fast and cheap to operate. However, the models are comparatively less accurate and physically weaker, although newer materials are being developed to address these issues. The EnvisionTEC machines selectively cure cross-sections of photopolymer using digital micro-mirror devices to project visible light. The Objet (now merged with Stratasys) machines aim to deliver the cost efficiencies of printing technology with the functionality and accuracy of SL.

Because of the technology-driven nature of RP companies and their products, the industry is awash with trade names, abbreviations and acronyms for the various processes, types of software, hardware and materials. Many of these are registered or recognised trademarks and these have been indicated where possible. A glossary of terms can be found in 'Glossary and explanatory notes'.

The most common RP processes are described later in this chapter. Each major RP process type is covered in principle and some detail. Although it is not practical to describe every aspect of every machine available, the sections should provide a good overview of the technologies, their pros and cons, and their appropriateness for various medical applications.

4.1.3   Layer-additive manufacturing

Rapid prototyping systems work by creating models as a series of contours or slices built in sequence, often referred to as layer manufacturing. The different RP systems vary in how they create the layers and in what material.

By convention, the axes X and Y represent the plane in which the layers are formed and the Z-axis is the build direction, usually referred to as the height. Consequently, the number of layers required for a given object is a function of the layer thickness and height of the model. The accuracy and resolution in the XY plane therefore depends on this mechanism. Most are accurate to fractions of a millimetre so that any geometry in the XY plane should be faithfully reproduced.

The layer thickness is dictated by the mechanical process that adds the build material and may vary according to the material being used. Layer thickness is usually in the order of 0.05–0.30 mm, although some printing-based technologies offer much thinner layers. This will lead to a stepped effect in geometry perpendicular to the XY plane.

As an example, consider a cube with two perpendicular holes through it, as shown in Figure 4.1. The hole in the top of the cube, as viewed from above, formed by the scanning mechanism in the XY plane will be formed perfectly (within the capability of the mechanism). However, the hole in the side of the cube, as viewed from the side, formed by the addition of layers will display a stepped effect (shown in Figure 4.2). Consequently when building objects, careful consideration should be given to the orientation so that the optimum features are formed in the XY plane and stepping is avoided. For example, a cylindrical shape should be built upright if the circular section is to be faithfully reproduced.

The extent of the stepping can be diminished by creating thinner layers, but because the overall build time tends to depend more on the layer addition process, a larger number of thinner layers lead to longer build times. Therefore, a compromise between surface finish and speed is established for each RP process.

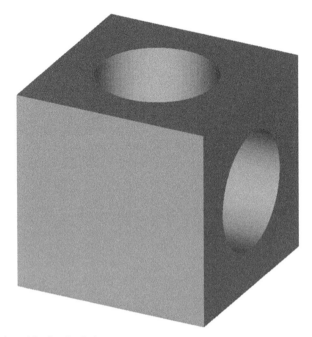

Figure 4.1 Cube with circular holes.

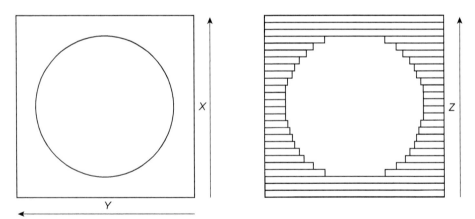

Figure 4.2 The stepped effect of layer manufacturing.

4.1.4 Boundary compensation

Because RP devices are additive processes, the method used to create the layers has a finite minimum size. In the case of the printing-based technologies, this is usually expressed as the pixel size that the printer is capable of reproducing. This is usually expressed in the same terms as paper printers in dots per inch. However, with laser-based technologies such as SL and laser sintering, the diameter of the laser beam provides the minimum capability. For maximum accuracy, the software that controls this

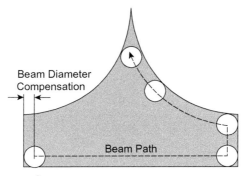

Figure 4.3 Beam compensation.

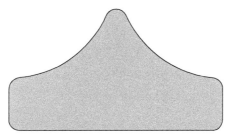

Figure 4.4 The effect of beam compensation.

process offsets the path of the laser by half of its diameter so that it draws a path within the boundary of the object. This is illustrated schematically in Figure 4.3. However, this means that small, thin features below the minimum size cannot be reproduced. A similar effect is shown by fused deposition modelling in which the minimum width of the deposited bead of molten plastic is compensated for in the same manner.

In practice, the size of offset is small, typically small fractions of a millimetre; however, the effect should be considered separately to dimensional accuracy. Although dimensional accuracy may remain good, large offset values will have a detrimental effect on the ability of the RP system to reproduce intricate detail, crisp edges and sharp corners, as illustrated in Figure 4.4.

4.1.5 Data input

All RP systems require a computer model of the object to be built. In all cases, software is used to take a 3D computer model of the object to be built and slice it into a vast number of thin cross-sections. These cross-sections are then translated into an appropriate format to direct the layer manufacturing process of the RP system.

Because there is a huge variety of CAD programs available and a number of RP systems, each with different software requirements, the industry required a standard translation format that could enable the RP process to build models designed in CAD. Stereolithography was the first RP process to market and employed a mathematically simple approximation of 3D CAD models, called the STL file, which is fully described in Chapter 3. The format describes models by closely approximating their shape with a surface made up of a large number of triangular facets. Because it was the first convenient transfer format to be offered by CAD software developers, the STL was adopted by other RP manufacturers and has since become a de facto standard in the industry. Consequently, despite some shortcomings, all RP machines can use the STL file. However, other formats are also available and a description of some of them is provided in Chapter 3.

In practice, RP software slices the STL file into a number of cross-sections. The software then creates control files that will instruct the RP machine as to how to construct each layer and how to deposit subsequent layer material. Depending on the sophistication of the process, there may be a degree of user interaction at this point that can help to optimise the build. The result is one or more files that are transferred to the RP machine itself. Typically, the build file is then checked and the machine is prepared for a new build.

Although the models may take many hours to build, the machines typically operate unattended, nonstop. Therefore, a build will frequently be started last thing in the working day and the machine will run overnight; often the model will be completed by the next morning.

All of the systems have some method of supporting the build in progress and this will require some degree of post-process cleaning and finishing after the model is built. Finishing is usually done by hand to remove residual material, remnants of support structures and the step effect. The level of finishing employed depends on the end use of the model.

RP technologies are constantly being developed and incrementally improved, so the descriptions here are intended to provide an overview of how the technologies work and compare their relative strengths and weaknesses. However, when assessing RP technologies from manufacturers or service providers, it is always advisable to obtain the latest specifications. RP manufacturer websites are the best source of such information, and the most popular are listed in the bibliography.

4.1.6   Basic principles of medical modelling: orientation

By definition, every medical model is unique and the characteristics of each model need to be considered carefully when selecting and using a particular RP system. Compared with engineering products that have been designed for manufacture, models of human anatomy may be much more challenging to prepare for a successful build, even for those experienced in the operation of their RP machines.

The following sections describe some of the most common RP processes and highlight their key technical considerations. However, some basic principles apply to nearly all RP technologies when building medical models. The most important consideration is the orientation of the build. The orientation of the build will have an influence on the surface finish of the model, the time it takes to build, the cost of the model, the amount of support required and the risk of build failure.

All of these factors are interdependent; the key to producing a high-quality medical model is a thorough understanding of them, correctly identifying the priorities and reaching a compromise solution that best meets the needs of the clinician. The effect of orientation on these factors is explored in the following text.

1. Build time and cost

   Because all RP processes work on a layer-by-layer basis, the builds consist of two repeated stages: drawing or creating the layer and recoating or depositing material for the next layer. Generally, the material deposition stage takes longer and often poses the greatest risk to build failure. Therefore, orienting a model such that it minimises the number of layers will reduce the build time.

   Generally, RP model costs are directly related to the build time. Therefore, the longer the build time, the more the model will cost. In many cases, the automatic option is to orient a model for minimum height and therefore minimum cost. However, as described in the following text, this may have an undesirable effect on the quality of the model.

2. Surface finish, model quality

   As described in Section 4.1.3, the layer-by-layer building process results in a stair-step effect on sloping or curved surfaces. Depending on the shape of the object, the orientation may have a great effect of the degree of stair stepping on the model surface. However, the mechanisms that create the layer geometry usually offer better resolution. When considering engineering parts, the most important feature is identified and the build is oriented to provide the optimum surface quality for that feature. However, human anatomy usually possesses curved surfaces in all directions and the optimum orientation may depend on other, more important factors. However, there are some obvious examples in which orientation makes a considerable difference to surface quality. The long bones of the arms and legs, for example, are essentially cylindrical in form. Orienting a model of a long bone so that it lies flat will minimise build time and cost, but the layers will be readily apparent in the model. Orienting the build in an upright sense will increase build time and cost considerably, but the layered effect would be drastically reduced in comparison, leading to much better model quality. Figure 4.5 illustrates the effect of stair stepping on a model of a proximal tibia (the relative thickness of the layers has been increased to exaggerate the effect). The upper model was built lying horizontally. Although this minimised build time, the thickness of the layers had a negative effect on the reproduction of the contours of the

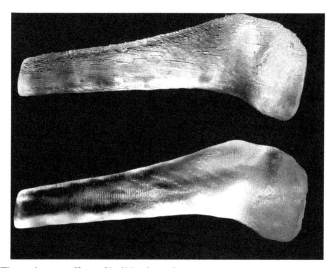

Figure 4.5 The stair-step effect of build orientation.

model. The lower model was built upright, and consequently the layer thickness has not had such a detrimental effect, at the cost of increased build time.

3. Support

All RP processes provide support to the model as it is built. Some processes build supports concurrently with the model whereas others employ the unused build material to support the part as it is built. Usually, parts that are supported by the unused material may be oriented to provide the minimum cost, or to fill the available build volume with the greatest number of parts. However, for those processes that construct supports or deposit a second supporting material, their effect on model quality and their subsequent removal should be considered.

Typically, the more supports or supporting materials that are present, the more time will be spent manually removing them, which can add considerably to the overall time to delivery and labour costs. Where supports are constructed from build material, the surface finish and quality will also be affected because most supports leave a witness mark on the surface after their removal. For example, the effect of support removal from SL models is shown in Case study 3 in Section 5.3. Another consideration when building medical models is the presence of internal cavities. The skull, for example, contains many such cavities, and supports within them may prove difficult or impossible to remove by hand. Some RP processes use soluble supports, which can be removed by immersing the model in a solvent or can melt away supporting materials with a lower melting point (typically wax). Although this still adds to the overall time to delivery, labour costs are reduced and supports inside cavities are easily removed. Orientation may be optimised to minimise support materials because they represent a considerable materials cost (and, of course, waste). In some circumstances, the support material may prove to be a significantly higher volume than the actual model. It is therefore advantageous to orient a model to minimise the amount of support, but not to the degree where the build process is threatened.

4. Risk of build failure

Although RP machines are generally reliable and able to operate unattended for long periods, the build process can be threatened by pushing parameters to their operational limits. As has been stated previously, building models of human anatomy poses challenges to RP owing to the highly complex nature of the forms being built. This makes the risk of build failure higher when attempting medical modelling compared with engineering parts. Often the risks to build failure depend highly on the specific RP process being used, but some general principles apply. The overall stability of the model is important in most RP processes. It therefore makes sense to orient the model so that it is most stable, i.e. wider at the bottom than the top. This will almost certainly also directly affect the supports required. Generally, the less support needed, the lower risk of build failure. The principle of stability and the subsequent effect on the supports is shown clearly in the example in Figure 4.6, where the orientation on the right offers a more stable build and a much reduced amount of support.

Economic factors are also important. Most RP parameters are set to provide the fastest and therefore cheapest build possible. However, setting parameters for speed usually increases risk of build failure. The complex nature of medical modelling means that parameters may have to be altered to lower risk, and consequently build time and cost may increase compared with an engineering model of similar size.

5. Data quality

Most orientation decisions are arrived at by considering the final model's shape and size, but the computer data used to build it may also influence the choice of build orientation. The choice of data format could limit the options available for orienting the model. For example, if a 2.5-dimensional format is used, such as an SLC file, the choice of orientation is fixed when the data are created. Usually these formats are created in the same orientation as the original scan data and this may provide a higher quality data file.

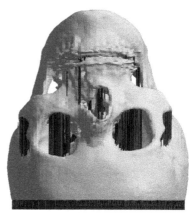

Figure 4.6 The effect of orientation on model support.

The advantages of the data format in terms of quality and efficiency may override other orientation considerations. The data formats and their various advantages and disadvantages are described in Chapter 3.

4.1.7 Basic principles of medical modelling: sectioning, separating and joining

When preparing an RP build, it is worth considering the nature of the model and its intended use before finalising the data. There are several considerations to address relating to the overall size of the model, whether it is one structure or a number of related structures and whether their positional relationship is important. For these reasons, it may become necessary to section, separate or join different parts to make the model or models required.

1. Sectioning
   Sectioning may be done when selecting the data from the original scan data or can be done by splitting the export data files, such as STL files. The most obvious reason for sectioning a model is to reduce the extents of the model to only those areas that are needed. This reduces build time and cost. However, it may also be used to split large models into a number of parts that can be accommodated within the build volume of a given RP machine. Sectioning may also be used to gain access to trapped volumes (internal cavities). This may be necessary to remove waste material or supports, but it may be because the internal anatomy is also interesting to the clinicians.

   When sectioning models it may be advisable to incorporate a stepped or keyed section so that the two separate parts are easily located when they are put back together. Many RP software packages provide functions to achieve this. An example is shown in Figure 4.7, in which a long bone has been cut into two pieces. The keyed section helps align the two pieces when they are joined together.

2. Separating
   As opposed to sectioning models at convenient locations, it may be desirable to separate different adjacent anatomical structures. Often when preparing data from the original scan data, different anatomical structures are sufficiently close to each other that the data become

Figure 4.7 A long bone model with a keyed section.

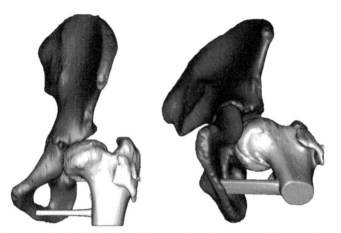

Figure 4.8 An example of a bridged model connecting the proximal femur to the pelvis.

a single object. For example, close joints can become closed, effectively creating a single object from two distinct bones. It may therefore be desirable to edit the data to separate the different anatomical structures so that they can be built individually. This may be so that the parts can be built separately to save time or cost, but it may also be because the clinicians wish to be able to articulate the two structures.

3. Joining

   Just as it may be desirable to separate adjacent anatomical structures, the opposite may be true. When building separate anatomical structures, their final use should be considered. If the spatial relationship between the structures is important, that will have to be created physically in the resultant model: for example, where bones have been fractured and displaced. In some cases, this may be achieved by leaving supports in place between the different parts. However, supports are not strong enough to achieve this reliably. Instead, it may be necessary to create bridges that join the separate parts together. When creating such bridges it is important to make them clearly artificial in appearance and locate them away from important areas so that they are not confused with the anatomy. An example of a bridge between two separate bones is shown in Figure 4.8.

4.1.8   Basic principles of medical modelling: trapped volumes

Compared with engineering parts, the presence of trapped volumes (internal cavities) can pose particular problems in medical modelling. As mentioned earlier, support materials need to be removed and this may prove difficult where physical access to the cavity is limited. Those processes with soluble or low melting point support materials have an advantage here, but they still need adequate access to drain the liquid materials. The trapped volume problem is particularly associated with SL, in which the effect can lead to build failure, because the resin in the trapped volume does not level correctly (see Section 4.2 for more detail on the trapped volume effect). However, the presence of trapped volumes can cause problems for many RP processes. To produce the unwanted effects of a trapped volume, the cavity does not need to be entirely closed. If the openings to such a cavity are small enough, the effect will be as bad as a fully closed cavity. Examples of cavities are the cranium, the sinuses in the skull and face, and the marrow space inside large bones.

The main problem with trapped volumes is removing the waste material and or supports from within them. Various techniques can be used to address trapped volumes. If the cavities are small, totally closed, and of no interest to the clinician, the data can be edited to fill in the cavity. It may be tempting to leave such cavities in place as they are fully closed; however, this poses risks. If the model is broken, sawn or drilled into at a later date, the unused material may leak out. This could be in the form of loose powder, solvent or liquid resin, which may damage the model or prove a nuisance (or, depending on the material being used, even a minor health risk) to the user or patient.

Where cavities are larger or deemed important, artificial openings can be created to enable the waste material to be removed manually or drained. When creating such openings, the location and shape should be created to make them obvious so that they are not confused with the anatomy. For example, drain holes should be made square so they are less likely to be confused with naturally occurring holes. Alternatively, as described earlier, the model can be sectioned into parts that are built separately to either enable access to the cavity or eliminate it altogether.

4.2 Stereolithography

4.2.1 Principle

Liquid resin is selectively cured to solid by ultraviolet (UV) light accurately positioned by a laser. The laser scans the layers onto the surface of the resin, with the first layers attached to a platform. Successive layers are cured by lowering this platform and applying an exact thickness of liquid resin.

4.2.2 Detail

The SL apparatus was developed and commercialised by 3D Systems, Inc., in the 1980s (the abbreviation SLA is 3D Systems' registered trademark). Models are made by curing a photopolymer liquid resin to solid using a UV laser. Models are built onto a platform that lowers by a layer thickness after each layer is produced. Wait states allow the liquid to flood over the previous layer and level out. Then a recoater blade will pass over the liquid, levelling the resin but also removing any bubbles or debris from the surface. The length of the wait states and speed of the recoater blade will depend on the viscosity of the resin. This method also means that there are problems with building objects with trapped volumes because the liquid in these areas is not in communication with the resin in the vat and does not level out, leading to build failure. In an attempt to remedy this, the recoater blade has a U-section that picks up a small amount of resin with a vacuum and deposits it over the previous layer.

Overhanging or unconnected areas have to be supported. Supports are generated by the build software and built along with the model. When a model is complete, excess resin is washed off using a solvent and the supports are removed. The model is then post-cured in a special apparatus by UV fluorescent tubes.

All lasers used in SL emit in the UV spectrum, and are therefore not visible to the naked eye. The laser represents a considerable cost and has a limited life, leading to comparatively high running costs.

The speed of the machine depends on how much energy the resin requires to initiate polymerisation, because the power of the laser is more or less constant; if more energy is required, the laser must travel slower. Material properties and accuracy also depend on the resin characteristics. As the material polymerises, there will be some degree of shrinkage; this can be compensated for in the build parameters but it may also lead to other problems, most notably curl. This was especially true early in the development of SL, when most systems used acrylate-based resins. These problems were partially eliminated by altering the build style, i.e. the way the laser scans the layers. The development of epoxy-based resins eliminated these problems because it shows low shrinkage, which gives accurate models although it requires more energy to polymerise and therefore builds slower. New materials are becoming available with physical properties increasingly similar to commonly used thermoplastics. In particular, much work has been done to improve the toughness of resins and reduce brittleness.

Typical SL medical models are shown in Figures 4.9 and 4.10. In medical modelling terms, SL is in many ways ideal. SL models show good accuracy and surface finish. The transparency of most SL materials enables internal details such as sinuses and nerve canals to be seen clearly. The fact that unused material remains liquid also means that it can be easily removed from internal spaces and voids. This is crucial when considering that most medical modelling is of the human skull, which possesses many such internal features as well as the cranium itself. These advantages can be clearly seen in the examples illustrated here. The solid, fully dense finished models lend themselves well to cleaning and sterilisation and the development of medical standard materials is an advantage.

A variety of resins are available that have been tested to internationally recognised standards (e.g. ISO 10993 and USP 23 Class VI; see Section 4.9), making parts suitable for handling in the operating theatre as a surgical aid or to be used as surgical templates and guides.

Some resins can also be selectively coloured. A single pass of the laser will solidify the resin; a subsequent high-power pass will cause the solidified resin to change colour. This allows internal features to be made visible through the thickness of the model. This can be used to show, for example, the roots of teeth in

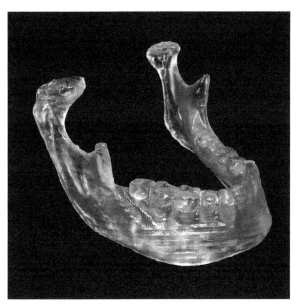

Figure 4.9 Stereolithography model of the mandible.

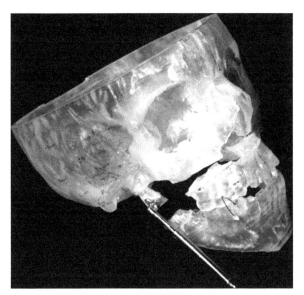

Figure 4.10 Stereolithography model used for planning maxillofacial surgery.

the jaws bones or tumors, as illustrated in Figures 4.11 and 4.12. These factors, combined with the inherent accuracy of SL, make it highly appropriate for medical modelling. The advantages and disadvantages of stereolithography are summarised in Table 4.1.

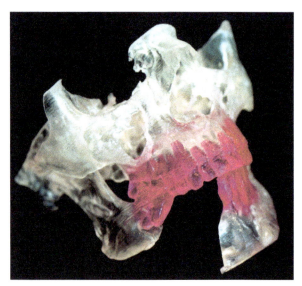

Figure 4.11 Selectively coloured model showing the roots of teeth.

Figure 4.12 Selectively coloured model showing a tumour.

Table 4.1 **Advantages and disadvantages of stereolithography**

Advantages	Disadvantages
Well-developed, reliable machines and software	High cost of machine and materials
Established sales, support and training	High maintenance costs
High accuracy, good surface finish	Resin handling requirements
Little material waste	Trapped volumes problematic
Medical standard materials	
Can be sterilised and selectively coloured	
Transparent models	

4.3   Digital light processing

4.3.1   Principle

Liquid resin is cured to solid by visible light accurately projected as a 2D mask using a digital micro-mirror device (DMD). The cross-sections are projected as images by the DMD onto the bottom of a shallow glass vat, curing the layers onto a rising build platform. Successive layers are cured by raising the build platform by an exact layer thickness and allowing resin to flow under the previous layer.

4.3.2   Detail

This process was developed by the German company EnvisionTEC GmbH (Digital Light Processing (DLP) is their trademark). Models are made by curing a photopolymer liquid resin to solid using visible light. In their Perfactory machine, models are built onto a platform that rises by a layer thickness after each layer is produced. Raising the build platform allows the liquid to flood into the gap between the transparent base of the vat and the model. The flat, transparent base means that layers are created at the exact layer thickness and are perfectly flat and level. Creating this fixed gap eliminates any of the settling or levelling procedures required in SL, which in turn accelerates the process. This process also means that the layer thickness can be reduced, leading to models built from thin layers, which improves surface finish. A range of machines are available from small desktop machines with a build envelope of only $40 \times 30 \times 100$ mm to larger free-standing machines with envelopes as large as $192 \times 120 \times 160$ mm. The smaller machines are typically used for dental or jewellery applications, whereas larger machines compete with SL. The most popular models are in the Perfactory range.

Because the cross-section is projected in one instant, the exposure time for each layer is fixed regardless of the geometry of any given cross-section. This provides rapid and predictable build rates. This method also means that there are no problems when building objects with trapped volumes.

As in SL (but upside-down on the Perfactory), overhanging or unconnected areas have to be supported. Supports are generated by the build software and built along with the model. When a model is complete, excess resin is washed off, sometimes using a solvent and the supports are removed. Unlike SL, the models do not require post-curing, although if required this can be achieved without special equipment because the resin will cure in daylight. Also unlike SL, there is no laser, which leads to considerable savings in maintenance and running costs. The projector bulbs require regular replacement.

The accuracy of the process is a function of the projected image of the cross-section. A DMD has a fixed pixel array (e.g. 1920×1200) but the image can be scaled up or down using optics. Therefore, the resolution of the model will be defined by the overall size of the build area divided by the number of pixels (i.e. pixel size) and the layer thickness. This resolution (pixel size) in the XY plane can range from as little as 0.016 to 0.069 mm. The layer thickness may range from 0.015 to 0.150 mm depending on the machine specification and optics used.

The materials used are similar to SL resins and produce models that are solid, reasonably tough and transparent, although usually amber or red in colour. A range of different colours is available for hearing aid manufacture. An example is shown in Figure 4.13. Materials have been developed and approved for specific medical application in dental technology and hearing aid manufacture. Application-specific versions of the smaller machines are available for a variety of hearing aid and dental appliance manufacture. An example of a hearing aid shell manufactured by DLP is shown in Figure 4.14. The advantages and disadvantages of DLP are summarised in Table 4.2.

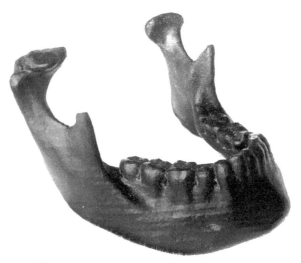

Figure 4.13 Perfactory® model of a mandible.

Figure 4.14 A hearing aid shell manufactured using DLP™.

Table 4.2 **Advantages and disadvantages of DLP^{TM}**

Advantages	Disadvantages
High accuracy, excellent surface finish	Resin handling requirements
Little material waste	Limited material choice
Low maintenance and running costs	Limited build size
Transparent models	

4.4 Fused deposition modelling

4.4.1 Principle

Thermoplastic material is fed in filament form to a heated extrusion head. Layers are made by molten material deposited as a fine bead. The build table lowers an exact amount and the next layer is deposited onto the previous layer bonding owing to partial melting.

4.4.2 Detail

This process was developed in the United States by Stratasys, Inc., and the technology is used in most of their machines, including the less expensive Dimension branded machines (Fused Deposition Modelling (FDM) is a Stratasys trademark). In recent years, the expiration of key patents and the RepRap open source movement (www.reprap.org) has led to a proliferation of cheap, small '3D printers'. These machines range in quality from home hobbyist's kits to professional standard machines. The use of low-cost machines in medical applications is possible, but care must be taken with materials properties, quality and accuracy. Models made on low-cost machines are likely to be inferior to models built on professional/industrial standard equipment and quality control and certification will be difficult to achieve for clinical use. Consequently, this section will focus on industry standard machines.

Models are made by extruding thermoplastic materials through a heated nozzle. The nozzle moves in the X- and Y-axes to produce layers; then the build table lowers by a layer thickness and the next layer is produced. Supports for overhangs are built up and are removed when the model is complete. The materials are thermoplastics such as acrylonitrile-butadiene-styrene, nylon, polycarbonate and polyphenylsulphone that are fed in the form of a filament from a spool. The models produced therefore can be handled directly and require no special cleaning or curing. The physical properties are close to injection-moulded plastic components and are therefore strong and durable compared with other RP models. The difficulty of removing the supports has been addressed by the use of a second support material that does not adhere to the build material, enabling the supports to be more easily removed. On some machines, the support material can be dissolved away in an agitated bath of water and detergent. This not only reduces labour but also reduces the risk of physical damage to delicate models.

The process also means that the machine is quiet and clean and is suitable for an office environment. Because the process is relatively simple, there are few moving

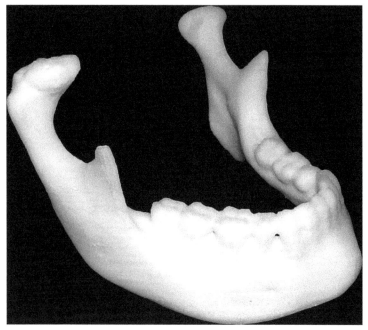

Figure 4.15 Fused deposition model of the mandible.

parts and the machine is therefore reliable and comparatively easy to maintain. Surface finish and accuracy are not as good as SL, for example, but the machines cost less than similar-sized SL machines.

As with SL, FDM models are hard and completely solid, which makes them suitable for sterilisation for use in the operating theatre. FDM models are opaque (usually white) which hides internal details that could be visible in an SL model. A typical FDM medical model is shown in Figure 4.15.

FDM is well suited to medical modelling. FDM models show reasonable dimensional accuracy and surface finish. The models are particularly tough and can withstand repeated handling, which makes them ideal for teaching models. The finished models lend themselves well to cleaning and sterilisation and the use of medically acceptable materials is an advantage.

Unlike SL, the unused material is solid and although it does not adhere to the model, it can be difficult and sometimes impossible to remove from the internal cavities often found in human anatomy. The option of soluble support materials has made removing supports from medical models much easier. A small number of materials are available that have been tested to internationally recognised standards including, for example PC-ISO, a polycarbonate that is particularly strong compared with most other RP materials.

The fact that the materials are opaque makes the identification of internal voids such as sinuses and nerve canals impossible. However, the white material tends to represent bones in a familiar manner and models of long bones and joints can be well modelled. These advantages can be clearly seen in the examples illustrated in Figures 4.16 and 4.17. The advantages and disadvantages of FDM are summarised in Table 4.3.

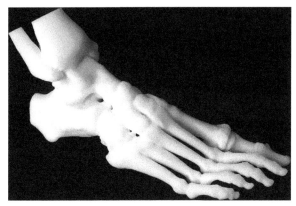

Figure 4.16 Fused deposition model of a human foot.

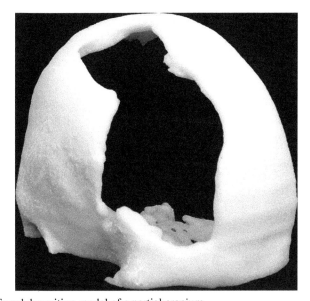

Figure 4.17 Fused deposition model of a partial cranium.

Table 4.3 **Advantages and disadvantages of FDM™**

Advantages	Disadvantages
Relatively cheap to buy and run	Poorer accuracy and surface finish compared with SL
Reliable	Small features difficult
Clean and safe process	Opaque material
Very strong, tough models	
Medical standard material	
Can be sterilised	

4.5   Laser sintering

4.5.1   Principle

Laser sintering (LS) is similar to SL except that it uses powders instead of liquid resins. A powerful laser locally fuses or sinters thin layers of fine particulate material. The build platform is lowered, fresh powder is applied and the next layer is scanned on top of the previous layer. Local melting also forms an interlayer bond.

4.5.2   Detail

Models are made by selectively sintering thermoplastic powder material using a laser. The materials are heated to near melting point and the laser scans the cross-sections, locally heating the powder enough to fuse the particles together. The build platform lowers each layer and fresh powder is spread across the build area by a roller. The inherent dangers of handling fine powders are controlled by purging the build volume with nitrogen gas. Models are supported by the unused powder. Overall, build times are comparatively slow to allow for heat-up, around 1.5 h, and cool-down of the powder, around 2 h. When completed, the model is dug out of the powder and bead blasted to remove any powder adhering to the model's surface. The machines are large and heavy and require water cooling, extraction and nitrogen supply. Consequently, the operating costs are considerable and therefore a high throughput is required to justify the purchase of the technology.

Although these machines are produced by two manufacturers, the SLS abbreviation is a 3D Systems registered trademark. EOS GmbH refer to their machines as laser sintering machines. Laser sintering machines typically use thermoplastic materials. The most commonly used glass-filled nylon material gives relatively low porosity, resulting in strong, robust models compared with SL, for example. Surface finish is relatively poor compared with SL, as may be expected from a powder. Accuracy is almost as good as SL and comparable with FDM. Specific materials allow LS models to be used as sacrificial patterns in investment casting. An elastomeric material is also available for prototyping flexible components, and aluminium-filled materials are available for enhanced physical properties. When installed it takes a while to set up the machine parameters, involving a certain amount of trial running. However, once correctly set up, the machines are relatively reliable.

Although reasonably accurate and tough, the materials remain porous. Although the materials themselves pose no inherent medical problems, the porosity makes them difficult to clean and sterilise effectively. Like most RP models, LS models are opaque, which may hide internal details. A typical LS medical model is shown in Figure 4.18.

In medical modelling terms, LS models have proven useful. Laser sintering models offer reasonable accuracy and surface finish and would be comparable to FDM models. The strength and toughness of the models are definite advantages but the powder nature of the material leads to a rough surface, which can trap dirt and grease in handling. Despite these issues, LS parts are now routinely used to make surgical guides (single use) as well as some end-use devices such as foot orthoses. The porosity of the

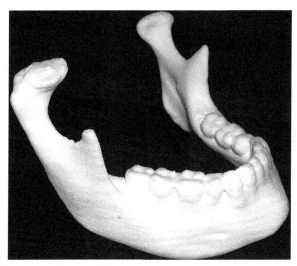

Figure 4.18 SLS model of the mandible.

Table 4.4 **Advantages and disadvantages of LS**

Advantages	Disadvantages
Strong, tough models Reasonably accurate Wide range of materials	High cost of machine and materials Large machine, nitrogen supply, water cooling Heat-up and cool-down time Poor surface finish Materials not suited to sterilisation

models poses some concerns where long-term use is concerned, especially if patient contact is being considered. Unused material remains as loose powder so it can be removed from internal spaces and voids, although this may sometimes be difficult. The advantages and disadvantages of LS are summarised in Table 4.4.

4.5.3 Laser melting

A number of manufacturers offer machines that can process metals. Although similar in principle to LS, these processes fully melt the powder to form fully dense metal parts. This is sometimes referred to as Selective Laser Melting (SLM, Realizer GmbH , now Renishaw, UK), Direct Metal Laser Sintering (DMLS, EOS GmbH), Laser Melting (Renishaw plc.) or Laser Cusing (Concept Laser GmbH). All are similar in principle, using high-power fiber lasers to melt fine metal powders in inert gas atmospheres. The full melting enables the production of solid, dense metal parts in a single process (i.e. not using binders or post-process furnace operations that have been previously used to make metal parts indirectly via LS). A variety of metals can be used, including stainless steels, cobalt-chrome and titanium. Although these processes are not suited to the production

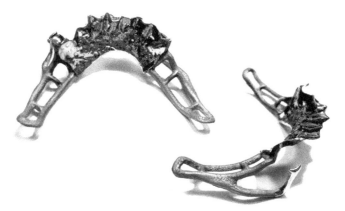

Figure 4.19 Partial removable denture framework (RPD) made using SLM.

Table 4.5 **Advantages and disadvantages of SLM™ and DMLS®**

Advantages	Disadvantages
Strong, tough metal parts	Relatively new technology
Accurate	High cost of machine and materials
Materials can be sterilised easily (using autoclave)	Inert gas supply required
Wide range of metal materials	

of anatomical models, they are being used to produce dental appliances, surgical guides, custom-fit implants and prostheses, as described in the case studies in Chapter 5. For example, Figure 4.19 shows a partial removable denture framework (RPD) made using SLM.

The process offered by Arcam (Arcam AB, Krokslätts Fabriker 30, SE-431 37 Mölndal, Sweden) also uses a powder bed but an electron beam directed by magnetic fields supplies the melting energy. Parts are commonly made from titanium but other metals are also being used. Compared with laser melting processes, the Arcam process typically has faster deposition rates and thicker layers. The build chamber is a vacuum rather than using an inert gas. The advantages and disadvantages of SLM and DMLS are summarised in Table 4.5.

4.6 Powder bed 3D printing

4.6.1 Principle

Powder material is deposited layer by layer and selectively bonded with adhesive printed onto the powder by heads, similar to those used in inkjet printers.

4.6.2 Detail

This process was originally developed at the Massachusetts Institute of Technology and commercialised by the US-based company, Z Corp, subsequently bought

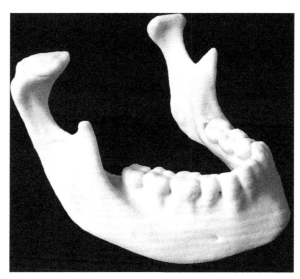

Figure 4.20 Powder bed 3D printed model of the mandible.

by 3D Systems. This specific process was referred to as 3D printing; however, in recent years the term '3D printing' has come to mean any additive manufacturing process; to distinguish this process, it is sometimes referred to as powder bed 3D printing. The machines use simple, relatively cheap powder materials such as starch and plaster, which are selectively bound by printing an adhesive. More recently, machines have become available that use plastic powders and solvent binders that can produce more robust models. The models have to be removed from the unused powder, which supports the build as it progresses. At this point, the models tend to be delicate and soft and require infiltration of a hardener material such as polyurethane or cyanoacrylate resin. The machines are comparatively inexpensive and the material and running costs are reasonable. The finished models are not as accurate and the surface finish not as good as SL, for example, but the advantage is that the machine builds extremely quickly compared with other processes. Combined with low running costs, this makes the models comparatively cheap. The finished models remain relatively weak compared with FDM or LS models. The machines and materials are safe if slightly messy.

The materials tend to be highly porous and therefore not well suited to sterilisation and no medical-grade materials are available. The models are opaque, which may hide internal details. A typical printed medical model is shown in Figure 4.20. However, machines are available that can produce full-colour models. This is achieved by introducing coloured inks into the adhesive print head (exactly like an inkjet printer) and using a white powder. Therefore, a wide range of colours is available. This may prove particularly useful in teaching models, in which different adjacent objects need to be identified. Figure 4.21 shows a coloured model used for surgical planning (Segments 1, 2 and 3 are different colours). The advantages and disadvantages of three-dimensional printing are summarised in Table 4.6.

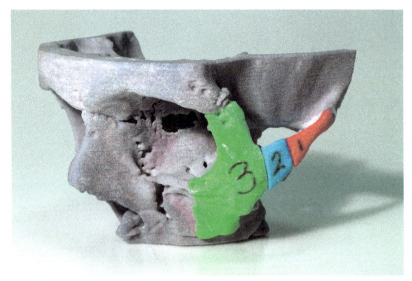

Figure 4.21 Colour model for surgical planning.

Table 4.6 **Advantages and disadvantages of three-dimensional printing**

Advantages	Disadvantages
Relatively affordable machines	Lower accuracy and surface finish
Fast build times	Sterilisation difficult
Easy to use	Relatively poor physical properties
Low maintenance and running costs	
Full-colour models possible	

4.7 Material jetting technology

4.7.1 Principle

Build material is deposited discretely by jetting heads, similar to those used in ink-jet printers. The material is ejected as a liquid, solidifying on contact with solid material or the build platform. The head moves in X- and Y-axes to build the layers. The build platform lowers by a layer thickness and material is deposited onto the previous layer.

4.7.2 Detail

Several companies produce small office-based machines that use deposition processes similar to those found in inkjet printers, such as Objet, Solidscape and 3D Systems. Machines such as the now obsolete 3D Systems ThermoJet® and those from Solidscape deposit wax-based materials in their liquid state that instantly cool and solidify to

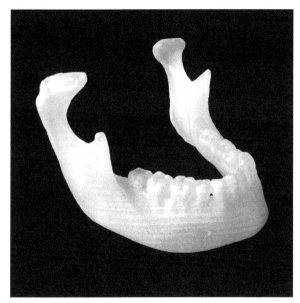

Figure 4.22 ProJet model of the mandible.

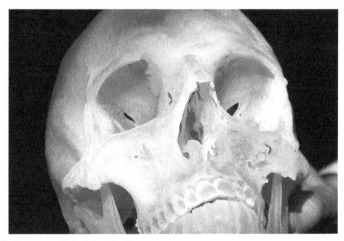

Figure 4.23 ProJet model of the skull.

produce the model, whereas the 3D Systems ProJet 3000 and 5000 machines and the Objet Polyjet machines use photo-polymerising resins that are simultaneously solidified by UV light as they are printed. ProJet models are shown in Figures 4.22 and 4.23 and an Objet model is shown in Figure 4.24.

Generally, the machines and materials are clean and safe and do not require extraction or special handling. ProJet machines surround the model with a secondary wax support material that is melted away from the completed model before being soaked in cooking oil and degreased. The CP range of ProJet machines fabricate in

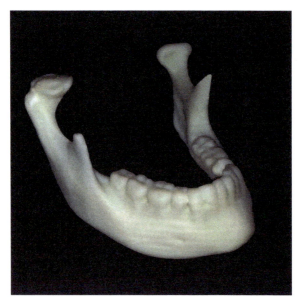

Figure 4.24 Objet model of the mandible.

a wax-based polymer and are the closest analogous machine to the now obsolete ThermoJet technology. Objet systems, however, deposit a gel-like support material that can be removed easily by hand or with brushes, although pressurised water jets are typically used. A caustic soda/hydrochloric acid (2% concentration) solution can also be used for model cleaning. The Solidscape machine uses a soluble support material.

As might be expected from technologies that rely on printing technologies, the resolution of these machines is often quoted in terms of dots per inch. Machine specifications vary, but generally the models are comparable to SL models.

One of the advantages of printing technologies is the ability to produce thin layers. This reduces the stepping effect and models are smooth compared with other layer-manufacturing techniques. Layer thicknesses of 0.030 mm are typical but can be as thin as 0.016 mm. Another advantage is that each layer is created during a pass of the print head regardless of the geometry of any given cross-section, which makes prediction of build times relatively simple and accurate.

With the exception of the Solidscape systems, most of the machines use at least two print heads, with the Objet Connex featuring up to eight print heads in total: four for distributing support material and the rest used to distribute model material. The print heads traverse across the build platform on a single block to strategically deposit photopolymer resin. The benefit of having large numbers of print heads is accelerated building speed. The Solidscape machine, however, uses a single print head, which makes these machines highly accurate, with thin build layers and hence excellent surface finish. However, this makes them

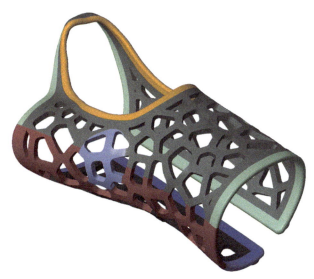

Figure 4.25 Multiple material CAD model.

extremely slow and limited to a small build envelope. They are, however, highly popular with the jewellery trade because of their small size and suitable casting materials.

In medical terms, the wax-like materials used in Solidscape and ProJet CP machines are extremely fragile and difficult to sterilise owing to their low melting points. The materials are opaque, which may hide internal detail, but the wax materials lend themselves well to casting processes and other maxillofacial and dental laboratory techniques. The Solidscape machine is capable of producing accurate models of extremely small objects. The relative accuracy of the Objet process and the transparency of the materials make them ideal for medical modelling and comparable in most respects to SL models. In addition, an Objet build material has been tested to an internationally recognised standard for plastics (USP 23 Class 6) and therefore may be sterilised and used in the operating theatre.

Uniquely, softer, rubber-like materials have also been developed for the Objet Connex process, and more recently by 3D Systems. These technologies enable multiple material builds. The Objet Connex machines, for example, can print two different build materials; typically, one will be a rigid plastic material and the other a soft, rubber-like material. In addition, intermediate materials are possible by printing differing combinations of the two materials. These intermediate materials are referred to as digital materials. This enables models to be made that exhibit a range of physical properties in a single build. To create multi-material models, a separate volume (STL file or shell) must be created to define each material region. This is illustrated in Figure 4.25, which shows a CAD model featuring multiple shells, and Figure 4.26, which shows an Objet Connex model featuring a range of soft black

Figure 4.26 Multiple material Objet model.

Table 4.7 **Advantages and disadvantages of jetting head technology**

Advantages	Disadvantages
Range of machine sizes and costs	Expensive materials
Easy to use	Expensive machines
Suitable for office environment	Replacing jetting heads can be expensive
Thin layers	
High resolution, good accuracy	
Transparent models possible	
Medically appropriate materials available	
Wide range of materials	
Multiple material builds	
Colour builds	

features, hard white features and intermediate grey features. More recently, technology has been developed to enable colour to be combined with the multiple materials. The advantages and disadvantages of jetting head technology are summarised in Table 4.7.

4.8 Laminated object manufacture

4.8.1 Principle

Inert, flat sheet materials are cut to the profile of a layer. Fresh material is bonded onto a previous layer and the next profile is cut. The layer thickness depends on the thickness of the sheet material.

4.8.2 Detail

In the now obsolete Helisys machines, models were made by cutting layers of paper with a laser using an *XY* plotter mechanism. Build time was relatively quick because only the perimeter of the layers was drawn. The paper was adhesive backed and the layers were adhered by a heated roller. The build table lowered, fresh paper was fed over the top of the build and heat bonded, the next layer was cut and so on. When the model was complete, the block was removed from the table and the waste material was broken away to reveal the model.

Currently, Mcor produces a more accessible laminated object manufacture (LOM)-based machine that uses normal office paper (e.g. A4 photocopier paper) to produce affordable models. Other machines have been produced that use PVC sheet material supplied on a roll that is precision cut with a blade and bonded with an adhesive. Laminated object manufacture materials undergo no state change, and therefore shrinkage is not a problem, although the height may alter slightly as built-in stresses are relieved. Paper models produced are good for handling and feel similar to wood.

In medical modelling terms, LOM has some uses. Laminated object manufacture models show reasonable accuracy and surface finish. However, the major advantage is the low cost of building large solid models. This makes the process ideal for producing models of large bones or soft tissues. These advantages can be clearly seen in the examples illustrated in Figures 4.27–4.29. The fact that unused material is solid means that it cannot be removed from internal spaces and voids. The advantages and disadvantages of LOM are summarised in Table 4.8.

4.9 Computer numerical controlled machining

Unlike RP techniques, milling has always been of limited use when producing shapes with undercuts, re-entrant features and internal voids. Yet, it proves economical and rapid when forming simple, solid shapes and forms. The principal advantages of machining are its availability and versatility. A large number of machines exist, from cheap desktop routers to large factory-based machine tools. Unlike most RP technologies, machining is not restricted to a limited range of materials. Almost any material can be machined, ranging from the hardest metals to soft foams. Computer numerical controlled machining (CNC) allows machining to be carried out under computer control based on CAD data. Depending on the machine and material configuration, CNC allows complex forms to be machined rapidly and accurately.

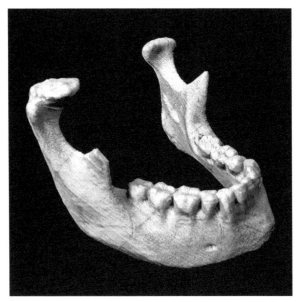

Figure 4.27 Laminated object manufacture model of a mandible.

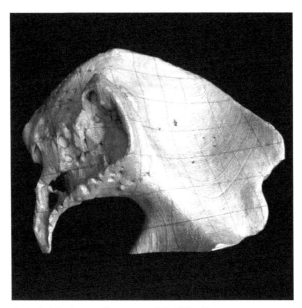

Figure 4.28 Laminated object manufacture model of the ileum.

The characteristics of machining make their use in medical modelling particularly suitable to producing larger models of the external anatomical topography. Typical applications may be for moulds and formers or custom-fit supports and wearable devices.

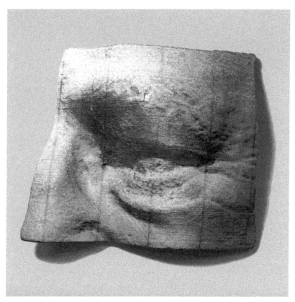

Figure 4.29 Laminated object manufacture of a partial face.

Table 4.8 **Advantages and disadvantages of LOM™**

Advantages	Disadvantages
Relatively cheap machines Cheap materials Clean, safe materials Good for large, thick, simple models	Labour-intensive post-processing Poor for hollow parts, thin walls and small features Waste material removal

The artificial hip model shown in Figure 4.30 was made to replicate the human body in impact testing of hip protection devices for the elderly. The model is made of dense, closed-cell foam and shaped using CNC machining to replicate not only the external anatomical topography, but also an internal pocket in which an artificial femur can be located. The advantages and disadvantages of CNC are summarised in Table 4.9.

4.10   Cleaning and sterilising medical models

4.10.1   Introduction

The application of AM/RP technologies in the medical sector has led technology and material vendors to develop processes and materials that meet prerequisite standards that can enable products to be used in direct contact with skin or as temporary or permanent implants. Common standards that developers cite are USP 23 Class VI

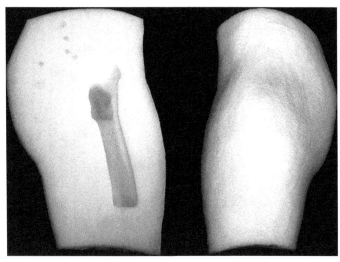

Figure 4.30 CNC machined foam model of the soft tissue of the hip.

Table 4.9 **Advantages and disadvantages of CNC**

Advantages	Disadvantages
Wide range of machine size and cost	Pre-processing time consuming
Wide range of materials	Setup time
Clean, safe materials	Clamping the part during cutting
Good for large, thick, simple models	Poor for thin walls and small features
Good accuracy possible	Poor for internal detail, hollow parts or undercuts

and/or specific parts of ISO 10993. However, although some materials have undergone testing to demonstrate levels of biocompatibility, it is the responsibility of the device manufacturer to ensure that it is fit for purpose and complies with the guidance set out in the Medical Device Directive (MDD). Cleaning and sterilising become increasingly important parts of the production process as the risk to the individual using the device or the patient receiving treatment increases. MDD classification is related to many factors, including the duration of exposure, the invasiveness, and whether it is passive or active. The higher the MDD classification, the more risk it poses to the health of the patient and the more rigorous the controls required in the specification and testing.

Compared with alternative manufacturing processes such as CNC machining, RP processes can involve the use of far fewer non-biological contaminants such as cutting fluids, oils and swarfs. However, RP materials themselves and the processing and post-processing of them provide significant opportunity to introduce contamination that has the potential to harm when used for medical purposes. Many RP manufacturing processes have prescribed methods of post-processing, but there is still significant opportunity for deviation in cleanliness, especially when dealing with complex, anatomical-based structures that include areas that are difficult to access with fluids and cleaning implements. The suitability of various RP processes for any given medical application will vary based on factors such as:

*   The geometry being produced (will it have any deep crevices or features that may prevent support structures being removed or adequate cleaning?)
*   The intended application (classification according to the MDD. This will dictate the necessary level of material testing, cleaning and sterilising to which parts should be subjected.)

When choosing an RP process for fabricating end-use medical devices, the ability to rigorously clean and/or sterilise parts should be at the forefront of consideration. Some processes, such as powder bed 3D printing, do not lend themselves well to cleaning and would not withstand most forms of sterilising. Other processes, such as SL, SLS, DLP, some jetting head technologies and most metal-based technologies, have a more demonstrated track record for producing end-use medical devices. The common characteristic of these processes is the availability of materials that have undergone prerequisite testing to USP/23 Class VI or ISO 10993 and ability to produce parts that can withstand rigorous cleaning (will not degrade with handing and scrubbing, for example). With the exception of SLS, processes that have a more demonstrated track record in medical applications also tend to be those capable of producing 100% dense components, which makes them easier to clean and sterilise. Further advantages are offered by processes that either do not require a support structure (such as SLS) or that use the same material for support structure and part (such as SL or DLP). This means that only one material requires validation. However, one drawback of using a support structure in the same material as the part is that they can be difficult to fully remove from hard to reach areas (such as sinuses on anatomical models) and leave witness marks. Jetting head technologies such as Objet and ProJet use support materials that are dissimilar and encase the part, further compounding the difficulty in fully removing residues or unwanted contamination.

Despite the wide range of medical applications in which RP technologies are used, there is a limited range of academic literature describing how parts should be cleaned and sterilised. The following sections will describe general procedures used to clean and sterilise parts that are employed as medical devices and will put this in the context of RP production.

4.10.2   Cleaning

The purpose of cleaning can generally be classified as twofold: the removal of non-biological (manufacturing) contamination and minimisation of biological contamination (bioburden) to controlled levels.

Post-processing and cleaning regimes are particular to each RP process. A limited number of material suppliers and equipment manufacturers have published guidelines that describe both build parameters and post-processing procedures that should be adhered to when cleaning for medical applications. However, it may also be necessary for the device manufacturer to adapt processes to suit the final application and demonstrate through testing that its processes meet the requirements defined in the MDD.

In general, any cleaning method should be undertaken in a controlled environment, where the risk of contamination or bioburden is reduced. This means carefully considering routes of contamination and implementing controls to avoid it. Further consideration must also be given to how the cleaning process may affect the mechanical and chemical properties of the RP material. For resin-based SL and DLP processes, solvents such as isopropyl alcohol are used to remove residual monomer.

Ultrasonic cleaning methods are commonly used to remove contaminants from metal components but are less effective on polymers. Typical cycles will include multiple-stage ultrasonic cleaning (with each stage gradually reducing the contamination) and passivation (for titanium implants), followed by a rinse and drying.

4.10.3 Sterilisation

The purpose of sterilisation is to remove all forms of microbial life present on a surface in a fluid, medicine or other compound. The aim is to achieve a high Sterility Assurance Level (SAL), a term used to describe the probability of a single unit being non-sterile after the sterilisation process. There are numerous methods of sterilisation, some more compatible with RP processes than others. Three common methods are:

• Autoclave
• Gamma radiation
• Ethylene oxide

Autoclaving is a widely used and highly accessible method of sterilisation available in the vast majority of hospitals. It involves the use of disinfecting/cleaning stages, combined with high-temperature steam and pressure to remove debris and sterilise the part. A typical cycle will involve disassembly and labelling of equipment, which is then placed on a metal tray; pressure washing in a detergent to remove debris; inspection; sorting according to the operating theatre lists; wrapping in a linen bag or packing in a special bag; labelling with a sterilisation indication marker; autoclaving on a specified cycle at 134 °C; and cooling. The entire cycle can take around 4 h. If at any point during the cycle there is an indication that the parts are not sterile, the cycle is repeated. The high temperatures and pressures may make autoclaving an unsuitable method for sterilising many polymer parts, especially if it is necessary to sterilise them multiple times.

Gamma sterilisation is a method commonly used to sterilise disposable medical devices and those that are sensitive to heat and steam. This method of sterilisation is typically found in medical device factories and is less common in hospitals. Irradiation can also have a detrimental effect on some polymer materials.

Ethylene oxide (EO or EtO) is used to sterilise objects sensitive to temperatures greater than 60 °C and/or radiation, such as plastics. It is commonly used for large-scale sterilisation of disposable devices but is less common in hospitals; it is often outsourced to private companies. One reason for this is that EtO gas is highly flammable, toxic and carcinogenic, with a potential to cause adverse reproductive effects. After initial cleaning and pre-conditioning phases, the sterilisation process can take around 3 h. This is followed by a stage to remove toxic residues.

In general, the physical and chemical effect of sterilisation processes on many RP materials, particularly polymers, is not well documented in the research literature.

Case Studies

Introduction

This chapter covers a wide range of applications of medical scanning, 3D CAD and RP in a variety of different medical disciplines and research projects. The chapter is made up of individual case studies based on the authors' research over the last 17 years. The chapter is divided into six sections covering the implementation of RP in medicine, surgical applications, maxillofacial prosthetic rehabilitation, orthotic rehabilitation, dental technology and finally the application of medical modelling techniques in the support of research projects. The case studies go into detail about the technologies and techniques employed based on the concepts introduced in Chapters 1 through 4. The majority of the work has been previously published in international peer-reviewed journals and conference proceedings and where this is the case full acknowledgements and citations to the original publications are provided including digital object identifiers where available.

The first section, Implementation, covers the issues to consider when collaborating between designers, engineers, RP service providers, radiographers and clinicians. The papers highlight possible problems and suggest ways of avoiding or overcoming them. The second section, Surgical applications, covers the use of medical modelling techniques in the support of surgical planning and the design and manufacture of surgical guides and implants. The third section, Maxillofacial prosthetic rehabilitation, focuses on the use of medical modelling techniques to support the provision of maxillofacial prosthetics (typically facial prostheses but also including burns conformers and breast prostheses). The fourth section, Orthotic rehabilitation, describes the use of scanning, CAD and RP in the development of custom-fitting, wearable orthotic devices; specifically wrist splints. The fifth section, Dental applications, describes the development of CAD and RP technologies for the design and manufacture of removable partial denture (RPD) frameworks, digital reproduction and the design and manufacture of custom-fitting sleep apnoea devices. Finally, the sixth section, Research applications, covers a variety of applications where medical modelling techniques have been used to support research projects including the validation of finite element analysis, the appearance of AM materials in CT images, texture reproduction, archaeology, bone surrogate materials and surgical training devices.

Medical Modelling. http://dx.doi.org/10.1016/B978-1-78242-300-3.00005-6

Implementation

5.1 Implementation case study 1: computed tomography guidelines for medical modelling using rapid prototyping techniques

Acknowledgements

The work described in this case study was first reported in the reference below and is reproduced here in part or in full with the permission of Elsevier Publishing.

Bibb, R., & Winder, J. (2010). A review of the issues surrounding three-dimensional computed tomography for medical modelling using rapid prototyping techniques. *Radiography, 16*, 78–83. http://dx.doi.org/10.1016/j.radi.2009.10.005.

5.1.1 Introduction

Medical modelling is the term for the production of highly accurate physical models of anatomy directly from three-dimensional (3D) medical image data using computer-controlled manufacturing machines commonly referred to as rapid prototyping (RP). Medical modelling involves acquiring 3D image data of human anatomy, processing the data to isolate individual tissues or organs of interest, optimising the data for the RP technology and finally building the physical model in an RP machine. These machines have been primarily developed to enable designers and engineers to build exact models of products that they have designed using computer-aided design software (CAD). Consequently, RP technologies have developed to produce models of very high accuracy as rapidly as possible.

Medical modelling has many applications and the most common has been in head and neck reconstruction including neurosurgery, craniofacial/maxillofacial surgery and manufacturing prosthetics and implants. Medical models are routinely used for diagnosis, communication and presurgical planning but also they are increasingly used in the design and manufacture of implants and prosthesis (Bibb, Bocca, & Evans, 2002; D'Urso et al., 2000; Eufinger & Wehmoller, 1998a; Evans, Eggbeer, & Bibb, 2004; Heissler et al., 1998a; Hughes, Page, Bibb, Taylor, & Revington, 2003a; Joffe et al., 1999a; Singare, Dichen, Bingheng, Zhenyu, & Yaxiong, 2005; Winder, Cooke, Gray, Fannin, & Fegan, 1999a), creating surgical guides (Bibb, Bocca, Sugar, & Evans, 2003; Di Giacomo, Cury, De Araujo, Sendyk, & Sendyk, 2005; Goffin, Van Brussel, Vander Sloten, Van Audekercke, & Smet, 2001; Goffin et al., 1999; Sarment, Al-Shammari, & Kazor, 2003; Sarment, Sukovic, & Clinthorne, 2003) and

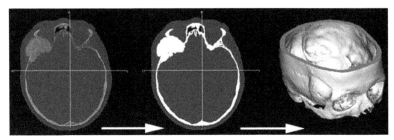

Figure 5.1 Steps from CT image to 3D reconstruction.

Figure 5.2 Example of a medical model.

making imaging phantoms and teaching. Clear demonstration of benefits published in case studies and review articles have led to increasing interest in medical modelling (Bibb & Brown, 2000a; Petzold, Zeilhofer, & Kalender, 1999a; Sanghera, Naique, Papaharilaou, & Amis, 2001). Hundreds of medical models are produced in the United Kingdom each year and the numbers are growing rapidly making it increasingly likely that radiographers will be asked to provide computed tomography (CT) images for medical modelling.

Usually, a surgeon or clinician who requires a medical model will request a 3D scan of the area of interest. The medical modelling process requires the relevant anatomy to be captured in a 3D format and although a number of medical imaging technologies have been successfully employed to make models, including magnetic resonance imaging (MRI) (Swann, 1996) and ultrasound (D'Urso & Thompson, 1998), volumetric

CT (3D CT) is by far the commonest imaging modality. The 3D medical image data is processed, mathematically modelled and subsequently transferred to a rapid prototype model provider for manufacture. Useful reviews of the RP methods and clinical applications are available (Bibb, 2006; Webb, 2000a). After acquisition and transfer, the images are imported into specialist RP software and techniques such as image thresholding and region growing are used to isolate the desired anatomical structure (Gonzalez & Woods, 2007, pp. 443–458). This data is then exported in a format called stereolithography (STL) that can be used by RP machines. The process is illustrated in Figure 5.1.

The software for the RP machine then slices the 3D data to produce a cross-section for each layer the machine will build. It is not possible to describe each RP technology in detail here but more information can be found in reference texts (Bibb, 2006). A typical medical model of the mid-face and mandible is shown in Figure 5.2. The model is multicoloured to help the clinician distinguish between different tissue types identified within the CT images.

This technical note aims to provide radiographers with recommendations that will enable them to provide good quality 3D CT images that can fulfil the requirements of medical modelling quickly and efficiently. There have been many case studies describing the use of medical models but comparatively few have addressed the issues encountered when attempting to use medical modelling (Sugar, Bibb, Morris, & Parkhouse, 2004; Winder & Bibb, 2005). Potential problem areas in data acquisition and transfer are addressed in this note and the guidelines described here aim to help avoid these errors.

5.1.2   CT guidelines for medical modelling

5.1.2.1   Anatomical coverage

An ideal CT acquisition should be free from image artefacts, have isotropic voxel resolution, high image contrast between the anatomy of interest and neighbouring tissues and low noise. It is clear that the series of axial images must begin and end either side of the anatomy of interest but it is important to begin and end the scan some distance either side of the region/anatomy of interest. It is better to include anatomy (perhaps a few centimetres either side up to a maximum of 5 cm) above and below the area of interest, the amount being dependent on the region being scanned. In some cases, anatomy beyond the area of direct interest is used to help form or shape a repair to a bone of soft-tissue defect (D'Urso et al., 2000; Vander Sloten, Van Audekercke, & Van der Perre, 2000; Verdonck, Poukens, Overveld, & Riediger, 2003; Winder, McRitchie, McKnight, & Cooke, 2005; Winder et al., 1999a). The data volume should be continuous as noncontinuous data may contain areas where the patient has shifted position slightly and the separate series will not align perfectly.

5.1.2.2   Patient arrangement, positioning and support

Movement during the CT acquisition will result in movement artefact and distort the image data, which will translate directly through to the finished model. Anything

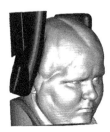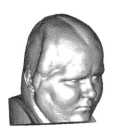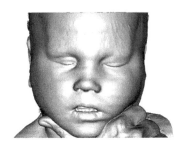

Figure 5.3 Soft-tissue deformations caused by support.

greater than a 1-mm movement may render a model unusable. Involuntary movement of the chest, neck, head or mouth can occur through breathing or swallowing.
Gated acquisition for respiratory motion compensation as well as cardiac compensation is becoming increasingly common in multislice CT scanning (Li et al., 2008)
and their implementation would lead to significant movement artefact reduction.
Other sources of movement like swallowing or talking should be controlled as far
as possible.

It is increasingly common for 3D CT scans to be used in the multidisciplinary
management of a patient, where the image data is not only used for diagnosis but
also for surgical planning, computer guided surgery, medical modelling and prostheses design. The 3D CT data may be used to represent both hard and soft tissue
internally but may also find subsequent application in tissue reconstruction or prosthetic rehabilitation. It is therefore important to consider the positioning and support
of soft tissues to eliminate or reduce unwanted deformation of soft tissue. The use
of positioning pads should be considered so that the anatomy of interest does not
become distorted prior to scanning. Patient immobilisation techniques such as vacuum pillows or simple foam pads may be used to support the patient, even to the
extent that the surgical position of the patient may be replicated within the scanner.
Examples of unwanted soft-tissue deformations caused during CT acquisition can
be seen in Figure 5.3, where the images were acquired for the soft tissue information. Note the use of a hand in supporting a child's chin in Figure 5.3. The use of the
data should be clarified with the referring clinician to ensure appropriate positioning
and support is used.

When using CT data for surgical planning it is often necessary for different
bones to be manufactured individually so that they can be moved independently to
simulate surgical techniques. It is common for example when planning maxillofacial surgery to perform osteotomies and move parts of the mandible or maxilla.
Often the patient is scanned with a closed bite, which causes the data for upper
and lower teeth to merge. This makes subsequent separation of the mandible and
maxilla using image-processing techniques very difficult. The effect of this overlap can be seen in Figure 5.4. For maxillofacial cases, it is recommended that a
slightly open bite or spacer be used that will enable the different jaws to be separately segmented.

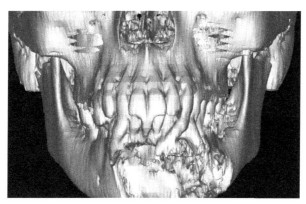

Figure 5.4 Overlap of the occlusion.

5.1.2.3   CT parameters

There is a very wide range of CT technology available in the Health Service (from single-slice helical to 64 slices or greater) and the variation in clinical practice throughout the United Kingdom and Ireland is very wide. It would not be appropriate for us to provide a definitive CT scanning protocol for any or all regions of the body, given that different centres will have different technology and levels of experience. It is recommended that radiographers use their routine 3D volumetric protocol for the given anatomical area also taking into consideration some of the issues that follow.

5.1.2.4   Slice thickness

To maximise the data acquired, the reconstructed slice thickness should be minimised. Some scanners can produce 0.5-mm slices, which gives excellent results but this must be balanced against increased X-ray dose. Typically, slice thicknesses of 0.5–1 mm produce acceptable results. A slice thickness of 2 mm may be adequate for larger structures such as the long bones or pelvis. A slice thickness greater than 2 mm will give poorer results and it is not recommended for rapid prototype models.

Consideration may also be given to collimation and overlap. In most circumstances, the scan distance and collimation should be the same. However, using a slice distance that is smaller than the collimation gives an overlap. When scanning for very thin sections of bone that lie in the axial plane, such as the orbital floor or palate, an overlap may give improved results (Hughes et al., 2003a).

5.1.2.5   Gantry tilt

Gantry tilt is commonly used in CT to provide the appropriate angle of slice relative to the anatomy of interest and also to reduce the radiation dose to the orbits in head scanning. However, for the purposes of medical modelling, gantry tilt should be set to zero (0°) because it does not significantly improve the quality of the acquired data and provides an opportunity for error when the service provider imports the images.

Large gantry tilt angles are clearly apparent on visual inspection of the data and may be corrected. However, small angles may not be easy to check visually and may remain undetected or be compensated for incorrectly. Even the use of automatic import of the medical image standard Digital Imaging and Communications in Medicine (DICOM) is no guarantee as, although the size of angle is included in the format, the direction is not. Failure to compensate for the direction correctly will lead to a distorted model, wasting time and money and potentially leading to errors in surgery or prosthesis manufacture.

5.1.2.6   X-ray scatter

Dense objects such as amalgam or gold fillings, braces, bridges, screws, plates and implants scatter X-rays resulting in a streaked appearance in the image. It is common practice in radiography to remove all metal to reduce artefacts where possible. However, where metal implants cannot be removed, the effects of these can be manually edited using medical imaging software to produce a better medical model. However, this does depend on the expertise of the service provider and consequently the accuracy of the model in the affected areas cannot be guaranteed.

5.1.2.7   Noise

Noise is a fundamental component of a CT image and is especially prevalent in dense tissues. Medical modelling depends on identifying smooth boundaries between tissues. Noise interrupts the boundary, resulting in poor 3D reconstructions and consequently poor medical models, typically appearing rough or porous. This is shown in Figure 5.5. Efforts should be made to reduce noise where medical models are required.

5.1.2.8   Image reconstruction kernels

During the reconstruction, digital filters (kernels) are applied which enhance or smooth the image depending on the clinical application. Typically, the options will

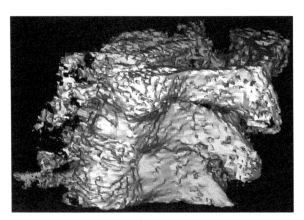

Figure 5.5 Porous appearance caused by noise.

range from 'sharp' to 'smooth'. Sharpening filters increase edge sharpness but at a cost of increasing image noise. Smoothing filters reduce noise content in images but also decrease edge sharpness. In general, when building medical models, smooth filters tend to give better results and are easier to work with. Although the smooth image contrast appears poor on screen (Windows computers can only display 256 shades of grey and the human eye can only perceive about 70 grey levels), density profiles show that the actual contrast is good and allows a lower threshold to be used.

5.1.2.9 Data transfer

3D medical image processing and medical RP software (e.g. Mimics, Materialise NV, Technologielaan 15, 3001 Leuven, Belgium; AnalyzeDirect, Inc., 7380 W 161st Street, Overland Park, KS, 66085 USA) require DICOM V3.0 data format and are usually sent to a medical modelling service provider on CD-ROM. In nearly all cases, only the reconstructed axial/transverse images are required as further image processing and modelling will be carried out by the service provider. CT images should be written without image compression or the automated viewing software. From a patient confidentiality point of view, the exclusion of the manufacturer's viewing software means that images on a lost or misplaced CD cannot be easily viewed without specialist knowledge and software. The images do not need to be 'windowed' before storage or transfer as access to the original DICOM images allows the service provider to view them with their own settings. Careful consideration should be given to patient confidentiality and data security and procedures should be agreed with the service provider to ensure all data is securely and ethically treated. For example, data should be sent by registered post, the service provider should be requested to store data in a locked cabinet, and access to any data should be password protected. Arrangements should also be made to return or destroy the CD on completion of the model.

5.1.3 Conclusion

Following the general guidelines presented here will help to improve the source volumetric CT image data used for medical RP and subsequently improve on model quality.

Acknowledgements

Figure 5.2 is reproduced with the permission of Queen Elizabeth Hospital Birmingham, Figure 5.3(a) was kindly provided by Dr Jules Poukens, University Hospital Maastricht, the Netherlands. Figures 5.3(b) and 5.4 were kindly provided by Carol Voigt, Brånemark Institute, South Africa.

References

Bibb, R. (2006). *Medical modelling: The application of advanced design and development technologies in medicine*. Cambridge: Woodhead.

Bibb, R., Bocca, A., & Evans, P. (2002). An appropriate approach to computer aided design and manufacture of cranioplasty plates. *Journal of Maxillofacial Prosthetics and Technology*, *5*(1), 28–31.

Bibb, R., Bocca, A., Sugar, A., & Evans, P. (2003). Planning osseointegrated implant sites using computer aided design and rapid prototyping. *Journal of Maxillofacial Prosthetics and Technology*, *6*(1), 1–4.

Bibb, R., & Brown, R. (2000a). The application of computer aided product development techniques in medical modelling. *Biomedical Sciences Instrumentation*, *36*, 319–324.

Di Giacomo, G. A. P., Cury, P. R., De Araujo, N. S., Sendyk, W. R., & Sendyk, C. L. (2005). Clinical application of stereolithographic surgical guides for implant placement: preliminary results. *Journal of Periodontology*, *76*(4), 503–507.

D'Urso, P. S., Earwaker, W. J., Barker, T. M., Redmond, M. J., Thompson, R. G., Effeney, D. J., et al. (2000). Custom cranioplasty using stereolithography and acrylic. *British Journal of Plastic Surgery*, *53*(3), 200–204.

D'Urso, P. S., & Thompson, R. G. (1998). Fetal biomodelling. *Australian and New Zealand Journal of Obstetrics and Gynaecology*, *38*(2), 205–207.

Eufinger, H., & Wehmoller, M. (1998a). Individual prefabricated titanium implants in reconstructive craniofacial surgery: clinical and technical aspects of the first 22 cases. *Plastic and Reconstructive Surgery*, *102*(2), 300–308.

Evans, P., Eggbeer, D., & Bibb, R. (2004). Orbital prosthesis wax pattern production using computer aided design and rapid prototyping techniques. *Journal of Maxillofacial Prosthetics and Technology*, *7*, 11–15.

Goffin, J., Van Brussel, K., Vander Sloten, J., Van Audekercke, R., & Smet, M. H. (2001). Three-dimensional computed tomography-based, personalized drill guide for posterior cervical stabilization at C1-C2. *Spine*, *26*(12), 1343–1347.

Goffin, J., Van Brussel, K., Vander Sloten, J., Van Audekercke, R., Smet, M. H., Marchal, G., et al. (1999). 3D-CT based, personalized drill guide for posterior transarticular screw fixation at C1-C2: technical note. *Neuro-Orthopedics*, *25*(1–2), 47–56.

Gonzalez, R. C., & Woods, R. E. (2007). *Digital image processing* (3rd ed.). Reading, MA: Addison Wesley.

Heissler, E., Fischer, F. S., Bolouri, S., Lehmann, T., Mathar, W., Gebhardt, A., et al. (1998a). Custom-made cast titanium implants produced with CAD/CAM for the reconstruction of cranium defects. *International Journal of Oral and Maxillofacial Surgery*, *27*(5), 334–338.

Hughes, C. W., Page, K., Bibb, R., Taylor, J., & Revington, P. (2003a). The custom-made titanium orbital floor prosthesis in reconstruction for orbital floor fractures. *British Journal of Oral and Maxillofacial Surgery*, *41*(1), 50–53.

Joffe, J., Harris, M., Kahugu, F., Nicoll, S., Linney, A., & Richards, R. (1999a). A prospective study of computer-aided design and manufacture of titanium plate for cranioplasty and its clinical outcome. *British Journal of Neurosurgery*, *13*(6), 576–580.

Li, G., Citrin, D., Camphausen, K., Mueller, B., Burman, C., Mychalczak, B., et al. (2008). Advances in 4D medical imaging and 4D radiation therapy. *Technology in Cancer Research and Treatment*, *7*(1), 67–81.

Petzold, R., Zeilhofer, H.-F., & Kalender, W. A. (1999a). Rapid prototyping technology in medicine – basics and applications. *Computerized Medical Imaging and Graphics*, *23*(5), 277–284.

Sanghera, B., Naique, S., Papaharilaou, Y., & Amis, A. (2001). Preliminary study of rapid prototype medical models. *Rapid Prototyping Journal, 7*(5), 275–284.

Sarment, D. P., Al-Shammari, K., & Kazor, C. E. (2003). Stereolithographic surgical templates for placement of dental implants in complex cases. *International Journal of Periodontics and Restorative, 23*(3), 287–295.

Sarment, D. P., Sukovic, P., & Clinthorne, N. (2003). Accuracy of implant placement with a stereolithographic surgical guide. *International Journal of Oral and Maxillofacial Implants, 18*(4), 571–577.

Singare, S., Dichen, L., Bingheng, L., Zhenyu, G., & Yaxiong, L. (2005). Customized design and manufacturing of chin implant based on rapid prototyping. *Rapid Prototyping Journal, 11*(2), 113–118.

Sugar, A., Bibb, R., Morris, C., & Parkhouse, J. (2004). The development of a collaborative medical modelling service: organisational and technical considerations. *British Journal of Oral and Maxillofacial Surgery, 42*, 323–330.

Swann, S. (1996). Integration of MRI and stereolithography to build medical models. *Rapid Prototyping Journal, 2*(4), 41–46.

Vander Sloten, J., Van Audekercke, R., & Van der Perre, G. (2000). Computer aided design of prostheses. *Industrial Ceramics, 20*(2), 109–112.

Verdonck, H. W. D., Poukens, J., Overveld, H. V., & Riediger, D. (2003). Computer-assisted maxillofacial prosthodontics: a new treatment protocol. *The International Journal of Prosthodontics, 16*(3), 326–328.

Webb, P. A. (2000a). A review of rapid prototyping (RP) techniques in the medical and biomedical sector. *Journal of Medical Engineering and Technology, 24*(4), 149–153.

Winder, R. J., & Bibb, R. (2005). Medical rapid prototyping technologies: state of the art and current limitations for application in oral and maxillofacial surgery. *Journal of Oral and Maxillofacial Surgery, 63*(7), 1006–1015.

Winder, J., Cooke, R. S., Gray, J., Fannin, T., & Fegan, T. (1999a). Medical rapid prototyping and 3D CT in the manufacture of custom made cranial titanium plates. *Journal of Medical Engineering and Technology, 23*(1), 26–28.

Winder, J., McRitchie, I., McKnight, W., & Cooke, S. (2005). Virtual surgical planning and CAD/CAM in the treatment of cranial defects. *Studies in Health Technology and Informatics, 111*, 599–601.

5.2   Implementation case study 2: the development of a collaborative medical modelling service – organisational and technical considerations

Acknowledgements

The work described in this case study was first reported in the reference below and is reproduced here in part or in full with the permission of the British Association of Oral and Maxillofacial Surgeons. An update has been added to reflect more recent developments in data transfer.

Sugar, A., Bibb, R., Morris, C., & Parkhouse, J. (2004). The development of a collaborative medical modelling service: organisational and technical considerations. *British Journal of Oral and Maxillofacial Surgery, 42*(4), 323–330. http://dx.doi.org/10.1016/j.bjoms.2004.02.025

This work was only been possible because of the enthusiasm and hard work of the entire collaborating team from the National Centre for Product Design & Development Research and Morriston Hospital. In particular, the authors gratefully acknowledge the work of Peter Evans and Alan Bocca of the Maxillofacial Unit and Dr E Wyn Jones, Rose Davies and Sian Bowen of the Radiology Department of Morriston Hospital. We are also grateful to our Neurosurgical colleague, Tim Buxton, who has supported this project physically and financially since its inception.

5.2.1   Introduction

Medical modelling has been shown to be a valuable aid in the diagnosis of medical conditions, the planning or surgery and production of prostheses (Greenfield & Hubbard, 1984, pp. 91–130; Jacobs, 1996a). RP techniques have shown significant advantages over previous milled models (Klein, Schneider, Alzen, Voy, & Gunther, 1992). However, the application of the technologies requires close and efficient collaboration between radiographers, surgeons, prosthetists and rapid prototyping service providers. As yet, no optimum method of achieving this has been demonstrated and often the process is ad hoc leading to potential errors and delays. Close cooperation between the Maxillofacial Unit of Morriston Hospital and the National Centre for Product Design & Development Research in several successful cases led to the desire to improve the process in terms of data transfer and clinical decision-making (Bibb & Brown, 2000b; Bibb, Brown et al., 2000; Bibb, Freeman et al., 2000). This case study describes the efforts of this hospital and RP service provider to develop an efficient and rapid collaboration that serves the clinical needs of the hospital.

5.2.2   Aims of medical modelling collaboration

For most medical models produced in the United Kingdom, the preparation of the RP build files from patient scan data is undertaken by a very small number of specialist suppliers. One of the principal reasons for this is that scan data is only normally available in proprietary formats on archive media, which cannot be read unless the appropriate hardware and software is purchased. This equipment differs for each manufacturer of medical scanner. Additional software is also required to segment the scan data and produce the RP build files. Many RP service providers are therefore unwilling to invest the large sums necessary to purchase this expensive software and equipment when the number of models made would make it difficult to recoup the investment.

Therefore, in most cases scan data on archive media (e.g. magnetic optical disk, DAT tape, CD-ROM) is posted from the radiology department to the specialist supplier. The supplier translates and segments the data from which the RP build files are created. The transfer of medical scan data is described in Section 2.6. This requires careful communication so that the correct parameters, tissues and extents are used. Once translation is complete, the RP build files are then returned to the RP service provider who builds and delivers the model to the hospital.

This procedure, schematically shown in Figure 5.6 and called the 'disconnected' procedure, is time-consuming and removes potentially important clinical decision-making opportunities from medical staff.

In an effort to improve this situation, the authors piloted a more collaborative approach. Although the RP generation software needed to be purchased, the use of direct communication eliminated the need to translate data from the archive media saving a considerable amount of time and money. This process involved sharing the necessary procedures between the medical staff and the RP service provider. Effectively, the tasks were accomplished by the most appropriate staff, improving decision-making opportunities whilst reducing cost and turnaround time. This approach was further accelerated by the use of electronic communication rather than routine mail.

Because of the modular nature of the software used, the Initial Integrated Procedure shown in Figure 5.7 was first attempted (Swaelens & Kruth, 1993). Experienced radiographers in a location accessible to the surgeons conducted the segmentation of the image data within the radiology department. The segmented data was then sent to the RP service provider who used the remaining software modules to prepare and manufacture the models. Once such a route was established, the turnaround time from initial scan to finished model was reduced from weeks to days.

This approach had both technical and organisational implications as described in the next section. Such considerations may limit the adoption of what appears at first sight to be the optimal route. The technical issues may all be overcome given sufficient resources but organisational conditions will dictate whether such resources are available or appropriate.

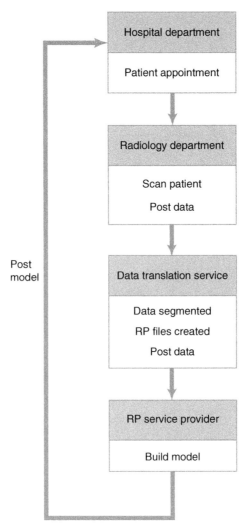

Figure 5.6 The 'disconnected' procedure.

5.2.3   Implementation

The software used in this study consists of one module (Mimics, Materialise NV, Leuven, Belgium) used for the import and segmentation of scan data. A separate module (CT Modeller) generates the rapid prototyping build files from data exported from Mimics. This allowed the segmentation software (Mimics) to be installed in both the radiology and the clinical departments while the other module (CT Modeller) remained at the RP service provider. The segmentation module was installed on well-specified PCs connected to the hospital network allowing direct access to the scan data. The basic principles of data segmentation are more fully described in Chapter 3.

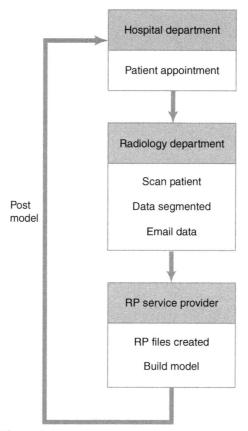

Figure 5.7 The initial integrated procedure.

When using this software, there is a convenient intermediate step between the different modules. The export file format from Mimics to CT Modeller (called a .3dd file) is highly compressed, typically in the region of 1 MB. Files this small are easily transferred as an attachment to an ordinary email. The use of existing email protocols including the firewalls at the hospital and the service provider eliminated security and network access issues.

After training and some initial trials, it became clear that the file-import, preparation and segmentation were too time-consuming to be undertaken by the radiographers. Although highly trained in the operation of the scanner, the radiographers were unfamiliar with other computer formats and network procedures. As the radiology department was already working at capacity, it was felt that this was not the best use of their time. It was originally envisaged that radiographers would be better able to segment the scan data. However, in practice the simplicity of the software allows accurate segmentation to be accomplished by any adequately trained user and is well within the competence of clinicians. Furthermore, clinicians have a much better idea of what they want to achieve by the segmentation.

It was also felt that the majority of the clinical decision-making should be the responsibility of the clinical department. This would eliminate any potential

misunderstandings between medical and technical staff. Therefore, a modified procedure was implemented (Figure 5.8, The Current Integrated Procedure). This procedure
involved the clinical department having the software to enable them to segment the
data and email the compressed file format (a 3D image) to the RP service provider.

This method also keeps as much of the workload (and therefore cost) within the
clinical department and minimises the workload transferred to radiology. The added
benefit of this method is the capability of the clinical department to produce 3D reconstructions on screen. The surgeon can then view and move these images at will rather

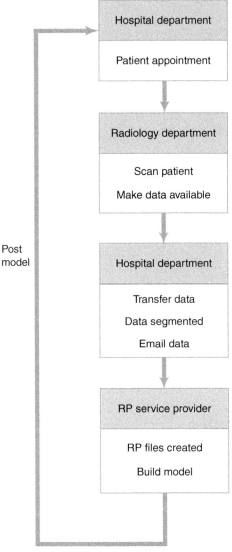

Figure 5.8 The current integrated procedure.

than relying on a selection of fixed views produced on film by the radiology department. Axial slices can be viewed simultaneously with coronal and sagittal reformats; and the slices can be run through in seconds. Areas of specific interest can be generated in three dimensions and viewed from any angle. The increased access to 2D and 3D data in various forms may also eliminate the need to produce a physical model in many cases, with a potential saving of time and money. The images produced can also be saved on hard disk, printed on plain paper for storage in patients' notes and exported into standard image software and from there, into slide presentation formats for teaching/lecturing/demonstrations. Hard copy can also be given to patients to help them understand their condition.

When following this procedure, the radiology department obtains the scans in the usual manner and saves the data into a secure directory on a PC hard-wired to the scanner. This PC enables easy storage of data in the radiology department in their preferred format. The PC was connected to the hospital's local Intranet, which makes the data available to the registered user in the maxillofacial unit using file transfer protocol (FTP) to another PC in the maxillofacial unit also connected to the intranet. Other registered users in other hospital departments can also be included if necessary. The data is then burned in the maxillofacial unit onto CD, saved on the PC's hard disk and imported into the specialised imaging software (Mimics®). Image manipulation can then be carried out by clinicians on screen in 3D. Sections can be mirror-imaged and manipulated and areas measured without interfering with the schedule of a busy radiology department. The file for RP model construction, if required, can then be emailed to the RP provider as an attachment.

Of course, when hospitals change their CT scanners, it is necessary to check that existing procedures still work and that the imaging software (e.g. Mimics) is configured appropriately to receive the new scan format. Current scanners almost invariably now comply with a standard digital format (DICOM) and the importation of such scans into Mimics is in fact made much easier in this format. When Morriston Hospital acquired a new multislice CT scanner in 2002, all its raw CT data was saved on a server within the hospital Intranet. The maxillofacial unit was then given direct access to this server through the Intranet and CT scans can be picked up directly and saved. This method has effectively replaced the FTP arrangement.

5.2.4 Discussion

5.2.4.1 Technical issues

The most important technical issues in implementing such a modelling process are:

- The export file format of the scan data
- The nature of the storage media
- The electronic transfer of the data

If the scan data media and/or format are proprietary, the appropriate software and hardware may have to be purchased by the collaborating parties. Many radiology departments are not accustomed to exporting data and may only have such archiving formats available as DAT or optical disk. In addition, some radiology departments are still not networked although this is likely to change over the next few years as teleradiology becomes more widespread. These data transport issues can be eliminated if the data can be directly transferred across computer networks. Electronic communication can be achieved in a number of ways and the method chosen should be economic and fast whilst maintaining security.

Problems may be encountered when attempting to transfer very large data sets (a full set of 3D scan data may be 50–300 MB) as network disruptions become more probable during long download times. Such problems may result in corrupted or incomplete data. The use of compression software may help but the data sets may still be large. However, evidence suggests that data is frequently transferred successfully in this manner. FTP is commonly used for transferring data across local and remote networks. To maintain security FTP sites can be protected with user names and passwords as can access to servers.

Although the National Health Service (NHS) in the United Kingdom has a dedicated network (for example, in Wales the local network is called Digital All Wales Network or DAWN) it is, in the short term, unlikely that external RP service providers would be allowed access to such a network. Maintaining security is understandably a high priority for NHS networks. Even if access were allowed, the network capacity is limited when compared to other high capacity networks, such as the Higher Education service 'JANET' (Joint Academic Network).

If the data files are small enough, they can be sent as an email attachment. In this case, security would be maintained through the email server's existing firewall. However, most email systems would become unsuitable if the file size exceeded 3 MB or the number of files sent per day proves excessive, as most servers set storage limits for users.

An ISDN line could be used but the security protocols for connecting to NHS networks may still apply in this case. If a PC is connected to a hospital intranet (and through it the Internet), connection simultaneously to an ISDN line could breach firewall regulations. The installation of ISDN lines within the hospital is likely to require some form of management approval and agreement concerning which budget will cover the cost of installation, line rental and call charges. Once in place, however, it should prove a reliable and secure transfer method.

Some consideration will have to be made concerning the amount of traffic the link will be expected to handle. It may well be that in the short term there will be insufficient traffic to warrant the investment in new networks. Alternatively, more commonly used high capacity storage systems such as CD-R, CD-RW, DAT, ZIP or JAZ drives could be used.

It is worth bearing in mind that the majority of medical models required will be for scheduled elective operations. In such cases, delay in communication of data can be anticipated and planned. However, in the long term, it is hoped that the service may be used to aid in the treatment of trauma victims in which case speed will be of great importance and a direct link may save crucial hours from the turnaround time.

A description of the commonly used formats and some of the issues that may be encountered in the effective transfer of medical scan data can be found in Section 2.6.

5.2.4.2   Organisational issues

It is self-evident that setting up a successful dedicated medical modelling collaboration between a hospital and an RP service provider requires the full cooperation of the surgical departments, the radiology department, the information technology department and the RP service provider. The implications for the different departments will be different depending on which technical solution is adopted. The chosen procedure should reflect the overall needs of the patient whilst taking due account of economic factors. Consequently, the organisational considerations relate to:

• Economics and budgeting
• Staff workload and responsibility

It is widely acknowledged that hospital departments in the United Kingdom usually lack resources and operate within strictly controlled budgets. It is therefore crucial to obtain the support and commitment of senior hospital management at the outset of a project such as this. The costs involved in setting up the service will have to be met but there is likely to be pressure within departments to pass on as much of the cost to other sources of funding thus minimising the impact on departmental budgets.

In terms of workload, it is likely that each of the departments will be running at capacity, which may lead to conflicts between departments when the workload is distributed. Departments will try to resist any increase in their budget requirements or workload. There may be economic reasons for transferring workload from one department to another.

Careful thought will have to be given as to who takes responsibility for the key decisions taken during the medical modelling process. It is likely that clinical departments will want to maintain the maximum amount of responsibility and control whilst minimising workload and expenditure. The only practical method of resolving this difficulty is to try different procedures and determine which one best meets the clinical need. It should then be a matter of using this sound knowledge to apply for the correct budgeting and resources from hospital management.

It is worth emphasising that the use of medical models may drastically reduce theatre time and its associated costs as well as improving the quality and accuracy of surgery. However, cost savings have to be seen in context. For elective cases in the British NHS, the operating time available to surgeons is finite and spare time will be taken up by other cases. The principal effect, therefore, in a public funded health care system will be on waiting lists and waiting times for surgery. Quality of patient care and cost-effectiveness of care will improve but the need for a 3D model will impose an additional charge on tight budgets. Clinical control of the process, however, can enable many of these problems to be minimised. For example, Figure 5.9 shows a 3D CT reconstruction of a skull with a fronto-zygomatic bone defect (left). A model was required to enable a cranioplasty plate to be constructed and to assist planning for the placement of craniofacial osseointegrated implants. By eliminating unwanted parts of the 3D image, a new 3D image (Figure 5.9, right) was created by the clinicians and emailed in the form of a .3dd file to the RP provider so that the model constructed could be much smaller and therefore very much cheaper.

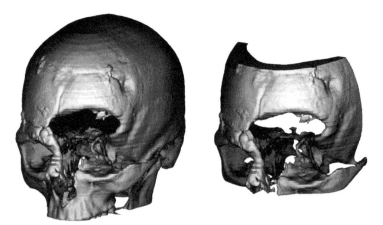

Figure 5.9 Whole 3D reconstruction (left) and reduced to the area of interest (right).

5.2.5   Conclusions

The clinical benefits of medical modelling have been documented and this work has demonstrated that a dedicated collaboration between the RP service provider and medical staff can greatly improve speed, efficiency and quality of service. Evidently, careful planning is necessary before embarking upon such collaboration. In particular, close cooperation between the rapid prototyping service provider, the radiology department, information technology departments, surgical teams and hospital management is of the utmost importance. Although it is inevitable that investment will be required, all of the long-term clinical and economic benefits should be clearly indicated to fund managers. Besides the principal goal of creating a medical modelling service, the setting up of convenient data formats and electronic communications can be exploited in many other ways. This can be seen in the number of hospitals embracing teleradiology, which allows consultants to view medical images from remote locations.

It should be anticipated that existing computer facilities might have to be upgraded with particular attention paid to hardware specification, communication and data transfer issues. In some cases, the scanner may also have to be networked or upgraded to allow the export of image data in a convenient format.

Although it is not in the form originally anticipated, the approach described here enabled a service to be established that provided the desired medical modelling service to the hospital in a fast, efficient and economic manner. The personnel most able to complete each task effectively handle the individual procedures. Data transfer is fast and secure without compromising either party's network security. Importantly, the bulk of the costs and decisions are kept within the surgical department that requires the medical model and that department is thus able to control the process and minimise those costs.

In the near future, it is likely that there will be many other factors driving hospitals to improve their networks and electronic communications. Issues regarding data format compatibility and file transfer will become more apparent and new standards applied to ensure the efficient and secure transfer of all kinds of clinical information.

Many of the issues encountered in this collaboration will be overcome by the overriding proliferation of computerised clinical information in all hospitals.

5.2.6 Update

Since the original publication of this work, direct access to hospital data through secure data connections are possible although strict data protection and security protocols have to be agreed and implemented. Whilst this enables faster transfer of data, eliminating the need for some of the steps detailed previously or the use of storage media and postage, the implications for clinical decision-making discussed in this article remain relevant.

References

Bibb, R., & Brown, R. (2000b). The application of computer aided product development techniques in medical modelling. *Biomedical Sciences Instrumentation, 36*, 319–324.

Bibb, R., Brown, R., Williamson, T., Sugar, A., Evans, P., & Bocca, A. (2000). The application of product development technologies in craniofacial reconstruction. In *Proceedings of the ninth European conference on rapid prototyping and manufacturing* (pp. 113–122). Athens, Greece.

Bibb, R., Freeman, P., Brown, R., Sugar, A., Evans, P., & Bocca, A. (2000). An investigation of three-dimensional scanning of human body surfaces and its use in the design and manufacture of prostheses. *Proceedings of the Institute of Mechanical Engineers Part H, Journal of Engineering in Medicine, 214*(6), 589–594.

Greenfield, G. B., & Hubbard, L. B. (1984). *Computers in radiology.* New York: Churchill-Livingstone.

Jacobs, P. F. (Ed.). (1996a). *Stereolithography and other RP&M technologies.* One SME Drive, Dearborn, MI: Society of Manufacturing Engineering.

Klein, H. M., Schneider, W., Alzen, G., Voy, E. D., & Gunther, R. W. (1992). Pediatric craniofacial surgery: comparison of milling and stereolithography for 3D model manufacturing. *Pediatric Radiology, 22*, 458–460.

Swaelens, B., & Kruth, J. P. (1993). Medical applications of rapid prototyping techniques. In *Proceedings of the fourth international conference on rapid prototyping* (pp. 107–120). Dayton, OH.

5.3   Implementation case study 3: medical rapid prototyping technologies – state of the art and current limitations for application in oral and maxillofacial surgery

Acknowledgements

The work described in this case study was first reported in the reference below and is reproduced here in part or in full with the permission of the American Association of Oral and Maxillofacial Surgeons.

Winder, R. J., & Bibb, R. (2005). Medical rapid prototyping technologies: state of the art and current limitations for application in oral and maxillofacial surgery. *Journal of Oral and Maxillofacial Surgery, 63*(7), 1006–1015. http://dx.doi.org/10.1016/j.joms.2005.03.016

5.3.1   Introduction

Medical rapid prototyping (MRP) is defined as the manufacture of dimensionally accurate physical models of human anatomy derived from medical image data using a variety of RP technologies. It has been applied to a range of medical specialities including oral and maxillofacial surgery (Anderl et al., 1994; Arvier et al., 1994; D'Urso et al., 1999; Eufinger & Wehmoller, 1998b; Gateno, Allen, Teichgraeber, & Messersmith, 2000; Hughes, Page, Bibb, Taylor, & Revington, 2003b; Sailer et al., 1998), dental implantology (Heckmann, Winder, Meyer, Weber, & Wichmann, 2001), neurosurgery (Heissler et al., 1998b; Winder, Cooke, Gray, Fannin, & Fegan, 1999b) and orthopaedics (Minns, Bibb, Banks, & Sutton, 2003; Munjal, Leopold, Kornreich, Shott, & Finn, 2000). The source of image data for 3D modelling is principally CT, although MRI and ultrasound have also been used. Medical models have been successfully built of hard tissue such as bone and soft tissues including blood vessels and nasal passages (Nakajima, 1995; Schwaderer et al., 2000). MRP was described originally by Mankovich, Cheeseman, and Stoker (1990). The development of the technique has been facilitated by improvements in medical imaging technology, computer hardware, 3D image processing software and the technology transfer of engineering methods into the field of surgical medicine.

The clinical application of medical models has been analysed in a European multi-centre study (Erben, Vitt, & Wulf, 2000). Results were collated from a questionnaire sent out to partners of the Phidias Network on each institution's use of MRP STL models.

The 172 responses indicated the following the following range applications.

• To aid production of a surgical implant
• To improve surgical planning

- To act as an orienting aid during surgery
- To enhance diagnostic quality
- Useful in preoperative simulation
- To achieve patient's agreement before surgery
- To prepare a template for resection

Further, it was noted that the diagnosis in which a medical model was employed were as follows: neoplasms (19.2%), congenital disease (20%), trauma (15%), dentofacial anomalies (28.9%) and others (16.9%). MRP is also being developed for use in dental implants. Greater accuracy was achieved with the use of rapid prototyped surgical guides for creating osteotomies in the jaw (Sarment, Sukovic, & Clinthorne, 2003) and a CAD/computed-aided manufacturing approach to the fabrication of partial dental frameworks has been developed (Williams, Bibb, & Rafik, 2004).

The creation of medical models requires a number of steps: the acquisition of high-quality volumetric (3D) image data of the anatomy to be modelled; 3D image processing to extract the region of interest from surrounding tissues; mathematical surface modelling of the anatomical surfaces; formatting of data for RP (this includes the creation of model support structures which support the model during building and are subsequently manually removed); model building; quality assurance of model quality and dimensional accuracy. These steps require significant expertise and knowledge in medical imaging, 3D medical image processing, CAD and manufacturing software and engineering processes. The production of reliable, high-quality models requires a team of specialists that may include medical imaging specialists, engineers and surgeons.

The purpose of this report is, first, to describe the range of RP technologies (including software and hardware) available for MRP; second, to compare their relative strengths and weaknesses; and third, to illustrate the range of pitfalls that we have experienced in the production of human anatomical models. The authors have a combined experience of 17 years working in the field of MRP and have direct experience of the technologies described later. The report begins with a description of 3D image acquisition and processing and computer modelling methods required, common medical RP techniques, followed by the discussion of model artefacts and manufacturing pitfalls. At the time of writing there was no suitable text describing MRP or its clinical applications; however, there are two useful review papers (Petzold, Zeilhofer, & Kalender, 1999b; Webb, 2000b).

5.3.2   3D image acquisition and processing for MRP

The modalities and general principles of acquiring medical scan data for RP are described in Chapter 2, but some of the more important observations resulting from a large number of actual cases are discussed here. The volumetric or 3D image data required for MRP models has certain particular requirements. Specialised CT scanning protocols are required to generate a volume of data that is isotropic in nature. This means that the three physical dimensions of the voxels (image volume elements) are equal or nearly equal. This has become achievable with the introduction of multislice CT scanners where in-plane pixel size is of the order of 0.5 mm and slice thickness as low as 1.0 mm (Fuchs, Kachelriess, & Kalender, 2000). Data interpolation is often required to convert the image data volume into an isotropic data set for mathematical modelling. Further image processing steps will be required to identify and separate out the anatomy (segmentation) for modelling from surrounding structures. Segmentation may be carried out by image thresholding, manual editing or auto-contouring to extract volumes of interest. Final delineation of the anatomy of interest may require 2D or 3D image editing to remove any unwanted details. Several software packages are available for data conditioning and image processing for MRP and include Analyze (Lenexa, KS, www.AnalyzeDirect.com), Mimics by Materialise (Leuven, Belgium, www.materialise.com) and Anatomics (Brisbane, Australia, www.anatomics.net). There is still a need for seamless and inexpensive software that provides a comprehensive range of data interpretation, image processing and model building techniques to interface with RP technology.

The first models created were of bone that was easily segmented in CT image data. Bone has a CT number range from approximately 200 to 2000. This range is unique to bone within the human body, as it did not numerically overlap with any other tissues. In many circumstances, a simple threshold value was obtained and applied to the data volume. All soft tissues outside the threshold range were deleted leaving only bone structures. Thresholding required the user to determine the CT number value that represented the edge of bone where it interfaced with soft tissue. Note that the choice of threshold may cause loss of information in areas where only thin bone is present.

In many circumstances, the volume of the body scanned is much larger than that actually required for model making. To reduce the model size, and therefore the cost, 3D image editing procedures may be employed. The most useful tool was a mouse-driven 3D volume editor that enabled the operator to delete or cut out sections of the data volume. The editing function deleted sections to the full depth of the data volume along the line of sight of the operator. Image editing reduced the overall model size which would also reduced RP build time. Clearer and less complex models may be generated making structures of interest more clearly visible. Other image processing functions such as smoothing, volume data mirroring, image addition and subtraction should be available for the production of models.

5.3.3   RP technologies

RP is a generic name given to a range of related technologies that may be used to fabricate physical objects directly from CAD data sources. RP enables design and manufacturing of models to be performed much more quickly than conventional manual methods of prototyping. In all aspects of manufacture, the speed of moving from concept to product is an important part of making a product commercially competitive. RP technologies enable an engineer to produce a working prototype of a CAD design for visualisation and testing purposes. There are a number of texts describing the development of RP technology and its applications (Jacobs, 1996b; Kai & Fai, 1997). Of the many RP processes that have been applied to medical modelling, the two

RP processes most extensively employed are stereolithography and fused deposition modelling.

5.3.3.1   Stereolithography

The following data provides some technical specifications of a specific type of SL machine that is in common use in medical modelling (3D systems SLA-250/40, 3D Systems, USA, www.3dsystems.com). However, a full description of the principles of SLA is provided in Chapter 4, Section 4.2.

- Laser beam diameter = 0.2–0.3 mm
- Laser scanning speed = 2.54 m/s
- Build platform = 250 × 250 × 250 mm
- Layer build thickness = 0.05–0.2 mm
- Minimum vertical platform movement = 0.0017 mm

These specifications indicate the precision of model building that is achievable with SL. The laser focus defines the in-plane resolution whilst the platform vertical increment defines the slice thickness at which the model is built. It should be noted that the imaging modality acquisition parameters are the limiting factors in model accuracy.

5.3.3.2   Fused deposition modelling

The technical specification of a commonly used fused deposition modelling (FDM)™ machine (Stratasys FDM-3000, Eden Prairie, MN, www.stratasys.com) used for models in the cases referred to here are as follows. A full description of the working principles of FDM™ is provided in Chapter 4, Section 4.4.

- Build envelope 254 × 254 × 254 mm
- Achievable accuracy of ±0.127 mm
- Road widths (extruded thermoplastic width) between 0.250 and 0.965 mm
- Layer thickness (extruded thermoplastic height) from 0.178 to 0.356 mm

These specifications indicate the precision of model building that is achievable with FDM™. It can be seen that the results are broadly similar to those achieved by SL. However, SL can achieve thinner layers and more precise control over the laser position compared to the deposition of plastic material in FDM™.

5.3.3.3   Computer controlled milling

Although generally not considered one of the many RP technologies, computerised numerically controlled (CNC) milling can successfully build some medical models (Joffe et al., 1999b). This technology was applied in the construction of custom titanium implants for cranioplasty. CNC milling uses a cutting tool, which traverses a block of material removing it on a layer-by-layer basis. Figure 5.10 shows a model of skull defect (only half the skull has been created). The complexity of models that can be achieved using CNC milling is limited as it only cuts on one side of the model data. If the model required has any internal features or complex surfaces facing a number

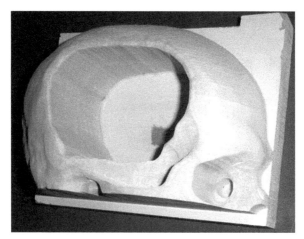

Figure 5.10 Half-skull model created by CNC milling demonstrating a large cranial defect.

of directions, then CNC milling would not be suitable. An overview of the principal differences between CNC milling and RP is provided in Chapter 4, Section 4.9.

5.3.3.4   Other RP technologies

Selective Laser Sintering (SLS®) locally heats a thermoplastic powder, which is fused by exposure to an infrared laser in a manner similar to SL. SLS® models do not require support structures and they are therefore cleaned relatively easily, thus saving labour costs. An example of the use of SLS® in medical modelling is described by Berry et al. (1997). Laminated object manufacturing (LOM) builds models from layers of paper cut using a laser, which are bonded together by heating. Inexpensive sheet materials make LOM very cost-effective for large volume models. However, the solid nature of the waste material means that it is not suited to models with internal voids or cavities often encountered in human anatomy. The SLS® and LOM processes are also described in more detail in Chapter 4, Sections 4.5 and 4.8, respectively.

5.3.3.5   Discussion of medical RP technologies

The main factors in choosing which RP technology are most appropriate for our clinical applications were as follows.

- Dimensional accuracy of the models
- Overall cost of the model
- Availability of technology
- Model building material

SL models are typically colourless to amber in colour, transparent and of sufficient accuracy to be suitable for MRP work. FDM™ models are typically made of white acrylonitrile-butadiene-styrene and attractive both in terms of appearance and material. It has been pointed out that medical models may be dimensionally accurate to

0.62 ± 0.35 mm (Choi et al., 2002). It should be noted that the limiting factor in model accuracy is the imaging technique rather than the RP technology employed. In general, CT and MRI typically acquire images slices, which have slice thickness of the order of 1.0–3.0 mm, which is much greater than the limiting build resolution of any of the RP technologies.

The potential benefits of exploiting RP techniques in surgical planning have been widely acknowledged and described. The process of producing accurate physical models directly from 3D scan data of an individual patient has proved particularly popular in head and neck reconstruction. In addition, most of the work done to date has concentrated on the use of 3D CT data as this produces excellent images of bone. However, the process is still not conducted in the large volumes associated with industrial RP and as such, practitioners applying these techniques to medicine often confront problems that are not encountered in industry. The small turnover associated with medical modelling also means that many manufacturers and vendors cannot justify investment in specific software, processes and materials for this sector. These characteristics combine to make medical modelling a challenging field of work with many potential pitfalls.

The authors' many years of practical experience in medical modelling has resulted in a knowledge base that has identified the problems that may be encountered, many of which are simple or procedural in nature. This paper aims to highlight some of these common problems, the effect they have on the resultant models and suggest methods that can be employed to avoid or minimise their occurrence or impact on the usefulness of the models produced.

5.3.4   Medical rapid prototyped model artefacts

Associated with all medical imaging modalities are unusual or unexpected image appearances referred to as artefacts. Some imaging modalities are prone to geometric distortion like MRI (Wang, Doddrell, & Cowin, 2004) and this should be accounted for in soft-tissue models manufactured from this source. CT does not suffer from the same distortion as MRI and models produced from this source have been proven to be dimensionally accurate (Barker, Earwaker, Frost, & Wakeley, 1994). In some circumstances, artefacts are easily recognisable and taken into account by the viewer whilst in other circumstances they can be problematic and difficult to explain. Artefacts present in the image data may subsequently be transferred to a medical model. In addition, because of the image processing steps and surface modelling required in the production of medical models, there is scope for the appearance of a wide range of artefacts. This section describes and illustrates some of the problems and pitfalls encountered in the production of medical models.

The procedures and potential problems associated with transferring and translating medical scan data are described in Chapter 2. However, the following observations serve to illustrate particular examples of some of the problems encountered by the authors. An example of some of the practical implications of transferring data from a hospital to an RP service provider is given in Case study 2 in Section 5.2.

5.3.4.1   CT data import errors

CT data consists of a series of pixel images of slices through the human body. When importing data the key characteristics that determine size and scale of the data are the pixel size and the slice thickness. The pixel size is calculated by dividing the field of view by the number of pixels. The field of view is a variable set by the radiographer at the time of scanning. The numbers of pixels in the x- and y-axes are typically 512×512 or 1024×1024. If there is a numerical error in any of these parameters whilst data is being translated from one data format to another, the model may be inadvertently scaled to an incorrect size. The slice thickness and any interslice gap must be known (although the interslice gap is not applicable in CT where images are reconstructed contiguously or overlapping). Numerical error in the slice thickness dimension will lead to inadvertent incorrect scaling in the third dimension. This distance is typically in the order of 1.5 mm but may be as small as 0.5 mm or as high as 5 mm. Smaller scan distances result in higher quality of the 3D reconstruction. The use of the internationally recognised DICOM (www.acrnema.org) standard for the format of medical images has largely eliminated these errors (Muller, Michoux, Bandon, & Geissbuhler, 2004).

5.3.4.2   CT gantry tilt distortion

A CT scanner typically operates with the X-ray tube and detector gantry perpendicular to the long axis of the patient (Z direction). The scan therefore produces the axial images that form the basis of 3D CT scans. However, in some cases the gantry may be inclined at an angle of up to $30°$. When a set of 2D slices is combined into an image volume for 3D modelling the gantry angle must be taken into account. With no gantry tilt, the slices are correctly aligned and they will produce an undistorted 3D volume. Slices acquired with a gantry tilt of $15°$ and converted into a data volume without the gantry tilt being taken into account may have a shear distortion arising from the misalignment of slices. At large angles, this is immediately visually apparent and can therefore be detected. However, at small angles it may not be so obvious. Building a model with a small, uncorrected gantry tilt angle could be easily done and result in significant geometrical inaccuracies in the resulting model. The use of the image transfer standard, DICOM, automatically provides the scan parameters including gantry tilt angle. However, the DICOM formal does not provide the direction of the angle and it cannot therefore be relied on to automatically correct gantry tilt. It is therefore advisable to avoid gantry tilt when acquiring a 3D CT image data set otherwise sophisticated mathematical algorithms are required to successfully correct the data. Figure 5.11 shows how a distorted 3D CT volume may be corrected using affine transformation to produce a data set with no distortion.

5.3.4.3   Model stair-step artefact

Two elements contribute to the stepped effect seen in medical models. One contribution is from the discrete layer thickness at which the model is built. This is a characteristic

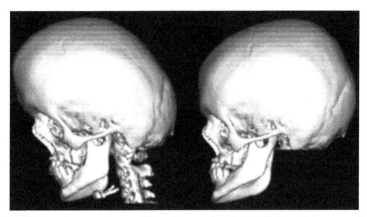

Figure 5.11 Effect of gantry tilt on a 3D surface rendered skull (left) and the same after correction (right).

of the particular RP process and material being used. Typically, these range from 0.1 to 0.3 mm. This affect can be minimised by selecting processes and parameters that minimise the build layer thickness. However, thinner layers result in longer build-times and increased costs, and an economic compromise is typically found for each RP process. As the layer thickness is typically an order of magnitude smaller than the scan distance of the CT images, it does not have an overriding effect on the quality of the model.

The second effect arises from the slice thickness of the acquired CT or MR images and any potential gap between them. The stair-step artefact is a common feature on conventional and single-slice helical CT scans where the slice thickness is near to an order of magnitude greater than the in-plane pixel size (Fleischmann et al., 2000). The artefact is manifest as a series of concentric axial rings around the model. The depth and size of these rings depends on the CT imaging protocol, but may be very slight where there is a thin slice used (e.g. 3-mm acquisition with 1-mm reconstruction interval). In thick-slice acquisitions (e.g. >3 mm with similar reconstruction interval to the slice thickness), the stair-step artefact will cause significant distortion to the model. Figure 5.12 shows a STL model of a full skull. The CT scan was performed on a conventional CT scanner with 5-mm slice thickness and no interpolation of the image data to create thin slices. Note there was significant stair-step artefact around the top of the skull and on the lower edge of the mandible. The stair-step artefact was most prominent on surfaces that were inclined to the data acquisition plane as is the case for 3D surface rendered images. This model was used for surgical planning and reconstruction but was limited in the use for obtaining physical measurements.

These effects can be countered to some degree by using interpolation between the original image data. The following images illustrate the difference between using no interpolation and using a cubic (natural curve) interpolation. Due to the natural appearance of the cubic curve, the resulting interpolated data results in a good, smooth and natural appearing surface.

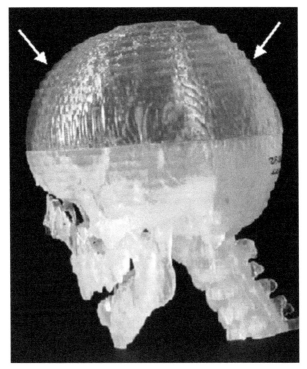

Figure 5.12 SL model with significant stair-step artefact.

5.3.4.4   Irregular surface resulting from support structures

Both SL and FDM™ required support structures during the build process. These were subsequently cleaned from the model manually although generally left a rough surface. This did not affect the overall accuracy of the model but contributed to a degradation of its aesthetic appearance. Figure 5.13 shows an SL model where surface roughness was attributed to the support structures. Models were easily cleaned using a light abrasive technique although this was felt unnecessary as the indentations were of submillimetre depth. It is unlikely that these structures would have a detrimental effect in surgical planning or implant design.

5.3.4.5   Irregular surface resulting from mathematical modelling

The mathematical modelling of a surface will introduce its own surface effects. The smoothness (governed by the size of the triangle mesh) of the model surface becomes poorer as the surface mesh becomes larger. A larger mesh resulted in a lower number of triangles, reduced computer file size and quicker rendering. A smaller mesh resulted in much better surface representation, much greater computer file size and slower rendering. Figure 5.14 shows irregular surface structures due to the mathematical modelling

Figure 5.13 Showing surface roughness on an SL model attributed to support structures.

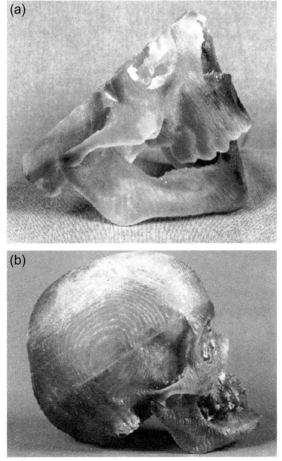

Figure 5.14 (a) Smooth surface resulting from a high-resolution meshing algorithm.
(b) Meshing contours visible resulting from a low-resolution meshing algorithm.

process (Karron, 1992). Figure 5.14(a) shows a model where the mesh structure is not readily apparent and (b) where the model contours are more clearly observed. In both cases, the surface produced was acceptable for its own clinical application. One could imagine that the mesh resolution used in model (b) would be unacceptable for smaller models where a fine detail would be masked.

These effects can be avoided by eliminating the creation of a 3D surface mesh and creating the RP build data directly from the CT image data. This essentially creates the two-and-a-half dimensional layer data for the RP machine from the CT images. Interpolation is used to create accurate intermediate layers between the CT images. This route not only eliminates surface modelling effects but also results in much smaller computer files and faster preparation.

5.3.4.6 Metal artefact

Metal artefact was present within CT scans of the maxilla and mandible because of the presence of metal within fillings of the teeth or the presence of dental implants. This was manifest as high signal intensities (in the form of scattered rays) around the upper and lower mandible. Figure 5.15 shows an FDM™ model with significant metal artefact around the teeth. These ray appearances extended from a couple of millimetres to over 1 cm in length. In some circumstances, the artefact may be reduced by software during CT image reconstruction (Mahnken et al., 2003). This artefact was plainly visible and added many superfluous structures to the medical models. Although, no significant geometric distortion was observed on models, large spikes were visible emanating from around the teeth which distorted the bone in the local area. The artefact may be removed by detailed slice-by-slice editing of the original CT images to produce a cleaner model. This process is very time-consuming and if not performed with great care can result in anatomy of interest being removed and the subsequent model becoming unusable.

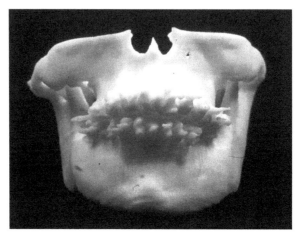

Figure 5.15 Metal artefact due to scattering of X-rays from metal within the teeth.

5.3.4.7 Movement artefact

CT scanning was prone to movement artefact if a patient was restless. This artefact was readily apparent in a model if the degree of movement was significantly large (i.e. greater than 1 mm). Figure 5.16 shows a mandible with a distinct artefact present. The patient moved slightly during the acquisition of a couple of images that left a bulge of 4-mm height extending right around the mandible. In addition, present in this model were concentric axial rings of about 3-mm thickness. These corresponded to the common stair-step observed in single slice helical CT scanning. Obviously, the degree of the movement during the scan determines the size of the movement artefact in the model. In the example shown, the artefact was felt not to be significant clinically as it did not interfere directly with the placement of a distraction device.

In another example where a model was being used for facial reconstruction, the patient moved whilst the scanner was acquiring data at the region where the surgery was to be performed. The movement artefact resulted in distortion of the model that the surgeon lost confidence in its physical integrity. During the scan, around the supra-orbital ridge, we believe the child rotated its head to look at a parent, which resulted in rotation of this part of the data, which was subsequently transferred to the model. In this case, the patient, a 7-year-old child, had to be rescanned under full general anaesthetic. It was interesting to note that the degree of the artefact was not noted until a physical model was produced. This indicates the need for good quality assurance of the original data set to ensure a useful model was produced.

5.3.4.8 Image threshold artefact

One of the simplest and commonest methods of tissue segmentation applied to the skull is CT number thresholding. A CT number range was identified by either region

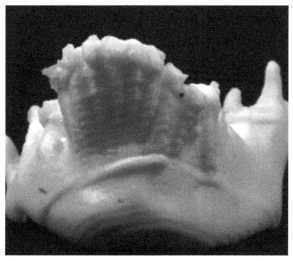

Figure 5.16 Distinct movement artefact on an FDM model of the mandible.

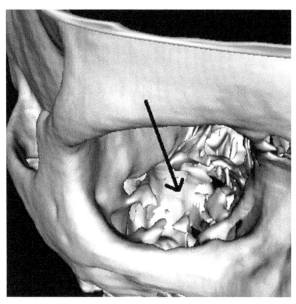

Figure 5.17 Removal of bone at the back of the orbit due to inappropriate choice of image threshold.

of interest pixel measurements or pixel intensity profiles, which was representative of bone. If the bone was particularly thin or the threshold inappropriately measured a continuous surface was unachievable. This left the model with a hole where the surface was not closed. In some cases, large areas of bone were removed completely, especially at the back of the orbit and around the cheekbones. Figure 5.17 illustrates bone deletion by data thresholding in the back of the orbit in this magnified surface shaded image. Anatomical detail is lost as the chosen threshold removed thin bone at the back of the orbit, as indicated by the black arrow. Adjusting the threshold to include bone in this case would have resulted in the inclusion of soft tissue that would have made the image more difficult to interpret. It is useful to specify what is required of a model clearly, so that an appropriate threshold can be chosen to preserve tissue of interest.

5.3.5   Conclusion

MRP models of human anatomy may be constructed from a number of image data sources and using a range of RP technologies. They are prone to artefacts from the imaging source, the method of manufacture and from the model cleaning process. It is important to ensure that high-quality source data is available to assure model quality. Clinicians requesting medical models should be aware of their physical accuracy and integrity which is generally dependent on the original imaging parameters and image processing rather than the method of manufacture that this is sufficient for its purpose. We have demonstrated a range of model artefact sources ranging from reading computer files to the removal of support structures and suggested ways to avoid or cure them. It is important that the source images be reviewed thoroughly; that robust image transfer and image processing procedures are in place and that the model building material is fit for the purpose for which it was intended. A multidisciplinary team approach to the manufacture of medical models with rigorous quality assurance is highly recommended.

5.3.6 Update

Whilst several newer RP technologies have become available since this article was first published, the sources of the majority of the potential errors involved in data acquisition and manipulation remain the same and the fundamental issues and problems discussed remain a concern regardless of which RP technology is used.

References

Anderl, H., Zur Nedden, D., Muhlbauer, W., Twerdy, K., Zanon, E., Wicke, K., et al. (1994). CT-guided stereolithography as a new tool in craniofacial surgery. *British Journal of Plastic Surgery*, *47*(1), 60–64.

Arvier, J. F., Barker, T. M., Yau, Y. Y., D'Urso, P. S., Atkinson, R. L., & McDermant, G. R. (1994). Maxillofacial biomodelling. *British Journal of Oral and Maxillofacial Surgery*, *32*, 276–283.

Barker, T. M., Earwaker, W. J., Frost, N., & Wakeley, G. (1994). Accuracy of stereolithographic models of human anatomy. *Australasian Radiology*, *38*(2), 106–111.

Berry, E., Brown, J. M., Connell, M., Craven, C. M., Efford, N. D., Radjenovic, A., et al. (1997). Preliminary experience with medical applications of rapid prototyping by selective laser sintering. *Medical Engineering and Physics*, *19*(1), 90–96.

Choi, J. Y., Choi, J. H., Kim, N. K., Kim, Y., Lee, J. K., Kim, M. K., et al. (2002). Analysis of errors in medical rapid prototyping models. *International Journal of Oral and Maxillofacial Surgery*, *31*(1), 23–32.

D'Urso, P. S., Barker, T. M., Earwaker, W. J., Bruce, L. J., Atkinson, R. L., Lanigan, M. W., et al. (1999). Stereolithographic biomodelling in cranio-maxillofacial surgery: a prospective trial. *Journal of Craniomaxillofacial Surgery*, *27*(1), 30–37.

Erben, C., Vitt, K. D., & Wulf, J. (2000). First statistical analysis of data collected in the Phidias validation study of stereolithography models. *Phidias Newsletter* (5), 6–7.

Eufinger, H., & Wehmoller, M. (1998b). Individual prefabricated titanium implants in reconstructive craniofacial surgery: clinical and technical aspects of the first 22 cases. *Plastic and Reconstructive Surgery*, *102*(2), 300–308.

Fleischmann, D., Rubin, G. D., Paik, D. S., Yen, S., Hilfiker, P., Beaulieu, C., et al. (2000). Stair-step artefacts with single versus multiple detector-row helical CT. *Radiology*, *216*, 185–196.

Fuchs, T., Kachelriess, M., & Kalender, W. A. (2000). Related articles, technical advances in multi-slice spiral CT. *European Journal of Radiology*, *36*(2), 69–73.

Gateno, J., Allen, M. E., Teichgraeber, J. F., & Messersmith, M. L. (2000). An in vitro study of the accuracy of a new protocol for planning distraction osteogenesis of the mandible. *Journal of Oral and Maxillofacial Surgery*, *58*(9), 985–990.

Heckmann, S. M., Winder, W., Meyer, M., Weber, H. P., & Wichmann, M. G. (2001). Overdenture attachment selection and the loading of implant and denture bearing area. Part 1: in vitro verification of stereolithographic model. *Clinical Oral Implants Research*, *12*(6), 617–623.

Heissler, E., Fischer, F., Bolouri, S., Lehmann, T., Mathar, W., Gebhardt, A., et al. (1998b). Aesthetic and reconstructive surgery – custom-made cast titanium implants produced with CAD/CAM for the reconstruction of cranium defects. *International Journal of Oral and Maxillofacial Surgery*, *27*(5), 334–338.

Hughes, C. W., Page, K., Bibb, R., Taylor, J., & Revington, P. (2003b). The custom made orbital floor prosthesis in reconstruction for orbital floor fractures. *British Journal of Oral and Maxillofacial Surgery*, *41*, 50–53.

Jacobs, P. (1996b). *Stereolithography and other rapid prototyping and manufacturing technologies*. Dearborn, MI: American Association of Engineers Press.

Joffe, J., Harris, M., Kahugu, F., Nicoll, S., Linney, A., & Richards, R. (1999b). A prospective study of computer-aided design and manufacture of titanium plate for cranioplasty and its clinical outcome. *British Journal of Neurosurgery*, *13*(6), 576–580.

Kai, C. C., & Fai, L. K. (1997). *Rapid prototyping: Principles & applications in manufacturing*. Singapore: John Wiley & Sons Ltd.

Karron, D. (1992). The 'spider web' algorithm for surface construction in noisy volume data. *SPIE Visualisation in Biomedical Computing*, *1808*, 462–476.

Mahnken, A. H., Raupach, R., Wildberger, J. E., Jung, B., Heussen, N., Flohr, T. G., et al. (2003). A new algorithm for metal artefact reduction in computed tomography: in vitro and in vivo evaluation after total hip replacement. *Investigative Radiology*, *38*(12), 769–775.

Mankovich, N. J., Cheeseman, A. M., & Stoker, N. G. (1990). The display of three-dimensional anatomy with stereolithographic models. *Journal of Digital Imaging*, *3*(3), 200–203.

Minns, R. J., Bibb, R., & Banks, R., Sutton, R. A. (2003). The use of a reconstructed three-dimensional solid model from CT to aid surgical management of a total knee arthroplasty: a case study. *Medical Engineering and Physics*, *25*, 523–526.

Muller, H., Michoux, N., Bandon, D., & Geissbuhler, A. (2004). A review of content-based image retrieval systems in medical applications-clinical benefits and future directions. *International Journal of Medical Informatics*, *73*(1), 1–23.

Munjal, S., Leopold, S. S., Kornreich, D., Shott, S., & Finn, F. A. (2000). CT generated 3D models for complex acetabluar reconstruction. *Journal of Arthroplasty*, *15*(5), 644–653.

Nakajima, T. (1995). Integrated life-sized solid model of bone and soft tissue: application for cleft lip and palate infants. *Plastic and Reconstructive Surgery*, *96*(5), 1020–1025.

Petzold, R., Zeilhofer, H., & Kalender, W. (1999b). Rapid prototyping technology in medicine-basics and applications. *Computerized Medical Imaging and Graphics*, *23*, 277–284.

Sailer, H. F., Haers, P. E., Zollikofer, C. P., Warnke, T., Carls, F. R., & Stucki, P. (1998). The value of stereolithographic models for preoperative diagnosis of craniofacial deformities and planning of surgical corrections. *International Journal of Oral and Maxillofacial Surgery*, *27*(5), 327–333.

Sarment, D. P., Sukovic, P., & Clinthorne, N. (2003). Accuracy of implant placement with a stereolithographic surgical guide. *International Journal of Oral and Maxillofacial Implants*, *18*(4), 571–577.

Schwaderer, E., Bode, A., Budach, W., Claussen, C. D., Danmmann, F., Kaus, T., et al. (2000). Soft-tissue stereolithography model as an aid to brachytherapy. *Medica Mundi*, *44*(1), 48–51.

Wang, D., Doddrell, D. M., & Cowin, G. (2004). A novel phantom and method for comprehensive 3-dimensional measurement and correction of geometric distortion in magnetic resonance imaging. *Magnetic Resonance Imaging*, *22*(4), 529–542.

Webb, P. A. (2000b). A review of rapid prototyping (RP) techniques in the medical and biomedical sector. *Journal of Medical Engineering and Technology*, *24*(4), 149–153.

Williams, R. J., Bibb, R., & Rafik, T. (2004). A technique for fabricating patterns for removable partial denture frameworks using digitized casts and electronic surveying. *Journal of Prosthetic Dentistry*, *91*(1), 85–88.

Winder, R. J., Cooke, R. S., Gray, J., Fannin, T., & Fegan, T. (1999b). Medical rapid prototyping and 3D CT in the manufacture of custom made cranial titanium plates. *Journal of Medical Engineering and Technology*, *23*(1), 26–28.

Surgical applications

5.4 Surgical applications case study 1: planning osseointegrated implants using computer-aided design and rapid prototyping

Acknowledgments

The work described in this case study was first reported in the references below and is reproduced here in part or in full with the permission of the Institute of Maxillofacial Prosthetics and Technologists and the Council of the Institute of Mechanical Engineers.

Bibb R, Bocca A, Sugar A, Evans P, "Planning Osseointegrated Implant Sites using Computer Aided Design and Rapid Prototyping", The Journal of Maxillofacial Prosthetics & Technology, 2003, Volume 6, pages 1-4.

Bibb R, Eggbeer D, Bocca A, Evans P, Sugar A, "Design and Manufacture of Drilling Guides for Osseointegrated Implants Using Rapid Prototyping Techniques", Proceedings of the 4th National Conference on Rapid & Virtual Prototyping and Applications, 2003, ISBN 1-86058-411-X, pages 3-11, Professional Engineering Publishing, London, UK.

5.4.1 Introduction

In recent years, rapid prototyping (RP) has been used to build highly accurate anatomical models from medical scan data. These models have proved to be a valuable aid in the planning of complex reconstructive surgery, particularly in maxillofacial and craniofacial cases. Typically, RP is used to create accurate models of internal skeletal structures on which operations can be accurately planned and rehearsed (Bibb & Brown, 2000; Bibb et al., 2000; D'Urso & Redmond, 2000; Klein, Schneider, Alzen, Voy, & Gunther, 1992). Such models have been successfully used for positioning osseointegrated implants. Osseointegrated implants are titanium screws attached directly to a patient's bone structure that pass through the skin to provide a rigid and firm fixture for dentures, hearing aids, and prostheses (Branemark & De Oliveira, 1997) (see medical explanatory note in Section 7.2.1 for an explanation of osseointegrated implants). The accuracy of the RP models allows the depth and quality of bone to be assessed, improving the selection of drilling sites before surgery. Although this process has dramatically improved the accuracy and reduced the theatre time of some surgical procedures, it incurs significant time and cost to produce the anatomical model. Although it uses RP technologies, this route does not fully exploit the potential advantages of computer-aided design (CAD).

To address this issue, it was decided to complete as much planning as possible in the virtual environment and only use RP to make small templates that would guide the surgeon in surgery. This route would allow clinicians to conduct all planning and explore many options without damaging an expensive RP model. To be successful, the approach would have to be simple to conduct and have low investment requirements.

5.4.2   Proposed approach

The approach would use three-dimensional (3D) computed tomography (CT) data to create virtual models of the elements necessary to plan the osseointegrated implants required to secure a prosthetic ear. The elements consisted of the soft tissue of the head, a copy of the remaining opposite ear, and the bone structure at the implant site. The simple and popular STL (Manners, 1993) format was chosen as the 3D representation of the entities. This format ensures easy access to a number of software options at a reasonable cost. In this case, a popular software package used in the RP industry was used (Magics version 7.2, 2003). The STL file format is more fully described in Chapter 3. The entities were created as STL format files from CT data using one of a number of specialist software packages available for creating STL files from CT data (Mimics® version 7.2, 2003).

The STL manipulation software was then used to mirror the copy of the ear and position it in an anatomically and aesthetically appropriate location. The software was then used to create cylinders representing the implants. These cylinders were positioned in the preferred location by the prosthetist observing a lateral view. Then, the bone quality at the implant sites could be assessed.

5.4.3   Scanning problems

Misalignment of the patient's head to one side or the other means that the optimum accuracy obtained in the axial plane during scanning is not axial to the patient. This means that entities will not be in alignment with software coordinate system. Although this is not a major issue, it makes control of the angles at which the entities meet more difficult.

Unintended displacement of the soft tissue during the CT scan results in poor representation of the anatomy. In this example, the ear of one of the patients had become folded over, resulting in a deformed anatomical entity (Figure 5.18). This made positioning the contralateral ear and the implants problematic.

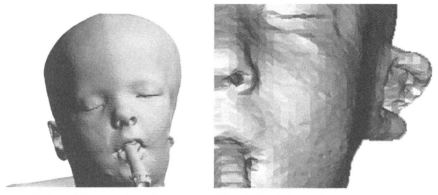

Figure 5.18 Problems resulting from poor position during CT scanning.

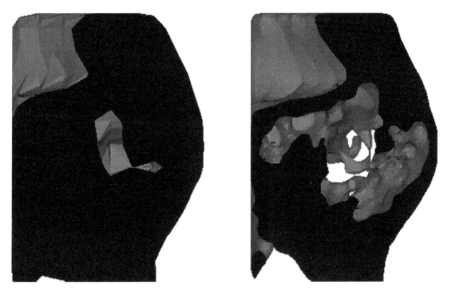

Figure 5.19 The effect of file size versus quality: low quality (left) and high quality (right).

5.4.4 Software problems

Initially, file size was a concern and all of the STL files were produced at a low resolution. This resulted in small STL files that could be handled easily and rapidly by the software. In visual terms, all of the entities appeared to be well represented by the low-quality STL files. However, when attempting one of the first cases, it was found that the difference in the representation of internal air cells in the highly pneumatised bone in the mastoid was dramatically altered by the resolution at which the STL file was produced. This led to the mistaken belief that this particular case had adequate bone thickness when in fact the bone was unusually thin (illustrated in Figure 5.19).

From this experience, it was decided to produce only the small amount of bone required for the implants but at the highest possible resolution. This resulted in only one entity with a large file size, which proved to be perfectly within the capabilities of a reasonable specification computer.

5.4.5   Illustrative case study

Three-dimensional CT data were used to create virtual models of the elements necessary to plan the osseointegrated implants required to secure a prosthetic ear. The elements consisted of the soft tissue of the head, a copy of the remaining opposite ear and the bone structure at the implant site. The surgeon and prosthetists then used 3D software to mirror the copy of the ear and position it relative to the head in an anatomically and aesthetically appropriate location (Figure 5.20).

Cylinders were created to represent the implants. These cylinders were positioned on the ear in the position preferred by the prosthetists observing a lateral view (Figure 5.21). The soft tissue entities were then removed to see where the implants intersected with the bone, as can be seen in Figure 5.21.

The bone quality at the implant sites was then assessed to check that they would be suitable for implants. Sectioning the virtual model enabled the quality and thickness of the bone to be measured accurately (Figure 5.22).

When the team was satisfied with the implant sites, a block was created that overlapped the implants and the surface of the skull (Figure 5.23). Then, by using a

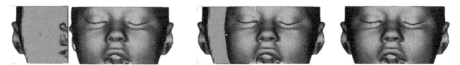

Figure 5.20 Positioning the contralateral ear.

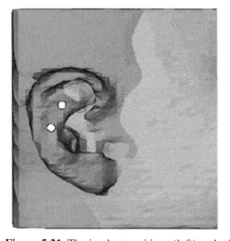
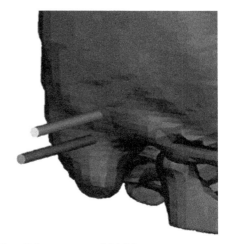

Figure 5.21 The implant positions (left) and with soft tissue removed (right).

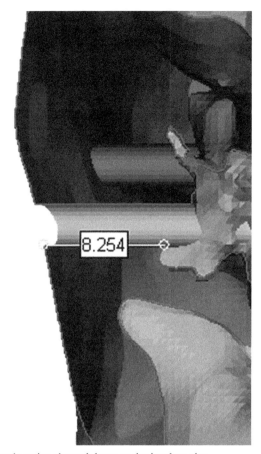

Figure 5.22 Measured section through bone at the implant site.

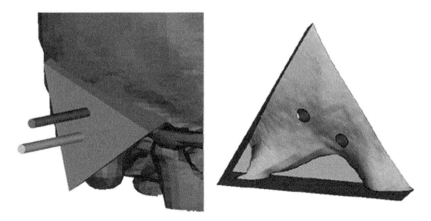

Figure 5.23 The block overlapping the bone surface and implants (left) final template design (right).

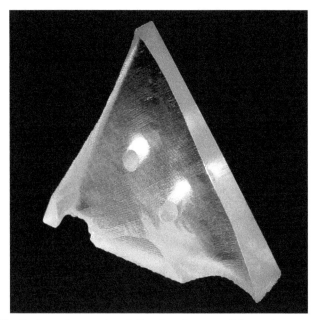

Figure 5.24 The final template produced directly using stereolithography.

Boolean operation, the skull and implant cylinders were subtracted from the block to create a template design (Figure 5.23).

This template, shown in Figure 5.24, was produced directly in a medically appropriate material by stereolithography (Jacobs, 1996; RenShape) (see Chapter 4, Section 4.2 for a full description of stereolithography). Because this was the only RP material that had been tested to a standard recognised by the Food and Drug Administration for patient contact in surgery meant that the stereolithography template could be used directly in surgery after sterilisation. Sterilisation presents no practical problems; appropriate methods include ethylene oxide (55 °C), formaldehyde, low-temperature steam (75 °C) and gamma irradiation. The template locates the anatomical features on the surface of the skull at the implant site and indicates the drilling sites to the surgeon. The mastoid and zygomatic process are exploited to provide positive anatomical features so that the template locates accurately and firmly.

5.4.6   Results

In surgery, the template fit accurately and securely to the area of the skull, as shown in Figure 5.25. Drilling was carried out and the bone thickness and quality were found to be as indicated by the data. The positions indicated by the template were much more accurate than those indicated by marks transferred from the soft tissue with ink and needle, by as much as 5 mm in one case. The team has now successfully carried out many similar cases using this approach, with equally positive results and considerable savings of time and money. The total time taken is 2.5 h if we consider the typical

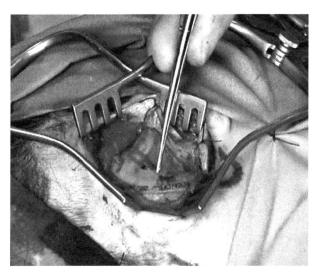

Figure 5.25 The template in position during surgery.

procedure for the traditional method as follows: carve a planning ear (30 min), take impression of defect site (15 min), create a template (30 min), planning (1 h) and mark the template (15 min). This involves at least one technician and one surgeon, and requires a patient appointment, which could represent a cost saving of approximately £250, depending on salaries and overhead.

5.4.7 Benefits and future development

The principle benefits resulting from this approach are reduced cost implications for planning activities. Once the entities are created from the CT data, they can be positioned and repositioned as many times as required. This allows different placement strategies to be performed and evaluated in 3D in a matter of minutes and with zero cost implication (other than time).

Once a plan has been agreed among clinicians, the implant sites themselves can be assessed for bone depth and quality (within the limits of the original CT scan). If they are unsatisfactory, these sites can be altered without incurring cost. When a final solution is achieved, the template model can be made in under 2 h and costs dramatically less than even a localised stereolithography model of the bone structure. The CT data used in these cases were taken at 1.5-mm slice distance and proved adequate. However, reducing this slice distance would increase the quality of the 3D entities.

Of course, the purchase and maintenance of the software required for this approach are significant and it can be anticipated that a high volume of cases would be required to justify this investment in isolation. However, it is the experience of the authors that such software has many useful applications in head and neck reconstruction as well as other medical specialities.

5.4.8   Update

The use of surgical guides has developed significantly in recent years with the development of specific software and service providers, particularly in orthopaedics. However, the considerations discussed in this chapter are still relevant when planning cases. The next chapter describes how the technique developed with more appropriate software and metal RP techniques.

References

Bibb, R., & Brown, R. (2000). The application of computer aided product development techniques in medical modelling. *Biomedical Sciences Instrumentation*, *36*, 319–324.

Bibb, R., Brown, R., Williamson, T., Sugar, A., Evans, P., & Bocca, A. (2000). The application of product development technologies in craniofacial reconstruction. In *Proceedings of the 9th European conference on rapid prototyping and manufacturing* (pp. 113–122). Athens, Greece.

Branemark, P., & De Oliveira, M. F. (Eds.). (1997). *Craniofacial prostheses, anaplastology and osseointegration* (pp. 101–110). Carol Stream, IL: Quintessence Publishing Co. Inc.

D'Urso, P. S., & Redmond, M. J. (2000). A method for the resection of cranial tumours and skull reconstruction. *British Journal of Neurosurgery*, *14*(6), 555–559.

Jacobs, P. F. (Ed.). (1996). *Stereolithography and other RP&M Technologies*. Dearborn, MI: Society of Manufacturing Engineering.

Klein, H. M., Schneider, W., Alzen, G., Voy, E. D., & Gunther, R. W. (1992). Pediatric craniofacial surgery: comparison of milling and stereolithography for 3D-model manufacturing. *Pediatric Radiology*, *22*, 458–460.

Magics version, 7. 2. (2003). Materialise NV, Technologielaan 15, 3001. Leuven, Belgium.

Manners, C. R. (1993). *STL file format*. Valencia, CA: 3D Systems Inc.

Mimics version, 7. 2. (2003). Materialise NV, Technologielaan 15, 3001. Leuven, Belgium.

RenShape H.-C. 9100R stereolithography material technical specification, Huntsman advanced materials, Duxford, Cambridge, UK.

5.5 Surgical applications case study 2: rapid manufacture of custom-fit surgical guides

Acknowledgments

The work described in this case study was first reported in the references below and is reproduced here with permission of Emerald Publishing.

Bibb R, Eggbeer D, Evans P, Bocca A, Sugar AW, "Rapid Manufacture of Custom Fitting Surgical Guides", Rapid Prototyping Journal 2009; 15(5): 346-354, ISSN: 1355-2546, http://dx.doi.org/10.1108/13552540910993879.

5.5.1 Introduction

Over the past decade, rapid prototyping (RP) techniques have been employed widely in maxillofacial surgery. However, they have concentrated on the reproduction of exact physical replicas of patients' skeletal anatomy that surgeons and prosthetists use to help plan reconstructive surgery and prosthetic rehabilitation (Bibb & Brown, 2000; Bibb et al., 2000; D'Urso et al., 2000; Eufinger & Wehmoller, 1998; Heissler et al., 1998; Hughes, Page, Bibb, Taylor, & Revington, 2003; Joffe et al., 1999; Knox et al., 2004; Petzold, Zeilhofer, & Kalender, 1999; Sanghera, Naique, Papaharilaou, & Amis, 2001; Webb, 2000; Winder, Cooke, Gray, Fannin, & Fegan, 1999).

Developments in this area are moving towards exploiting advanced design and fabrication technologies to design and produce implants, patterns or templates that enable the fabrication of custom-fit prostheses without requiring a model of the anatomy to be made (Bibb, Bocca, & Evans, 2002; Eggbeer, Evans, & Bibb, 2004; Evans, Eggbeer, & Bibb, 2004; Singare, Dichen, Bingheng, Zhenyu, & Yaxiong, 2005; Vander Sloten, Van Audekercke, & Van der Perre, 2000). However, there is also growing desire by clinicians to conduct more surgical planning using three-dimensional (3D) computer software. Several approaches have been undertaken in the application of computer-aided surgical planning, but the problem of transferring the computer-aided plan from the computer to the operating theatre remains. Two solutions exist to transfer the computer plan to the operating theatre: navigation systems and surgical guides. The use of navigation systems is a specialist field in itself and will not be described here. However, research presented by Poukens, Verdonck and de Cubber in 2005 suggested that navigation and the use of surgical guides are both accurate enough for surgical purposes (Poukens, Verdonck, & de Cubber, 2005). Rapid prototyping technologies provide a potential method of producing custom-fit surgical guides, depending on the nature of the planning software used. Previous work on the application of RP technologies in the manufacture of surgical guides has concentrated on the production of drilling guides for oral and extra-oral osseointegrated implants (Bibb, Bocca, Sugar, & Evans, 2003; Bibb, Eggbeer, Bocca, Evans, & Sugar, 2003; Goffin, Van Brussel, Vander Sloten, Van Audekercke, & Smet, 2001; Goffin et al., 1999; Sarment, Al-Shammari, & Kazor, 2003; Sarment, Sukovic, & Clinthorne, 2003; SurgiGuides, 2005; Van Brussel et al., 2004). This chapter describes one drilling guide case but also reports on two cases involving the use of surgical guides for osteotomies (saw cuts through bone), which has not been previously reported.

To be appropriate for the manufacture of surgical guides, RP processes have to be accurate, robust, rigid and able to withstand sterilisation. Because of these requirements, most surgical guides have been produced using Stereolithography (SL) and Laser Sintering (LS). (Stereolithography and LS are described in detail in Sections 4.2 and 4.5, respectively.) However, the use of SL in particular has necessitated local reinforcement of the guides using titanium or stainless-steel tubes to prevent inadvertent damage from drill bits and the use of low-temperature sterilisation methods such formaldehyde or ethylene oxide.

The recent availability of systems capable of directly producing fully dense solid parts in functional metals and alloys has provided an opportunity to develop surgical guides that exploit the advantages of RP while addressing the deficiencies of previous SL and LS guides. The ability to produce end use parts in functional materials means that processes such as these may be considered rapid manufacturing (RM) processes. Surgical guides produced directly in hardwearing, corrosion-resistant metals require no local reinforcement, can be autoclaved along with other surgical instruments, and are unlikely to be inadvertently damaged during surgery. The use of metals also enables surgical guides to be made much smaller or thinner while retaining sufficient rigidity. This benefits surgery because incisions can then be made smaller and the surgeon's visibility and access are improved.

5.5.2 Methods

To date, three surgical guides have been designed and produced as described here and subsequently used in surgery. The first case was a drilling guide for osseointegrated implants to secure a prosthetic ear. The remaining two cases were for osteotomy cutting guides for the correction of facial deformity. This section describes the general approach to the planning, rapid design and manufacture of surgical guides. The following section describes an individual case in which the approach was successfully employed for an osteotomy.

5.5.2.1 Step 1: three-dimensional CT scanning

The patients were scanned using 3D computed tomography (CT) to produce 3D computer models of the skull (see Chapter 2, Section 2.2). The CT data were exported in DICOM format, which was then imported into medical data transfer software (Mimics, Materialise NV, Technologielaan 15, 3001 Leuven, Belgium, www.materialise.com).

This software was used to generate the highest possible quality STL data files of the patient's anatomy using techniques described in Chapter 3. The STL files were then imported into the CAD software.

5.5.2.2 Step 2: computer-aided surgical planning and design of the surgical guide

The CAD package used in this study (FreeForm®, Geomagic, 3D Systems, USA, www.sensable.com) was selected for its capability of designing complex, arbitrary, but well-defined shapes that are required when designing custom appliances and devices that must fit human anatomy. The software has tools analogous to those used in physical sculpting and enables a manner of working that mimics that of the max-illofacial prosthetist working in the laboratory. The software uses a haptic interface (Phantom® Desktop™ haptic interface, Geomagic, 3D Systems, USA) that incor-porates positioning in three-dimensional space and allows rotation and translation in all axes, transferring hand movements into the virtual environment. It also allows the operator to feel the object being worked on in the software. The combination of tools and force feedback sensations mimic working on a physical object and allows shapes to be designed and modified in an arbitrary manner. The software also allows the import of scan data to create reference objects or bucks onto which objects may be designed.

The data of the patients' anatomy were imported into the software. The surgery was then planned and simulated by using the software tools to position prostheses and implants or cut the skeletal anatomy and move the pieces, as they would be in surgery. When the clinicians were satisfied with the surgical plan, the surgical guides were designed to interface with the local anatomy.

In general, the surgical guides were designed by selecting the anatomical surface in the region of the surgery (drilling or osteotomy) and offsetting it to create a struc-ture 1–2 mm thick. The positions of the drilling holes or cuts were then transferred to this piece by repeating the planned surgical procedure through it. Other features could then be added, such as embossed patient names, orientation markers or han-dles. When the design was completed to the clinicians' satisfaction, the human anatomy data were subtracted from the surgical guide as a Boolean operation. This left the surgical guide with the fitting surface as a perfect fit with the anatomical surface. Typically, the curvature and extent of the fitting surface provide accurate location when it is fitted to the patient. The final design was then exported as a high-quality STL file for rapid manufacture by Selective Laser Melting (SLM™). (SLM™ is described in Chapter 4).

5.5.2.3 Step 3: rapid manufacture

To build surgical guides successfully on the SLM™ machine (MCP Tooling Technol-ogies Ltd., now Renishaw, UK) adequate supports had to be created using Magics software (Version 9.5, Materialise NV). The purpose of the supports was to provide a firm base for the part to be built onto while separating the part from the substrate plate.

In addition, the supports conduct heat away from the material as it melts and solidifies during the build process. Inadequate supports result in incomplete parts or heat-induced curl, which leads to build failure as the curled part interferes with or obstructs the powder-recoating mechanism.

Recent developments in support design have resulted in supports with small contact points, which have improved the ease with which supports can be removed from parts. However, the parts were all oriented such that the amount of support necessary was minimised and avoided the fitting surface of the guide. This meant that the most important surfaces of the resultant part would not be affected or damaged by the supports or their removal.

The part and its support were sliced and hatched using the SLM™ Realizer software at a layer thickness of 0.050 mm. The material used was 316L stainless-steel spherical powder with a maximum particle size of 0.045 mm (particle size range, 0.005–0.045 mm) and a mean particle size of approximately 0.025 mm (Sandvik Osprey Ltd., Red Jacket Works, Milland Road, Neath, SA11 1NJ, United Kingdom, www.smt.sandvik.com/osprey). The laser had a maximum scan speed of 300 mm/s and a beam diameter 0.150–0.200 mm.

5.5.2.4   Step 4: finishing

Initially, supporting structures were removed using a Dremel handheld power tool using a reinforced cutting wheel (Dremel, Reinforced Cutting Disc, ref. Number 426). However, more recently, improved design of the supports has eliminated the need for cutting tools because the supports contact the part at a sharp point that can easily be broken away from the part.

The SLM™ parts described here were well formed with little evidence of the stair-stepping effect (resulting from the thin layers used) but showed a fine surface roughness. This roughness was easily removed by bead blasting to leave a smooth, matte finish surface. The parts were then sent to the hospital for cleaning and sterilisation by autoclave.

5.5.3   Case study

Although surgical guides have been produced using RP techniques for some years, the application of surgical guides for osteotomies had not been previously attempted. This case was the first attempt at such a guide. The surgery performed in this particular case involved distraction osteogenesis to correct deformity resulting from cleft palate. A description of distraction osteogenesis is given in medical explanatory note in Section 7.2.3. This required a Le Fort 1 osteotomy, which is a cut across the maxilla above the roots of the teeth but under the nose, to separate and move the upper jaw in relation to the rest of the skull. The maxilla is then gradually moved in relation to the rest of the skull, usually forward, by mounting it onto two devices that use precision screw

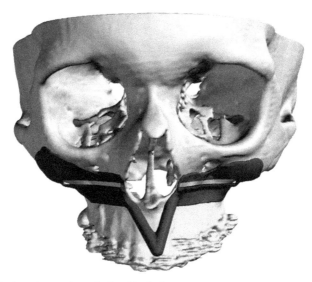

Figure 5.26 Patient data and surgical guide design.

Figure 5.27 Finished surgical guide design.

threads to advance the position by a small increment each day. The small increment causes the bone to grow gradually so that the shape of the face can be altered. When the desired position is reached, the bone is allowed to heal completely to give a strong and reshaped skeletal anatomy.

In this case, it was also the intention to include the drilling holes for the distraction devices as well as a slot for the osteotomy. The slot was then made sufficiently wide to allow the saw blade to move freely and sufficient irrigation during cutting. The lower edge of the slot is made flat and parallel to the direction of the cut to provide a reference surface on which the flat saw blade rests. As the cut is in two places on either side of the maxilla, the software design tools are used to join the two parts together into one device (see Figures 5.26 and 5.27). However, it was discovered that there was no way to simulate the bending of the distractor attachment plates using the software. Although it was theoretically possible to design and manufacture custom-fit plates and

Figure 5.28 Surgical guide and supports.

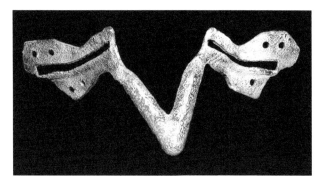

Figure 5.29 Finished surgical guide.

laser weld them to the distraction devices, the manufacturer of the devices would not allow the modifications.

The guide was therefore finalised, supported and built as described earlier. The support structure can be seen in Figure 5.28. The drilling positions for the distraction devices were planned on an SL model of the patient. The SLM™ surgical guide was then fit to the model and the drilling sites were transferred to it. The final guide is shown in Figure 5.29.

5.5.4   Results

The surgical guides were approved by clinicians before going into surgery and all were deemed satisfactory for surgical use. All of the guides used in surgery so far have displayed good accuracy and fit the patients' anatomy firmly and securely, as expected. The quality of the surgical guide fit for this particular case was assessed by an experienced maxillofacial prosthetist. This was achieved by fitting it to an SL model of the patient's facial skeletal anatomy, where it was found to show excellent fit. Figure 5.30 shows the guide in situ in surgery.

No problems were experienced sterilising or using the guides. The guides all resulted in some timesaving in surgery, particularly for the individual case described here. The surgical outcomes were good and turned out as planned.

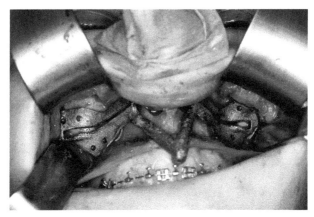

Figure 5.30 Guide being fitted to the patient during surgery.

5.5.5   Discussion

The surgical guides described here were all deemed successful and contributed to the successful transfer of computer-aided planning to surgery. The drilling guide was successful because it was thinner, more rigid, more hardwearing and easier to sterilise than previous reported attempts that used SLA (Bibb et al., 2003; SurgiGuides, 2005). As can be seen in Figure 5.31, the incorporation of embossed orientation markers and patient names was also beneficial and could help prevent errors in surgery (the patient name has been deliberately obscured to respect confidentiality).

However, the osteotomy guides proved challenging. There was no previous experience and there were no publications to build on; nevertheless, given the experimental nature of the two cases undertaken, the results were encouraging. The design of the guides will be significantly better in future cases based on the findings of these cases. The individual case described here illustrates examples of design improvement that resulted from this research. These improvements include better positioning of handles, having smaller fitting surfaces, and avoiding potential weaknesses. For example, in the case described here, the thin areas at the ends of the slots proved a potential weakness and the fitting surface was larger than necessary and was reduced by the maxillofacial prosthetist in the laboratory, as indicated in Figure 5.32.

However, the more fundamental problem of using the approach described here to include bending and fitting distractor plates will be addressed in future research by exploring other software applications and techniques.

5.5.6   Conclusions

Selective Laser Melting has been shown to be a viable rapid manufacturing method for the direct manufacture of surgical guides for both drilling and cutting. Stainless-steel

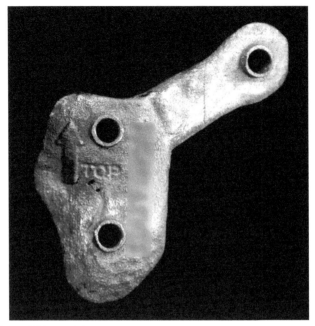

Figure 5.31 SLM drilling guide.

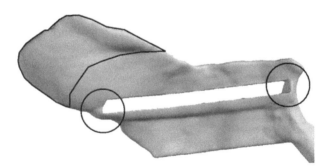

Figure 5.32 Areas for design improvement.

parts produced using the SLM process result in surgical guides that are comparable to previous experience with surgical guides produced using other RP processes in terms of accuracy, quality of fit and function, yet they display superior rigidity and good wear resistance and are easy to sterilise.

5.5.7   Update

This case has been followed by many more successful cases following similar procedures and the techniques have been developed and applied to the cases in Sections 5.9 and 5.10.

References

Bibb, R., Bocca, A., & Evans, P. (2002). An appropriate approach to computer aided design and manufacture of cranioplasty plates. *The Journal of Maxillofacial Prosthetics and Technology*, 5, 28–31.

Bibb, R., Bocca, A., Sugar, A., & Evans, P. (2003). Planning osseointegrated implant sites using computer aided design and rapid prototyping. *The Journal of Maxillofacial Prosthetics and Technology*, 6(1), 1–4.

Bibb, R., & Brown, R. (2000). The application of computer aided product development techniques in medical modelling. *Biomedical Sciences Instrumentation*, 36, 319–324.

Bibb, R., Brown, R., Williamson, T., Sugar, A., Evans, P., & Bocca, A. (2000). The application of product development technologies in craniofacial reconstruction. In *Proceedings of the 9th European conference on rapid prototyping and manufacturing* (pp. 113–122). Athens, Greece.

Bibb, R., Eggbeer, D., Bocca, A., Evans, P., & Sugar, A. (2003). Design and manufacture of drilling guides for osseointegrated implants using rapid prototyping techniques. In *Proceedings of the 4th national conference on rapid and virtual prototyping and applications* (pp. 3–12). London, UK: Professional Engineering Publishing, ISBN: 1-86058-411-X.

D'Urso, P. S., Earwaker, W. J., Barker, T. M., Redmond, M. J., Thompson, R. G., Effeney, D. J., et al. (2000). Custom cranioplasty using stereolithography and acrylic. *British Journal of Plastic Surgery*, 53(3), 200–204.

Eggbeer, D., Evans, P., & Bibb, R. (2004). The appropriate application of computer aided design and manufacture techniques in silicone facial prosthetics. In *Proceedings of the 5th national conference on rapid design, prototyping and manufacturing* (pp. 45–52). London, UK: Professional Engineering Publishing, ISBN: 1860584659.

Eufinger, H., & Wehmoller, M. (1998). Individual prefabricated titanium implants in reconstructive craniofacial surgery: clinical and technical aspects of the first 22 cases. *Plastic and Reconstructive Surgery*, 102(2), 300–308.

Evans, P., Eggbeer, D., & Bibb, R. (2004). Orbital prosthesis wax pattern production using computer aided design and rapid prototyping techniques. *The Journal of Maxillofacial Prosthetics and Technology*, 7, 11–15.

Goffin, J., Van Brussel, K., Vander Sloten, J., Van Audekercke, R., & Smet, M. H. (2001). Three-dimensional computed tomography-based, personalized drill guide for posterior cervical stabilization at C1-C2. *Spine*, 26(12), 1343–1347.

Goffin, J., Van Brussel, K., Vander Sloten, J., Van Audekercke, R., Smet, M. H., Marchal, G., et al. (1999). 3D-CT based, personalized drill guide for posterior transarticular screw fixation at C1-C2: technical note. *Neuro-Orthopedics*, 25(1–2), 47–56.

Heissler, E., Fischer, F. S., Bolouri, S., Lehmann, T., Mathar, W., Gebhardt, A., et al. (1998). Custom-made cast titanium implants produced with CAD/CAM for the reconstruction of cranium defects. *International Journal of Oral and Maxillofacial Surgery*, 27(5), 334–338.

Hughes, C. W., Page, K., Bibb, R., Taylor, J., & Revington, P. (2003). The custom-made titanium orbital floor prosthesis in reconstruction for orbital floor fractures. *British Journal of Oral and Maxillofacial Surgery*, 41, 50–53.

Joffe, J., Harris, M., Kahugu, F., Nicoll, S., Linney, A., & Richards, R. (1999). A prospective study of computer-aided design and manufacture of titanium plate for cranioplasty and its clinical outcome. *British Journal of Neurosurgery*, 13(6), 576–580.

Knox, J., Sugar, A. W., Bibb, R., Kau, C. H., Evans, P., Bocca, A., et al. (2004). The use of 3D technology in the multidisciplinary management of facial disproportion. In *Proceedings of the 6th international symposium on computer methods in biomechanics and biomedical engineering*. Cardiff, UK: Published on CD-ROM by First Numerics Ltd, ISBN: 0-9549670-0-3.

Petzold, R., Zeilhofer, H.-F., & Kalender, W. A. (1999). Rapid prototyping technology in medicine – basics and applications. *Computerized Medical Imaging and Graphics*, *23*(5), 277–284.

Poukens, J., Verdonck, H., & de Cubber, J. (2005). Stereolithographic surgical guides versus navigation assisted placement of extra-oral implants (oral presentation). In *2nd international conference on advanced digital technology in head and neck reconstruction, Banff, Canada* (p. 61). Abstract.

Sanghera, B., Naique, S., Papaharilaou, Y., & Amis, A. (2001). Preliminary study of rapid prototype medical models. *Rapid Prototyping Journal*, *7*(5), 275–284.

Sarment, D. P., Al-Shammari, K., & Kazor, C. E. (2003). Stereolithographic surgical templates for placement of dental implants in complex cases. *International Journal of Periodontics and Restorative Dentistry*, *23*(3), 287–295.

Sarment, D. P., Sukovic, P., & Clinthorne, N. (2003). Accuracy of implant placement with a stereolithographic surgical guide. *International Journal of Oral and Maxillofacial Implants*, *18*(4), 571–577.

Singare, S., Dichen, L., Bingheng, L., Zhenyu, G., & Yaxiong, L. (2005). Customized design and manufacturing of chin implant based on rapid prototyping. *Rapid Prototyping Journal*, *11*(2), 113–118.

SurgiGuides. (2005). Company information, Materialise Dental NV. Technologielaan 15, 3001: Leuven, Belgium.

Van Brussel, K., Haex, B., Vander Sloten, J., Van Audekercke, R., Goffin, J., Lauweryns, P., et al. (2004). Personalised drill guides in orthopaedic surgery with knife-edge support technique. In *Proceedings of the 6th international symposium on computer methods in biomechanics and biomedical engineering*. Cardiff, UK: Published on CD-ROM by First Numerics Ltd, ISBN: 0-9549670-0-3.

Vander Sloten, J., Van Audekercke, R., & Van der Perre, G. (2000). Computer aided design of prostheses. *Industrial Ceramics*, *20*(2), 109–112.

Webb, P. A. (2000). A review of rapid prototyping (RP) techniques in the medical and biomedical sector. *Journal of Medical Engineering and Technology*, *24*(4), 149–153.

Winder, J., Cooke, R. S., Gray, J., Fannin, T., & Fegan, T. (1999). Medical rapid prototyping and 3D CT in the manufacture of custom made cranial titanium plates. *Journal of Medical Engineering and Technology*, *23*(1), 26–28.

5.6   Surgical applications case study 3: use of a reconstructed three-dimensional solid model from computed tomography to aid in the surgical management of a total knee arthroplasty

Acknowledgments

The work described in this case study was first reported in the reference below and is reproduced here in part or in full with the permission of the Institute of Engineering and Physics in Medicine.

Minns RJ, Bibb R, Banks R, Sutton RA, "The use of a reconstructed three-dimensional solid model from CT to aid the surgical management of a total knee arthroplasty: a case study", Medical Engineering & Physics, 2003, Volume 25, Issue Number 6, pages 523–6. http://dx.doi.org/10.1016/S1350-4533(03)00050-X

5.6.1   Introduction

Reconstructing the knee with the aid of a prosthesis in patients with gross degenerative changes and large bone loss presents many challenges to the orthopaedic surgeon. Therefore, any aid to preoperative planning would enhance the outcome of this form of surgery. Plane radiographs give little insight into bone geometry in all three dimensions, especially the geometry of the cortex at the potential plane of resection.

The use of three-dimensional (3D) reconstructed images from computed tomography (CT) for the assessment and planning of complex hip pathologies has been investigated and is reported in the literature (Barmeir, Dubowitz, & Roffman, 1982; Bautsch, Johnson, & Seeger, 1994; Migaud, Corbet, Assaker, Kulik, & Duquennoy, 1997; Roach, Hobatho, Baker, & Ashman, 1997; Van Dijk, Smit, Jiya, & Wuisman, 2001). It has been shown to be helpful in the planning of surgery. More recently, the production of physical models of bone-deficient or dysplastic acetabulum using data generated from CT scans has been used in the computer-aided design and manufacture of implants (Munjal, Leopold, Kornreich, Shott, & Finn, 2000). The successful production of custom-made femoral components in total hip replacements has been also been reported (Bert, 1996; McCarthy, Bono, & O'Donnell, 1997; Robinson & Clark, 1996), as well as the use of 3D models in complex craniofacial surgery. However, their use in the reconstruction of complex bone shapes around the knee has not been reported.

5.6.2   Materials and methods

The patient was a 60-year-old woman with a history of Rheumatoid Arthritis (RA) that first presented at age 15 years, with deformity of the fingers. She had a

Benjamin double osteotomy (see explanatory note in Section 7.2.4) of the left knee at age 27 years because of the potential of subluxation (dislocation of the kneecap), and a synovectomy (surgery to remove inflamed joint tissue) of the right knee at 35 years. The left knee progressively became more varus (abnormally positioned towards the midline) and unstable and at age 59 years she became wheelchair bound with a grossly unstable, and deformed left knee. She was considered for total knee replacement; owing to the gross deformity of the joint, CT was performed. The scan was in the horizontal plane with the tibial axis aligned at right angles to the scanning plane on the machine's couch with soft firm padding (the tibia is the shinbone). Slices in the horizontal plane at 1.5-mm intervals were taken to 30 mm below the joint line, producing 20 sections. The data were stored on a magnetic/optical disk in DICOM format for processing and converting into the appropriate file system to produce a 3D model in the computer, and consequently a solid model to scale. The whole process to generate the solid model is shown in Figure 5.33.

The model of the knee was created using stereolithography apparatus (SLA®). The SLA® process is described fully in Chapter 4, Section 4.2. Preparation from CT data to machine-build files took less than 30 min and the SLA® machine produced the model in less than 4 h.

Aligning the tibia during the CT scan was advantageous to the planning of the resection of the proximal tibia in three ways. First, because the optimal surgical cut through the tibia is perpendicular to the tibial axis, this aligned approach means that the CT images could be visually inspected as sections through the tibia parallel to the plane of the intended surgical cut.

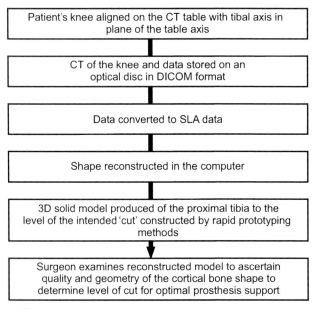

Figure 5.33 Workflow.

Second, it helps maintain accuracy when building the physical model. Computed tomography data are captured as a series of planar images with a gap between them typically of 1.5 mm, whereas the stereolithography process builds models at a layer thickness typically of 0.15 mm. Therefore, interpolation is used to create intermediate sections between the original images. A cubic interpolation ensured that the intermediate sections were anatomically accurate and natural. These sections then directly drive the rapid prototyping machine that builds the physical model. Aligning the CT images ensures that the layers from which the model is built correspond exactly to the plane of the planned surgical cut. This is a significant aid when viewing the stereolithography model because it ensures that the layers visible in the finished model could be used to guide a perfectly level cut.

Third, this 2.5D format (called SLC) can be more accurate and less memory-intensive than 3D approximation formats such as the commonly used triangular faceted STL file. These formats are described in more detail in Chapter 3.

Initial assessment of the CT sections suggested that the cut should be made 15 mm below the lateral joint line. A solid model of the proximal 15 mm of the tibia was produced showing the cortical bone geometry to assess the size and shape of the supportive bone available after a cut at this level, and compared with the undersurface shape and area of the tibial component. The model suggested that the bone shape and distribution would support a size small Minns™ meniscal-bearing tibial component; the thickness of the cut bone removed by the oscillating saw was assumed to be 3 mm.

In surgery, after preparation of the distal femur, the tibial cut was made as planned 15 mm below the lateral joint line, orthogonal to a line from the center of the knee to the center of the ankle in the sagittal plane. The CT-generated model accurately represented the clinical findings at the time of surgery, as confirmed when the removed bone was compared with the stereolithography model (Figure 5.34), and a Minns™ meniscal-bearing total knee was implanted as planned (Corin Medical Ltd., Cirencester, England) (Shaw, Minns, Epstein, & Sutton, 1997). A pair of 4-mm-thick meniscal bearings were most appropriate in this case and the remainder of the procedure was carried out without incident.

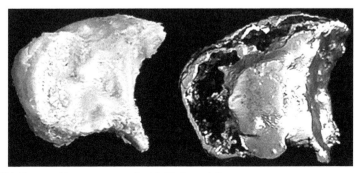

Figure 5.34 Resected bone compared with SLA model.

5.6.3   Postoperative management and follow-up

Clinically, the postoperative period was uneventful; however, on the following day a postoperative X-ray revealed a crack fracture of the shaft of the tibia that was thought to have occurred at some point during the operation and was believed to be most probably caused by the marked rheumatoid arthritis and disuse, because no traumatic event had been noted.

This altered the normal postoperative regimen of early mobilization and the patient was treated with 6 weeks in a full leg-length orthosis before mobilization thereafter. The patient went home on the 10th postoperative day mobilizing non–weight bearing with a walking frame and using a wheelchair. At 6 weeks, the fracture had united and the patient was allowed to mobilise. At 12 weeks, there was good range of knee movement. The patient walked without support and was pain-free with much improved gait; she was delighted with the result.

5.6.4   Discussion

The technique described earlier provides detailed information about the bone morphology in the region of the intended surgery and facilitated prediction of the level of transection of the deformed tibia at a level best suited to supporting the prosthesis. In addition, it provided a model with which the planned surgery could be carried out before ever reaching the patient. This degree of preoperative information is extremely useful, and if there had been insufficient bone at the proposed level of the tibial cut, the size of any required wedge could have been accurately predicted.

The fracture of the tibia may have been caused by an inability of the bone to resist ordinary preoperative handling as a result of the long-standing arthritis and subsequent disuse, or possibly inadvertent cortical contact of one of the trephines used to prepare the proximal tibia.

The authors recognise that there although it is small, there is an increased radiation dose to the patient associated with the use of a CT scan in this technique compared with plain radiographs normally used. Because the knee is an extremity, a scan does not put any other radiosensitive structures in the field. In addition to the images produced, there is also the opportunity, as was done in this case, for 3D reconstruction virtually and as a physical model, which can be compared with the proposed prosthesis preoperatively.

The technique can be used on any tissue that can be clearly distinguished in either CT or magnetic resonance images.

We do not suggest this technique be required for any joint replacement other than straightforward; however, this case demonstrates the value of this technique in knee arthroplasty (surgical knee joint repair) in cases with complex anatomy where the bone shape and quality are difficult to predict from plain films. The creation of a 3D model facilitates preoperative planning in difficult cases and provided valuable information before surgery.

References

Barmeir, E., Dubowitz, B., & Roffman, M. (1982). Computed tomography in the assessment and planning of complicated total hip replacement. *Acta Orthopaedica Scandinavica*, *53*, 597.

Bautsch, T. L., Johnson, E. E., & Seeger, L. L. (1994). True three-dimensional stereographic display of 3-D reconstructed CT scans of the pelvis and acetabulum. *Clinical Orthopaedics*, *305*, 138.

Bert, J. M. (1996). Custom total hip arthroplasty. *Journal of Arthroplasty*, *11*, 905.

McCarthy, J. C., Bono, J. V., & O'Donnell, P. J. (1997). Custom and modular components in primary total hip replacement. *Clinical Orthopaedics*, *344*, 162.

Migaud, H., Corbet, B., Assaker, C., Kulik, J. F., & Duquennoy, A. (1997). Value of a synthetic osseus model obtained by stereo-lithography for preoperative planning: correction of a complex femoral deformity caused by fibrous dysplasia. *Revue d'orthopedie et de chirurgie de l'appareil moteur*, *83*(2), 156.

Munjal, S., Leopold, S. S., Kornreich, D., Shott, S., & Finn, H. A. (2000). CT-generated 3-dimensional models for complex acetabular reconstruction. *Journal of Arthroplasty*, *15*, 644.

Roach, J. W., Hobatho, M. C., Baker, K. J., & Ashman, R. B. (1997). Three-dimensional computer analysis of complex acetabular insufficiency. *Journal of Pediatric Orthopedics*, *17*, 158.

Robinson, R. P., & Clark, J. E. (1996). Uncemented press-fit total hip arthroplasty using the identifit custom-molding technique; a prospective minimum 2-year follow-up study. *Journal of Arthroplasty*, *11*, 247.

Shaw, N. J., Minns, R. J., Epstein, H. P., & Sutton, R. A. (1997). Early results of the minns meniscal bearing total knee prosthesis. *The Knee*, *4*, 185.

Van Dijk, M., Smit, T. H., Jiya, T. U., & Wuisman, P. I. (2001). Polyurethane real-size models used in planning complex spinal surgery. *Spine*, *26*(17), 1920.

5.7   Surgical applications case study 4: custom-made titanium orbital floor prosthesis in reconstruction for orbital floor fractures

Acknowledgments

The work described in this case study was first reported in the reference below and is reproduced here in part or in full with the permission of the British Association of Oral and Maxillofacial Surgeons.

Hughes CW, Page K, Bibb R, Taylor J, Revington P, "The custom-made titanium orbital floor prosthesis in reconstruction for orbital floor fractures", British Journal of Oral and Maxillofacial Surgery, 2003, Volume 41, Issue Number 1, pages 50–3. http://dx.doi.org/10.1016/S0266435602002498

No financial support was given. The National Centre for Product Design and Development Research supplied the stereolithography model used in making the prosthesis.

5.7.1   Introduction

Few anatomical sites of such diminutive size have attracted so much variation in treatment as the orbital floor (the bottom of the eye socket) and its related fractures. The range of implant material in reconstruction after blowout fracture of the orbit is extensive and the decision as to which material is used remains a matter of debate (Courtney, Thomas, & Whitfield, 2000).

Autologous materials (those derived from human tissues) offer clear advantages; cartilage, calvarial bone, antral bone, rib and ilium have been described (Courtney et al., 2000). These grafts offer uncertain longevity and result in tissue damage at the donor site. Artificial materials such as Silastic® (Dow Corning Corporation, Auburn Plant, 5300 11 Mile Road, Auburn, MI 48611, USA) have the longest track record but a well-documented complication rate related in particular to extrusion of the graft (Morriston, Sanderson, & Moos, 1995). Other artificial materials such as polyethylene sheeting (Medpor®, Porex Surgical Products, Group USA, Porex Surgical, Inc., 15 Dart Road, Newnan, GA 30265-1017, USA) are reported to give satisfactory results (Rubin, Bilyk, & Shore, 1994) and newer resorbable materials such as polydioxanone are another option (Iizuka, Mikkonen, Paukku, & Lindqvist, 1991). The role of bioactive glass has been reported but its use is limited by the size of the defect (Kinnunen, Aitasalo, Pollonen, & Varpula, 2000). Titanium is an inert and widely used material (Dietz et al., 2001; Park, Kim, & Yoon, 2001) but in its preformed presentation can be cumbersome to use on the orbital floor, and if it has to be removed it can present an operative challenge.

Continued development in computer-aided diagnosis, management and construction of stereolithographic models offers unparalleled reproduction of anatomical detail (Bibb & Brown, 2000; Bouyssie, Bouyssie, Sharrock, & Duran, 1997). This technology is described in relation to planning in trauma surgery (Kermer, Lindner, Friede, Wagner, & Millesi, 1998) and ablative surgery for malignancies of the head and neck (surgery to remove cancer) (D'Urso et al., 1999; Kermer et al., 1998). Construction of custom-made orbital floor implants is possible (Hoffmann et al., 1998; Holk, Boyd, Ng, & Mauffray, 1999), although the material of choice is debated.

We describe a simple technique for construction of custom-made titanium orbital floor implants using easily available laboratory techniques combined with stereolithography models. We estimate the implant construction cost at around £300. This is largely accounted for by the cost of producing the model, which varies between £200 and £300, depending on the height of orbital contour required for the model. Construction of the implant takes about 2 h of a maxillofacial technician's time and the medical-grade titanium sheet costs only a few pounds. This compares favourably with some of the newer alloplastic materials. This cost would drop substantially with greater use of the technique, and when reduced operating time is taken into account, the cost comparison is more favourable.

5.7.2 Technique

5.7.2.1 Imaging

The detail given here is specific to this case; a more general overview of computed tomography (CT) scanning is given in Chapter 2, Section 2.2. Scanning protocols are observed to minimise the dose of ionising radiation to orbital tissues (Ionising Radiation medic, 2000). Maximum detail can be obtained scanning with a 0.5-mm collimation, but the 77% increase in dosage compared with using a 1-mm collimation may not be justified. We use a Siemens Somatom Plus4 Volume Zoom scanner with these settings: 140 kV, 120 mAs, 1 mm collimation, 3.5 feed per rotation, 0.75 rotation time, giving a displayed dose of 45 mGy/100 mAs. Data are reconstructed using 1-mm slices with 0.5-mm increments (50% overlap) and smooth kernel. Sharp reconstruction kernels normally associated with CT imaging of bony anatomy introduce an artificial enhancement of the edge. If used as part of a three-dimensional (3D) volume based on selection of specific Hounsfield values, the enhancement artefact will be included with the bony detail, thus degrading the image. The data obtained can be used to construct sharp multi-plane reformats for bony detail and 3D imaging for both hardcopy imaging and stereo viewing by the surgeons.

5.7.2.2 Model construction and stereolithography apparatus® (SLA)

Computed tomography scans are typically taken in the axial plane at intervals exceeding 1 mm. This means that thin bone that lies predominantly in the axial plane may fall between consecutive scans and therefore not be present in the data or 3D model created from it. To overcome this, scans were taken using a smooth kernel at a slice distance of 1 mm but with a 0.5-mm overlap, as described earlier. This improves the resolution of the data in these thin areas. The detail created is exceptionally good. The CT data were then segmented to select the desired tissue type, compact bone, using methods described in Chapter 3.

The production of models using stereolithography is described fully in Chapter 4, Section 4.2. In this case, to maintain the greatest level of accuracy, an epoxy resin was chosen (RenShape SL5220, Huntsman Advanced Materials, Everslaan 45, B-3078 Everberg, Belgium). This type of resin shows almost no shrinkage during the photo-polymerisation process and therefore can produce models with excellent accuracy.

5.7.2.3   Construction of prosthesis

From stereolithography models, the orbital defect is easily seen and assessed. The orbital defect is then filled with wax to reproduce a contour similar to the opposite side and an impression is taken of both orbital cavities using silicone putty impression material. The orbital injury side is then reproduced by pouring a hard plaster/stone model. The defect has been filled and therefore appears in its proposed reconstructed form. Using pressure flasks usually employed in the construction of dentures, a layer of 0.5-mm medical-grade titanium is swaged onto the stone/plaster model of the orbital floor, producing an exact replica of the proposed orbital floor and rim contour. The titanium sheet may then be trimmed to allow sufficient overlap and the positioning of a flange to fix the screws. The prosthesis is polished and sterilised for use according to local protocol for titanium implants.

5.7.3   Case report

A 54-year-old man sustained a blowout fracture of the left orbital floor and presented with diplopia (double vision) and restriction of upward gaze. Coronal plane CT scanning demonstrated the fracture (Figure 5.35). A stereolithography model was constructed that showed the trapdoor of the fractured orbital floor well (Figure 5.36). The model was then used to construct a plaster cast of the orbital defect. A medical-grade titanium prosthesis was constructed from this working cast (Figure 5.37). The prosthesis was packaged and sterilised by the hospital central sterile supplies department according to standard protocol for titanium medical implants.

The approach to the orbital floor was by a subciliary incision (through the lower eyelid) and the defect was exposed. Herniation and entrapment of the periglobar fat were released and the defect was prepared in a standard way (this means that the damaged layer of fat that surrounds the eyeball was repaired and put back in the correct position). The prosthesis fitted perfectly and was stabilised with 1.3-mm titanium screws from a standard plating kit (Figure 5.38). Forced duction was confirmed as normal (this is a test to check that the eye can rotate upward freely). Postoperative recovery was uneventful and radiographs revealed the prosthesis to be correctly positioned (Figures 5.39 and 5.40). At follow-up, complete return to normal range of ocular movement was found with resolution of the diplopia and no evidence of complications.

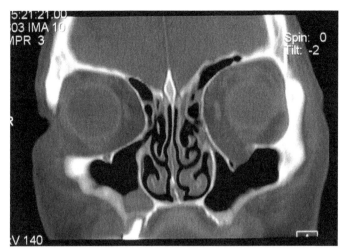

Figure 5.35 Coronal CT scan demonstrating classic orbital blowout fracture.

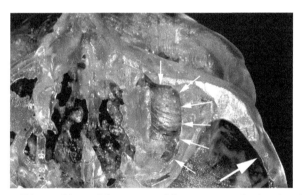

Figure 5.36 Stereolithographic model constructed from epoxy resin showing the trapdoor defect in the left orbital floor, viewed from below as if in the maxillary antrum looking up (lateral margins of the defect indicated with small arrows, under surface of zygomatic arch indicated with single large arrow).

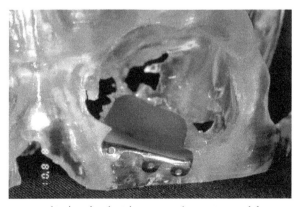

Figure 5.37 The custom titanium implant is seen on the master model.

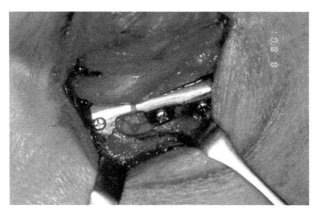

Figure 5.38 The implant inserted and fixed with 1.3-mm screws. The fit is precise.

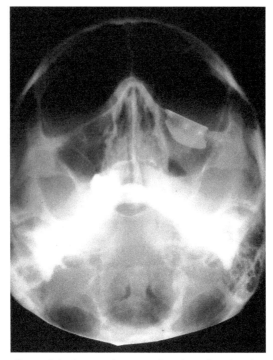

Figure 5.39 Plain radiograph in the anterior–posterior plane showing the position of the implant postoperatively.

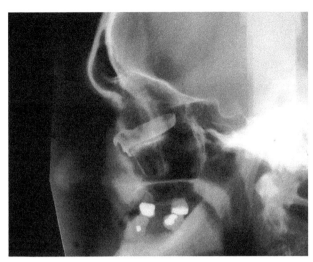

Figure 5.40 Plain radiograph in the lateral plane showing the position of the implant postoperatively.

5.7.4 Conclusion

We think that this technique has much to offer because of its simplicity and the reliability of titanium as a prosthetic material. The laboratory techniques are simple and readily available in most maxillofacial laboratories. The models require offsite production but their use is particularly valid in cases where defects may be complicated in three dimensions and where operating time should be reduced to a minimum. The cost of construction of models will drop substantially if numbers increase, and the technique may offer a financially viable alternative to current orbital floor prostheses.

References

Bibb, R., & Brown, R. (2000). The application of computer aided product development techniques in medical modelling. *Biomedical Sciences Instrumentation, 36*, 319–324.

Bouyssie, J. F., Bouyssie, S., Sharrock, P., & Duran, D. (1997). Stereolithographic models derived from x-ray computed tomography: reproduction accuracy. *Surgical and Radiologic Anatomy, 3*, 193–199.

Courtney, D. J., Thomas, S., & Whitfield, P. H. (2000). Isolated orbital blow out fractures: survey and review. *British Journal of Oral and Maxillofacial Surgery, 38*, 496–503.

Dietz, A., Ziegler, C. M., Dacho, A., Althof, F., Conradt, C., Kolling, G., et al. (2001). Effectiveness of a new perforated 0.15 mm poly-p-dioxanon-foil versus titanium dynamic mesh in reconstruction of the orbital floor. *Journal of Craniomaxillofacial Surgery, 2*, 82–88.

D'Urso, P. S., Barker, T. M., Earwaker, W. J., Bruce, L. J., Atkinson, R. L., Lanigan, M. W., et al. (1999). Stereolithographic biomodelling in cranio-maxillofacial surgery: a prospective trial. *Journal of Craniomaxillofacial Surgery, 1*, 30–37.

Hoffmann, J., Cornelius, C. P., Groten, M., Probster, L., Pfannenberg, C., & Schwenzer, N. (1998). Orbital reconstruction with individually copy-milled ceramic implants. *Plastic Reconstructive Surgery*, *3*, 604–612.

Holk, D. E., Boyd, E. M., Jr., Ng, J., & Mauffray, R. O. (1999). Benefits of stereolithography in orbital reconstruction. *Opthalmology*, *6*, 1214–1218.

Iizuka, T., Mikkonen, P., Paukku, P., & Lindqvist, C. (1991). Reconstruction of orbital floor with polydioxanone plate. *International Journal of Oral and Maxillofacial Surgery*, *20*(2), 83–87.

Ionising radiation (medical exposure) regulations. (2000). UK Government Department of Health Publication.

Kermer, C., Rasse, M., Lagogiannis, G., Undt, G., Wagner, A., & Millesi, W. (1998). Colour stereolithography for planning complex maxillofacial tumour surgery. *Journal of Craniomaxillofacial Surgery*, *6*, 360–362.

Kermer, C., Lindner, A., Friede, I., Wagner, A., & Millesi, W. (1998). Preoperative stereolithographic model planning for primary reconstruction in craniomaxillofacial trauma surgery. *Journal of Craniomaxillofacial Surgery*, *3*, 136–139.

Kinnunen, I., Aitasalo, K., Pollonen, M., & Varpula, M. (2000). Reconstruction of orbital floor fractures using bioactive glass. *Journal of Craniomaxillofacial Surgery*, *4*, 229–234.

Morriston, A. D., Sanderson, R., & Moos, K. F. (1995). The use of silastic as an orbital implant for reconstruction of orbital wall defects: review of 311 cases treated over 20 years. *Journal of Oral and Maxillofacial Surgery*, *53*, 412–417.

Park, H. S., Kim, Y. K., & Yoon, C. H. (2001). Various applications of titanium mesh screen implant to orbital wall fractures. *Journal of Craniofacial Surgery*, *6*, 555–560.

Rubin, P. A. D., Bilyk, J. R., & Shore, J. W. (1994). Orbital reconstruction using porous polyethylene sheets. *Opthalmology*, *101*, 1697–1708.

5.8 Surgical applications case study 5: use of three-dimensional technology in the multidisciplinary management of facial disproportion

Acknowledgments

The work described in this case study was first reported in the reference below and is reproduced here in part or in full with the permission of First Numerics Ltd.

Knox J, Sugar AW, Bibb R, Kau CH, Evans P, Bocca A, Hartles F, "The Use of 3D Technology in the Multidisciplinary Management of Facial Disproportion" Proceedings of the 6th International Symposium on Computer Methods in Biomechanics & Biomedical Engineering, Madrid, Spain, February 2004, ISBN: 0-9549670-0-3 (Published on CD-ROM by First Numerics Ltd. Cardiff, UK)

5.8.1 Introduction

Coordinated orthodontic/surgical treatment, which allows the predictable management of dentofacial disproportion, is largely a development of the latter third of the twentieth century. Traditionally, the diagnosis, treatment planning and postoperative evaluation of patients requiring such treatment have relied heavily on the use of cephalometric analysis. This has enabled the two-dimensional (2D) quantification of dental and skeletal relationships both before and after treatment, with reference to normative data in tabulated or template form (Ackerman, 1975; Broadbent, Broadbent, & Golden, 1975; McNamara, 1984).

However, the recent development of 3D measuring techniques has allowed a more clinically valid quantification of deformity and assessment of surgical outcomes (Ayoub et al., 1996, 1998; Ji, Zhang, Schwartz, Stile, & Lineaweaver, 2002; Khambay et al., 2002; Marmulla, Hassfeld, Luth, & Muhling, 2003; McCance, Moss, Wright, Linney, & James, 1992, McCance, Moss, Fright, Linney, & James, 1997). Computed tomography (CT), magnetic resonance imaging (MRI) and finite element analysis (FEA) have been employed in surgical planning and the visualization of treatment objectives (Bibb & Brown, 2000; Bibb et al., 2000; Gladilin, Zachow, Deuflhard, & Hege, 2002a,b; Gladilin, Zachow, Deuflhard, & Hege, 2003a,b; Nkenke et al., 2003; Xia et al., 2000, 2001). This presentation demonstrates the successful use of 3D tomography, surface laser scans and rapid prototyping in the surgical management and postoperative evaluation of a patient presenting with maxillary hypoplasia, who underwent surgical maxillary distraction.

5.8.2 Materials and methods

Three-dimensional virtual hard tissue images, shown in Figure 5.41, were constructed using Mimics® software (Materialise N.V., Technolgielaan 15, 3001 Leuven, Belgium)

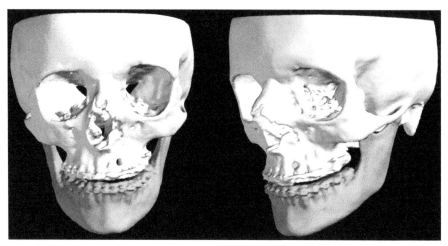

Figure 5.41 Virtual hard tissue images.

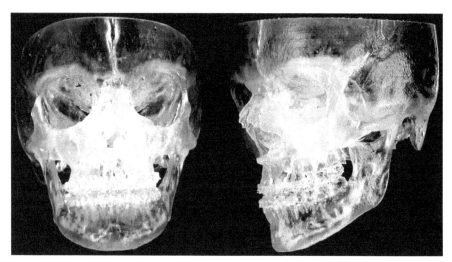

Figure 5.42 Stereolithography models.

from 0.5-mm slice CT DICOM data sets to identify tissue type. Upper and lower tissue density thresholds on the CT image were defined and the areas between the slices were interpolated to improve resolution. The data were then prepared for medical modelling using stereolithography. In addition, STL files were generated so that the same data could be imported into the Geomagic FreeForm® Plus software (Geomagic, 3D Systems, USA). Preparation of data for file transfer and medical modelling is described in detail in Chapter 3.

A stereolithography model was then constructed and used to visualise skeletal discrepancy, simulate surgical movements and adapt surgical distractors, as shown in Figures 5.42 and 5.43. The stereolithography process is described in detail in

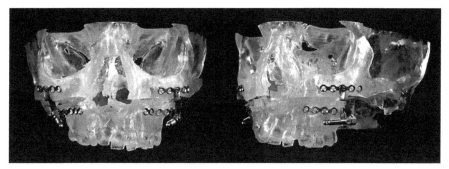

Figure 5.43 Surgical simulation and placement of distractors on stereolithography models.

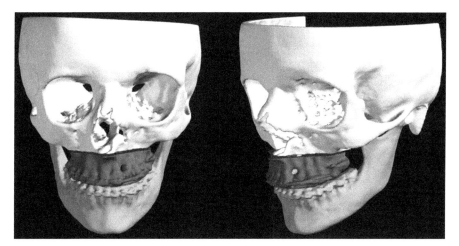

Figure 5.44 Virtual surgical simulations showing maxillary advancement at Le Fort 1 level.

Chapter 4, Section 4.2. FreeForm® software was used to produce a digital clay model allowing further simulation of surgical movements (see Figure 5.44).

Three-dimensional facial soft tissue images (see Figure 5.45) were captured before and after surgery using two high-resolution Minolta Vivid VI900 3D cameras operating as a stereo-pair. The scanners were controlled with multi-scan software (Cebas Computer GmbH, Lilienthalstrasse19, 69214 Eppelheim, Germany) and data coordinates were saved in the Minolta Vivid file format (called a.vvd). The scan data were transferred to a reverse modelling software package (Rapidform 2004, INUS Technology, Inc., SBC, Ludwig-Erhard-Strasse, 30-34, D-65760 Eschborn, Germany) for analysis.

5.8.3   Results

Surgical distraction of the maxilla at Le Fort 1 level was successfully completed (Le Fort 1 is a cut across the maxilla above the roots of the teeth but under the nose, to separate and move the upper jaw in relation to the rest of the skull). The

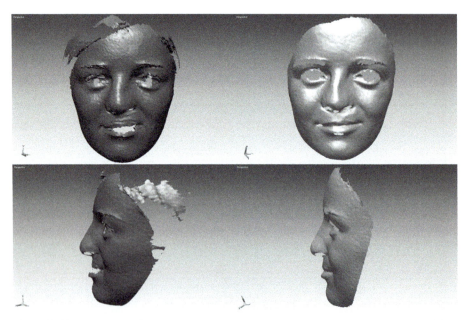

Figure 5.45 Three-dimensional facial images preoperatively (left) and postoperatively (right).

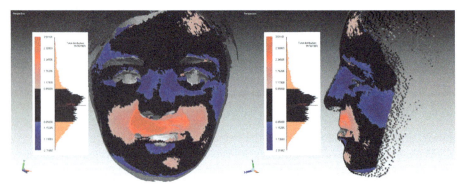

Figure 5.46 Merged pre- and postoperative facial scans demonstrating the magnitude of change in soft tissue morphology.

distractors were activated by 7 mm on the right and left sides, resulting in an equivalent advancement of the tooth-bearing portion of the maxilla. A description of distraction osteogenesis is given in medical explanatory note in Section 7.2.1. Superimposition of surface scans allowed quantification of soft tissue changes (Figure 5.46). Black areas indicate changes within 0.80 mm or less, which could be attributed to error of the technique. The blue areas demonstrate negative changes of 0.85–2.30 mm. The red areas demonstrate positive changes of 0.85–3.51 mm.

5.8.4 Discussion

Changes in maxillary prominence and lip relationship can be appreciated comparing pre- and postoperative scans in Figure 5.45. The magnitude of the soft tissue changes is demonstrated in Figure 5.46. Here, the primary effect of the maxillary distraction at Le Fort 1 level is advancement of the upper lip and paranasal areas of 1.7–3.5 mm (red areas). The small advancement (pink) demonstrated in the frontal region is an artefact introduced by overlying hair. The small advancement demonstrated in the left chin is probably the result of a change in lip relationship and a slight change in facial expression. The blue areas in Figure 5.46 demonstrate a reduction in lower lip prominence owing to the change in lip relationship caused by the maxillary advancement. The change demonstrated in the submandibular region and upper mid-face is suggested to be result from a reduction in body mass index.

References

Ackerman, R. J. (1975). The Michigan school study norms expressed in template form. *American Journal of Orthodontics*, *75*, 282–290.

Ayoub, A. F., Siebert, P., Moos, K. F., Wray, D., Urquhart, C., & Niblett, T. B. (1998). A vision-based three-dimensional capture system for maxillofacial assessment and surgical planning. *British Journal of Oral and Maxillofacial Surgery*, *36*(5), 353–357.

Ayoub, A. F., Wray, D., Moos, K. F., Siebert, P., Jin, J., Niblett, T. B., et al. (1996). Three-dimensional modeling for modern diagnosis and planning in maxillofacial surgery. *International Journal of Adult Orthodontics and Orthognathic Surgery*, *11*(3), 225–233.

Bibb, R., & Brown, R. (2000). The application of computer aided product development techniques in medical modeling. *Biomedical Sciences Instrumentation*, *36*, 319–324.

Bibb, R., Freeman, P., Brown, R., Sugar, A., Evans, P., & Bocca, A. (2000). An investigation of three-dimensional scanning of human body surfaces and its use in the design and manufacture of prostheses. *Proceedings of the institution of mechanical engineers Part H: Journal of Engineering in Medicine*, *214*(6), 589–594.

Broadbent, B. H., Sr, Broadbent, B. H., Jr, & Golden, W. H. (1975). *Bolton standards of dentofacial development and growth*. St. Louis: Mosby.

Gladilin, E., Zachow, S., Deuflhard, P., & Hege, H.-C. (2002a). A non-linear soft tissue model for craniofacial surgery simulations. In *Proceedings of the modeling & simulation for computer-aided medicine and surgery (MS4CMS), INRIA, Paris, France.*

Gladilin, E., Zachow, S., Deuflhard, P., & Hege, H.-C. (2002b). Biomechanical modeling of individual facial emotion expressions. In *Proceedings of visualization, imaging, and image processing (VIIP).*

Gladilin, E., Zachow, S., Deuflhard, P., & Hege, H.-C. (2003a). On constitutive modeling of soft tissue for the long term prediction of cranio-maxillofacial surgery outcome. In *Proceedings of computer assisted radiology and surgery (CARS), London* (pp. 343–348).

Gladilin, E., Zachow, S., Deuflhard, P., & Hege, H. C. (2003b). Realistic prediction of individual facial emotion expressions for craniofacial surgery simulations. In *Proceedings of SPIE medical imaging conference, San Diego, USA* (Vol. 5029) (pp. 520–527).

Ji, Y., Zhang, F., Schwartz, J., Stile, F., & Lineaweaver, W. C. (2002). Assessment of facial tissue expansion with three-dimensional digitizer scanning. *Journal of Craniofacial Surgery*,
13(5), 687–692.

Khambay, B., Nebel, J. C., Bowman, J., Walker, F., Hadley, D. M., & Ayoub, A. (2002). 3D stereophotogrammetric image superimposition onto 3D CT scan images: the future of orthognathic surgery. *International Journal of Adult Orthodontics and Orthognathic Surgery*,
17(4), 331–341.

Marmulla, R., Hassfeld, S., Luth, T., & Muhling, J. (2003). Laser-scan-based navigation in
cranio-maxillofacial surgery. *Journal Craniomaxillofacial Surgery*, *31*(5), 267–277.

McCance, A. M., Moss, J. P., Fright, W. R., Linney, A. D., & James, D. R. (1997).
Three-dimensional analysis techniques – Part 1: three-dimensional soft-tissue analysis of
24 adult cleft palate patients following Le Fort I maxillary advancement: a preliminary
report. *Cleft Palate-Craniofacial Journal*, *34*(1), 36–45.

McCance, A. M., Moss, J. P., Wright, W. R., Linney, A. D., & James, D. R. (1992). A three-
dimensional soft tissue analysis of 16 skeletal class III patients following bimaxillary
surgery. *British Journal of Oral and Maxillofacial Surgery*, *30*(4), 221–232.

McNamara, J. A., Jr (1984). A method of cephalometric evaluation. *American Journal of Orthodontics*, *86*(6), 449–469.

Nkenke, E., Langer, A., Laboureux, X., Benz, M., Maier, T., Kramer, M., et al. (2003). Validation of in vivo assessment of facial soft-tissue volume changes and clinical application
in midfacial distraction: a technical report. *Plastic and Reconstructive Surgery*, *112*(2),
367–380.

Xia, J., Ip, H. H., Samman, N., Wong, H. T., Gateno, J., Wang, D., et al. (2001). Three-
dimensional virtual-reality surgical planning and soft-tissue prediction for orthognathic
surgery. *IEEE Transactions on Information Technology in Biomedicine*, *5*(2), 97–107.

Xia, J., Samman, N., Yeung, R. W., Shen, S. G., Wang, D., Ip, H. H., et al. (2000). Three-
dimensional virtual reality surgical planning and simulation workbench for orthognathic
surgery. *International Journal of Adult Orthodontics and Orthognathic Surgery*, *15*(4),
265–282.

5.9   Surgical applications case study 6: appropriate approach to computer-aided design and manufacture of reconstructive implants

Acknowledgments

All of the cases described were undertaken as part of a multidisciplinary team.

Planning for case 1 was undertaken by Peter Evans and Adrian Sugar at Morriston Hospital Swansea and the surgery was undertaken by Adrian Sugar. Details of this case were presented as a poster at the 2009 Institute of Maxillofacial Prosthetists and Technologists Congress.

Planning for cases 2 and 3 were undertaken by Sean Peel and Dominic Eggbeer at PDR, Cardiff Metropolitan University and Satyajeet Bhatia at the University Hospital, Wales, Cardiff. Surgery for case studies 2 and 3 were undertaken by Satyajeet Bhatia.

Case 4 was planned by Adrian Sugar and Peter Evans at Morriston Hospital with input from Sean Peel from PDR, Cardiff Metropolitan University. Sean Peel designed the guides and implants. Adrian Sugar undertook the surgery.

5.9.1   Introduction

Advances in metal additive manufacturing (AM) processes have enabled complex forms designed in computer-assisted design (CAD) to be manufactured in biocompatible materials such as titanium. The layer-additive nature of AM offers a greater degree of design freedom over processes such as computer numerically controlled (CNC) machining; this represents an opportunity in the production of complex implants. There are a growing number of case studies in which AM technologies have been used to produce patient-specific implants in craniomaxillofacial and orthopaedic applications (Mazzoli, Germani, & Raffaeli, 2009; Murr, Gaytan, Martinez, Medina, & Wicker, 2012; Salmi et al., 2012; Wehmöller, Warnke, Zilian, & Eufinger, 2005), but the techniques are still not yet used routinely in most hospitals.

Chapter 5, Section 5.13, demonstrates an evolution of techniques for cranioplasty implant production, but this section will provide an overview of other maxillofacial cases in which CAD/AM technologies were used to produce more complex patient-specific implants.

5.9.2   Case 1: orbital rim augmentation implant

This study reports a case involving the design, manufacture and implantation of a custom-fit titanium implant to improve the orbital symmetry of a patient with a facial cleft. The 25-year-old woman had a repaired Tessier number 3 facial cleft and required

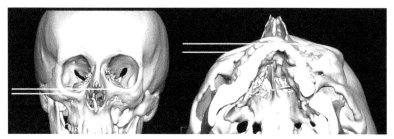

Figure 5.47 Preoperative CT showing defect on the patient's right orbit.

correction of severe right vertical orbital dystopia and enophthalmos, augmentation of the anterior zygomatico-maxillary region, revision of the facial cleft skin scar and lip revision. The patient's right eye was nonseeing, relatively immobile, turned inward and more than 1 cm lower than the left eye. The skin was tight over the anterior maxilla and body of the zygomatic bone where the augmentation would take place, so advancement was necessary to achieve tension-free skin cover over the augmentation when closing the facial cleft scar. The patient's major concern was about her appearance.

Evaluation of the computed tomography (CT) scan demonstrated that the whole orbit was not displaced inferiorly but the floor was, causing the eye and orbital contents to be drawn downward by the orbital floor (Figure 5.47). After much debate and discussion, it was decided that complete freeing of the orbital soft tissues by 360° periorbital dissection through a coronal flap approach followed by augmentation of the deficient orbital floor and zygomatico-maxillary region through opening up of the previous facial cleft scar was appropriate. It was hoped this would improve facial symmetry, eye position and overall appearance by helping to correct the dystopia and bone deficiency. It was recognised that repositioning the medial and lateral canthal ligaments would also be required. The ophthalmologists preferred to attempt to correct the severe squint at a later stage.

Various autogenous and alloplastic augmentation materials were considered. It was decided that the ideal would be a custom-fit solid titanium implant that would neither distort nor resorb. Additive manufacturing was deemed an appropriate route to production of the implant because of its complex shape, which would be impossible to achieve by swaging titanium. The dimensions of the implant also made it impossible cast a titanium implant at a reasonable cost, quality and time scale. Although CNC machining of custom-fit maxillofacial implants has been demonstrated (Wehmöller et al., 2005; Weihe et al., 2000), the complexity of shape involved in this example would have presented difficulty in clamping the work piece, required multiple process steps and involved a great deal of programming time to achieve the desired result. In addition, machining this implant would have required the purchase of an appropriately oversized billet of solid titanium which would have been expensive to purchase. The dimensions of the required billet would have been in the order of $5 \times 5 \times 5$ cm (125 cm^3) to machine an implant with a volume of just 2.52 cm^3, meaning that as much as 98% of the billet would have been removed as waste.

The decision to use AM necessitated CAD of the implant. A variety of AM processes are available that can produce parts from a range of materials from soft thermoplastics to high-performance alloys. The principles of AM technologies are described in Chapter 4 and in a range of texts (Chua, Leong, & Lim, 2010; Gibson, Rosen, & Stucker, 2009; Hopkinson, Hague, & Dickens, 2005; Noorani, 2005).

5.9.2.1 Materials and methods

Stage 1: three-dimensional data acquisition and transfer

A CT scan was taken with 1-mm slice thickness, 0.5-mm increments and 0.424-mm pixel size on a Toshiba Aquilion Multislice scanner. The data were imported into Mimics® software (Mimics version 13, Materialise N.V., Technolgielaan 15, Leuven 3001, Belgium), which was used to segment the bony anatomy, maxillary sinus volume and eye globes into separate volumes. The software also allowed measurements to be taken to quantify the difference in globe height and protrusion. The 3D reconstructed data of the bony anatomy were exported as high-quality STL files employing the smoothing parameters available when using Mimics. The STL files were imported into CAD software (FreeForm® Modelling Plus, Geomagic, 3D Systems, USA) to undertake the implant design. Use of this software has been reported in the design of other custom-fit medical devices including removable partial dentures and facial prosthetics (Bibb, Eggbeer, & Williams, 2006, 2009, 2010; Eggbeer, Bibb, & Williams, 2005, 2006; Williams, Bibb, Eggbeer, & Collis, 2006) and is described in other case studies in this book. The fill holes option was used to ensure the STL file was one single bound volume with no voids.

Stage 2: implant design

The design was conducted by a maxillofacial prosthetist who had several years of experience using the FreeForm® software in medical applications. On importing the patient data, the clay edge sharpness was set to 0.316 mm. This is how the quality of the model is controlled within FreeForm®. The patient anatomy data were imported using the buck setting, which prevents the data from being altered in any way, thus protecting the anatomy data from accidental modification. The basic implant shape was generated by selecting and copying an area of the patient's unaffected orbit corresponding to the defect area on the contralateral side and laterally inverting (mirroring) it about the mid-sagittal plane. Measurement tools and reference planes were used to establish the correct contours and position of the implant to create the best possible orbital symmetry. The positioned form was then expanded to overlap the protected model of the anatomy and final shaping undertaken to create a smooth shape that effectively restored the shape of the affected anatomy.

Two 3-mm-diameter screw holes were incorporated into the design according to the ideal anchor sites that eliminated potentially difficult drilling operations later. The model

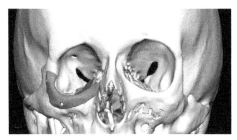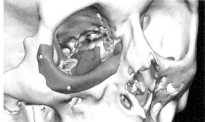

Figure 5.48 The design in FreeForm® CAD.

edge sharpness was refined by 0.1 mm to smooth off sharp edges. The design in FreeForm®
is shown in Figure 5.48. The original anatomy was then removed using a Boolean subtraction that resulted in a fitting surface that precisely matched the defect site. The final implant
design was then exported as an STL file. A stereolithography model of the patient's bony
anatomy was also created to enable the titanium implant fit to be validated before surgery.

Stage 3: additive manufacture

The implant was fabricated in Ti6Al4V alloy using direct metal laser sintering (EOS Materials data sheet, 2152). This AM process uses a fibre laser to selectively melt fine metal
powder particles together to form solid layers. Once each layer is complete, the build platform is lowered by one layer thickness and fresh powder is spread over it. The next layer
is then melted and also fused to the previous layer. The process repeats until the complete
metal object is created. The process also concurrently builds support structures that anchor
the part being built to the build platform and support overhanging features. The machine
used to produce the implant in this case was an EOSint M270 (EOS GmbH, Electro Optical Systems, Robert-Stirling-Ring 1, D-82152 Krailling, München, Germany).

The part was built using 0.03-mm layers and it took 4 h 47 min to build, with additional time and labour required to remove the supports and remove loose powder from
the surface by bead blasting. Abrasive finishing was undertaken in the hospital maxillofacial prosthetics laboratory to achieve a smooth, polished surface. The implant was
then passivated in concentrated nitric acid and autoclave sterilised.

Stage 4: fitting and surgery

The finished implant was test-fitted to the stereolithography model of the patient's
bony anatomy to check dimensional accuracy. The implant was found to be a precise
fit with secure location as shown in Figure 5.49.

At surgery, the orbital periosteum was raised through 360° around the right orbit via
a coronal incision. The previous facial scar was opened from the lower eyelid to and
including the upper lip. The orbital floor was identified and freed posteriorly and the anterior zygomatic bone and maxilla were exposed subperiosteally as required. The implant
was tried in and was observed to be a perfect fit. It was fixed to the orbital rims with two
1.5-mm-diameter titanium screws, as shown in Figure 5.50. The canthal ligaments were
then identified and repositioned superiorly to a screw on the right and a mini-plate on the

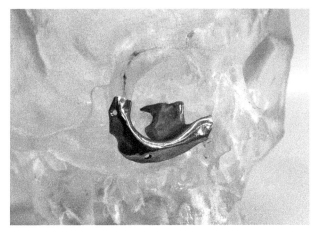

Figure 5.49 Additive manufacturing titanium implant test fitted to the stereolithography model.

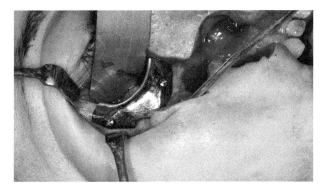

Figure 5.50 Implant fitted in surgery.

left. The lip scar was revised and the skin and other soft tissues were advanced over the implant. These tissues were then closed in layers and the scar was revised.

Results and conclusions from Case 1

The initial post-surgical outcome was good, resulting in a successful reconstruction and much improved aesthetics. Postoperative CT scans showed the implant to be securely fixed in its intended position, as shown in Figure 5.51. There were no complications after a 3-month review and subsequent reviews.

   A custom-fit, patient-specific implant was designed using an entirely digital process and fabricated employing AM in an appropriate, biocompatible titanium alloy. The prosthesis fit both the stereolithography model and the patient precisely, providing secure and positive location. However, because of the complex organic form, no convenient datum points were available and consequently no quantification of dimensional accuracy was undertaken to compare the implant with the CAD geometry. Whereas this case demonstrated that the digital design and AM process are capable of producing an appropriate

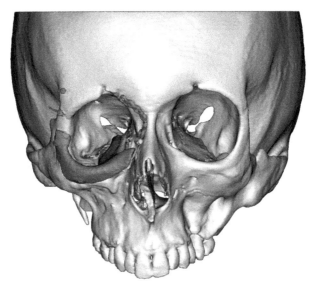

Figure 5.51 Postoperative views.

and accurate titanium implant, further case studies and technical experiments are proposed to further evaluate the potential of the digital process, addressing design capabilities, the capability of the AM process, material physical properties and process limitations.

5.9.3   Case 2: orbital floor implant incorporating placement guide

The first case study demonstrated the potential of a CAD/AM approach in congenital cases involving the orbit. The second study illustrates how the same approach can be used in trauma reconstructive surgery. There has been limited discussion on the application of CAD/AM in orbital floor reconstruction in the research literature (Salmi et al., 2012), but this is an area where a CAD/AM approach could improve the predictability of procedures.

The patient presented with enophthalmos (posterior displacement of the eye) of the right eye resulting from an orbital floor blowout fracture. In this case, trauma to the mid-face had caused the floor of the orbit to fracture and displace downward in a trapdoor fashion. The goal of surgical intervention was to correct the volume of the orbit by reconstructing the orbital floor with a bespoke implant, contoured to match the unaffected side. Orbital floor reconstructions can be challenging owing to the sensitive nature of the orbital anatomy, the proximity of the implant to the optic nerve and the confined space in which the procedure must take place. A CAD/AM approach was considered an appropriate route to ensure that the design was sufficiently small to minimise the necessary surgical incision and provide the best possible fit and contour.

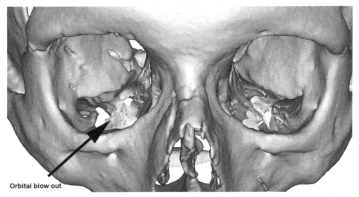

Orbital blow out

Figure 5.52 Three-dimensional representation of the orbital trauma in Mimics®.

5.9.3.1   Materials and methods

Stage 1: data segmentation

The same Mimics® CT segmentation techniques as described in the first case study and in Chapter 3, Section 3.2, were undertaken to obtain an accurate 3D representation of the orbital region that showed the defect, illustrated in Figure 5.52.

The mask of the original anatomy was duplicated. The edit mask tool was used to manually draw in holes in the orbital floor of the unaffected side, thereby creating a thickness that could be mirrored across the midsagittal plane to the damaged side. Both the original anatomy and version with the orbital floor drawn in were exported as high-quality STL files.

Stage 2: defect reconstruction

The STL files were imported into FreeForm® Modelling Plus as buck models (to protect them from accidental material removal) for implant design. An area of the left orbit incorporating the drawn-in portion was selected, copied and mirrored around the midsagittal plane. This was then repositioned by eye to provide the basic reconstruction shape for the damaged right side. Once the basic position had been established, the mirrored portion was converted to clay and combined into the original anatomy buck model for shape refinement. Various tools in the sculpt clay pallet were used to smooth and blend the clay into the buck, creating a seamless contour upon which the implant form could be designed. Measurement tools and planes were used to check the symmetry of the final contour.

Stage 3: implant and positioning guide design

The reconstructed orbital floor model was selected and converted to buck. An outline of the proposed orbital floor plate was drawn directly onto the buck surface using the draw curve tool. The layer tool was used to create an even 0.5-mm thickness of clay that represented the implant shape, which was subsequently verified by the prescribing

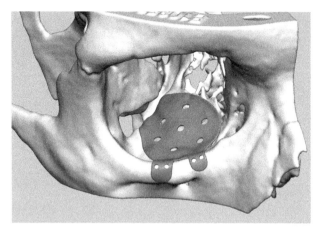

Figure 5.53 Completed orbital floor implant.

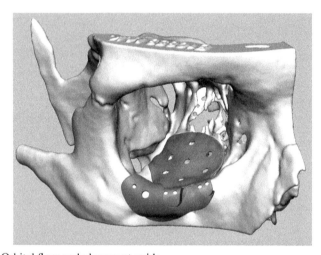

Figure 5.54 Orbital floor and placement guide.

surgeon. The implant size was designed to be as small as possible while still providing sufficient support to the orbital content. The buck was then removed to leave just the implant plate design. Then, 3-mm-diameter holes to allow fluid drainage beneath the eye globe were added to the plate using the carve with ball tool in random locations. Four 1.6-mm-diameter holes (intended for 1.5-mm-diameter screws, leaving 0.1-mm clearance) were then added on the tabs extending over the infra-orbital rim. The completed orbital floor implant design is shown in Figure 5.53.

To enable more accurate placement of the orbital implant, which was designed to minimise the bulk of material around the infra-orbital rim, the prescribing surgeon requested the design and production of a custom-fitting guide. This was designed as a transient use device to engage with a larger area around the infra-orbital rim, giving a stable and positive location. The contours were also designed to allow the implant to slot into the correct position and be secured (Figure 5.54).

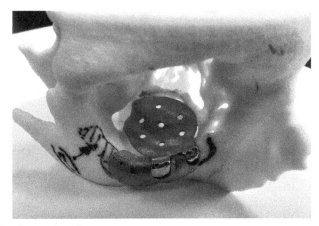

Figure 5.55 Implant and guide.

Stage 4: implant and guide fabrication

Additive manufacturing was chosen for the same reasons as in Case study 1 in Section 5.1. Both the implant and guide were fabricated in Ti6Al4V alloy using laser melting technology (described in Chapter 4, Section 4.5.3) by LayerWise (Leuven, Belgium). The implant screw holes were inspected under a 40× digital microscope, which highlighted the need to ream them to ensure the screws would fit without being forced. Both the guide and implant were then cleaned with detergent, ultrasonic-cleaned, rinsed and dried. The completed implant and guide are shown on a reference model in Figure 5.55.

Stage 5: fitting and surgery

The goal of the surgical approach was to minimise the size of the scar and insert the plate without causing surrounding tissue damage. A trans-conjunctival with lateral lid swing approach provided the access required to place the guide and locate the implant. The guide positioned adequately on the infra-orbital rim without the need to use screws. The orbital plate was carefully inserted beneath the eye and located into the corresponding cutout features in the guide, providing confidence that it was correctly located on the orbital rim. Two standard 1.5-mm Synthes Cortex screws (Synthes GmbH) were used to retain the plate.

Results and conclusions from case 2

This case demonstrated the potential of using a CAD/AM approach in orbital trauma reconstruction. The procedure was undertaken without complication and subsequent reviews demonstrated success in achieving improved vision.

In undertaking this case, small potential improvements with the process were also identified. The nature of AM in metal results in a relatively rough surface finish compared with pressed titanium and also meant that the small screw hole features, were not fully round or smooth. This meant that although clearance was left in the design, it was necessary to ream the holes, which represented a small degree of process inefficiency. In this case, it was also deemed unnecessary to countersink the screw holes, but the potential to incorporate this in future designs was noted. This would help reduce the prominence of screw heads into overlaying soft tissue.

5.9.4   Case 3: multipart reconstruction

The first two case studies demonstrated relatively simple cases involving small, localised implants for a congenital and a trauma case. The third case illustrates how a CAD/AM approach can be effectively used in a larger case that requires multiple components.

The patient presented with an extensive, cancerous tumour extending around the left temporal region of the head and involving the superior aspects of the orbital bones. Urgent, multidisciplinary surgery was required to remove the tumour and provide immediate reconstruction that would restore anatomical contours and enable further treatments, including radiotherapy, to be delivered effectively. This represented a significant challenge because the fast-growing nature of the tumour meant that it was difficult to predict how much bone and soft tissue would need to be removed. A CAD/AM approach was considered an appropriate route to fabricating reconstructive plates, but the extra challenge of not knowing how much tissue would require removal meant the design had to incorporate a degree of flexibility by not relying on an exact fit within the defect in predetermine areas.

5.9.4.1   Materials and methods

Stage 1: data segmentation

The same Mimics® segmentation techniques as previously described were used to create a 3D CAD model of the bony anatomy. In addition to this, the tumour margin was segmented based on operator interpretation and consultation with the prescribing surgeon. Segmenting the tumour provided the best method to determine how much tissue would require removal and, therefore, the extent of the reconstruction implants. Figure 5.56 shows the segmented 3D models of the anatomy and tumour extents. The bony anatomy and tumour were exported as high-quality STL files.

Stage 2: defect reconstruction

The STL files were imported into FreeForm® Modelling Plus as clay to enable virtual resection of the tumour area and smoothing of the margin. A Boolean subtraction of the tumour from the bony anatomy was undertaken, leaving a roughly resected clay model. The resection margins were smoothed and cleaned up by manually carving, being careful not to extend too far into unaffected anatomy where the implant

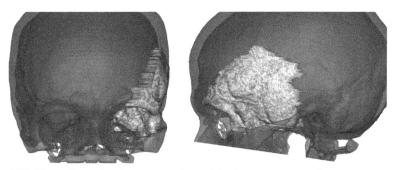

Figure 5.56 Three-dimensional reconstructions of the bony anatomy and tumour extent.

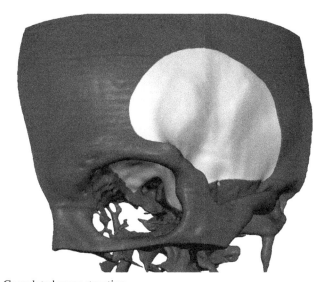

Figure 5.57 Completed reconstruction.

would locate. Once the computer-aided resection had been completed, an area of the patient's right temporal bony anatomy was selected, copied and pasted as a new piece of clay. This was mirrored around the midsagittal plane, repositioned by eye to form the basis of the reconstruction, and reduced in extent to approximately match the required shape. The original anatomy with tumour removed was converted into a buck model to protect it from carving and the reconstruction was then combined into it. A variety of carving and manipulation tools were then used to blend the reconstruction into the anatomy, creating a seamless join. The completed reconstruction is shown in Figure 5.57. This was duplicated and one version was turned into a buck model.

Stage 3: implant design

A three-part implant was decided as the optimum reconstruction method. This consisted of a two-part orbital implant to reconstruct the superior and lateral walls of the orbit and a temporal cranioplasty. Because the actual extent of the tumour could not be accurately predicted, it was decided that the cranioplasty needed to sit on top of the reconstruction rather than inside the surgically created resection. The orbital plates were, however, designed to sit inside the resection because it was considered that their extent and thickness would have compromised the patient's vision in the left eye.

All design work was undertaken in FreeForm® (Version 2014). Owing to the detailed, low-thickness nature of the proposed design, it was necessary to refine the clay coarseness to 0.18 mm. The temporal cranioplasty plate was designed by drawing the margin directly onto the reconstruction surface of the entirely buck model and using the Emboss Clay, Raise function to create a plate with a thickness of 0.5 mm on top of the surface. Localised fixation tabs were then added using the layer tool.

The orbital implants were more complicated because the intention was to set them inside the proposed surgical resection. The reconstructed model (with buck original anatomy and clay reconstruction) was duplicated again. A margin line extending beyond and enclosing the resection boundary was drawn on the surface of the model. The Emboss Clay, Lower function was used, but this time to lower the clay inside the margin line by 0.5 mm. This has the effect of removing only the clay reconstruction while leaving the buck anatomy unmodified. The model was then converted to buck and a duplication was made. One version of the modified model was then used to Boolean subtract material from the unmodified reconstruction model, leaving just a thin plate structure representing the contours of the orbital reconstruction. This basic form was then joined to the duplicated version of modified orbital reconstruction model for further refinement. Fixation tabs that extended onto the surface of the orbital rims were created in the same way as the cranioplasty, with the superior tab designed to interlock with corresponding contours of the cranioplasty plate to provide a location reference. Given the large extent of the orbital reconstruction, it was decided to split the plate into two separate pieces by drawing the margin on the surface of the clay and using the profile tool to create a 1-mm groove.

With further smoothing and refinement, design of the orbital and temporal implants was completed while still as layers on top of the buck reconstructed models. Once complete, the buck models were removed to leave just the implant designs, which were then refined to 0.15-mm clay sharpness to improve edge smoothness and enable further levels of detail to be added.

Computer-aided design representations of 1.5-mm Synthes Cortex screws were positioned to sit with the countersink protruding into the surface of the design. Because of the minimal plate thickness, a small portion of the screw countersink remained protruding above the outer surface. The plate thickness was therefore locally increased around the countersinks to avoid sharp edges that could cause soft tissue damage. Boolean subtraction was then used to subtract the screws from the implants.

To allow more efficient delivery of radiotherapy, reduce the bulk of the implant and make them less prone to distortion during the proposed AM fabrication process, it was decided to perforate the plate implant designs. A diamond pattern was drawn on a flat plane orientated approximately parallel with the plate surfaces and duplicated in a pattern. An unperforated plate margin was left, with diamond holes lying outside the margin deleted. The extrude, cut tool was then used to create the perforations.

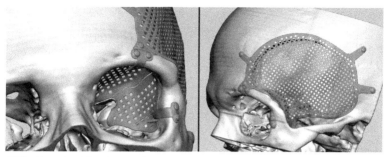

Figure 5.58 Completed implant designs.

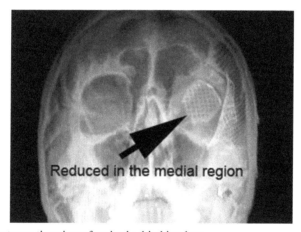

Figure 5.59 Postoperative view of revised orbital implant.

A final clay refinement operation was undertaken to smooth the implant designs, which can be seen in Figure 5.58.

The patient had combined craniofacial resection and reconstruction with neurosurgeons, maxillofacial surgeons and an ophthalmologist.

Results and conclusions from case 3

The case incorporated lessons learned from previous experience to provide a successful outcome. The designed-in flexibility of the temporal implant reduced reliance on the planned tumour excision but still enabled accurate aesthetic reconstruction of the anatomy. The interlocking features of the temporal and orbital roof plates gave extra confidence that they were located correctly. Vast number of areas for improvement were also noted. The tumour excision was less extensive than planned; therefore, it was necessary to use only the orbital roof part of the two-part orbital reconstruction plates. The orbital roof implant was also reduced in size during the operation by cutting with bone cutter, which although possible, was not an easy task given the strength of the material. Figure 5.59 shows a postoperative X-ray looking upward at the orbital implant, highlighting the area of reduction.

Based on feedback from the ophthalmic surgeon, it was suggested that it is unnecessary to extend farther back than the equator of the orbital globe when reconstructing the roof and lateral walls of the orbit. Subsequent implants of a similar nature have incorporated the feedback from the case illustrated.

5.9.5 Case 4: posttraumatic zygomatic osteotomy and orbital floor reconstruction

As the previous case studies have illustrated, the use of computer-aided planning, CAD and AM can help provide a more predicable surgical outcome. Previous application of CAD/AM to zygomatic osteotomy has been limited to translating a virtual plan into surgery using a polymer repositioning guide and a single moved bone piece (Herlin, Koppe, Béziat, & Gleizal, 2011). Case study 7 in Section 5.10, Chapter 5 notes the challenging nature of zygomatic osteotomies, especially when compounded by bone remodelling around the fracture sites. This clinical case demonstrates an evolution of the techniques described in that case study by employing anatomical models, digital reconstruction, virtual planning and CAD/AM end-use devices—a cutting guide, repositioning guide, custom zygomatic implant and custom orbital floor implant.

The patient's primary concern was appearance and improved vision in the left eye. The surgical goal was to improve facial symmetry by osteotomising bones that had been fractured, displaced and set in the incorrect position. A further goal was to improve the patient's vision by reconstructing the orbital floor and therefore restoring eye position during the same procedure.

5.9.5.1 Materials and methods

The complexity of this case meant that extensive, multidisciplinary planning was involved, including the development of multiple concept solutions. In the interest of brevity, only the chosen approach is described.

Stage 1: data segmentation

The patient underwent a CT scan with a 0.5-mm slice thickness using a Toshiba Aquilion Multislice scanner. The DICOM format data was imported into Mimics® (version 15) and the same segmentation techniques as previously described were used to create a 3D CAD model of the bony anatomy. In addition, a separate mask of the bony anatomy was created and the existing metal reconstruction plates from the primary reconstruction were erased manually on each CT slice. This was necessary because the surgical plan involved removing existing reconstruction plates to ensure the proposed guides and reconstructive implants would fit correctly. Figure 5.60 shows the extent of the bone displacement.

Once completed, the original anatomy with metal removed mask was exported in the STL file format. This file was fabricated using stereolithography (SLA 250-50,

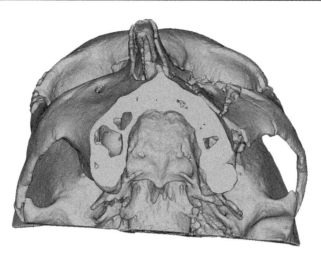

Figure 5.60 Inferior views showing the injury extents.

3D Systems, USA) to aid as a reference model to communicate the procedure to the surgical team and patient and assist in visualising the plan.

Stage 2: surgical planning and device design

FreeForm® Modelling Plus (version 13, Geomagic, 3D Systems, USA) provided the basis for collaborative surgical planning and device design sessions between surgeon, the assisting prosthetist and the design engineer. The STL file of the bony anatomy with existing reconstruction plates removed was imported as a buck reference model. A portion of the patient's right facial anatomy covering the entire orbit and parts of the zygomatic, maxilla temporal bones was selected, copied and pasted as a new clay piece. This was mirrored around the midsagittal plane and positioned using the surgeon's best judgement to create a target symmetrical reference.

Options for how best to achieve the optimal aesthetic outcome were then discussed, with the conclusion that is necessary to perform osteotomies on three separate sections of bone. The locations for cuts were defined by drawing representing polyline curves directly onto the anatomy. These curves would provide the basis for modelling precise groves onto the cutting guide design. The layer tool was used to create a 2.5-mm thickness of clay on top of the anatomy over the areas required for surgical cuts. The curves were then used to create precise groves and ridges corresponding to the proposed saw blade thickness (with 0.3 mm clearance) and angles required. Further refinement of the guide design was then undertaken using sculpting and refinement tools. Once the design was agreed, the buck anatomy upon which the clay design had been layered was subtracted, leaving a precise-fitting surface. Computer-aided design representations of standard Synthes 1.5-mm screws were then positioned at key points required to locate the guide securely. These were initially left as a visual reference to check the location in relation to the other planned positioning guide and implant screws. Figure 5.61 shows the design.

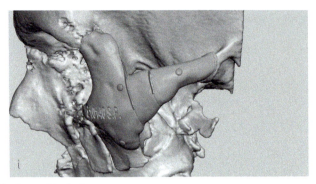

Figure 5.61 Cutting guide design.

Figure 5.62 Planned osteotomy illustrated in colour.

For osteotomy planning, a duplicate of the original anatomy was made and transformed into a modifiable clay piece. The three separate bone sections were then selected using the Select with Ball function, cut and pasted as new movable pieces. The pieces were then repositioned based on the surgeon's preference using the target symmetrical reference as a guide. Areas where the repositioned bone pieces overlapped the static anatomy were noted, to highlight where trimming would be required. With the new positions agreed, the individual pieces were set as unmodifiable buck parts to protect their condition for the remainder of the process. During the process, each bone fragment was assigned a bright, contrasting colour and number to assist in communication between clinical and engineering specialities. When a model of the planned reposition was built, a colour Z Corp 510 (3D Systems, USA) machine was selected for its ability to maintain the virtual colouring-in on the physical model. Similarly, a model was built with the separated bone fragments as loose, individual pieces, to facilitate procedure rehearsal (Figure 5.62).

To translate the planned osteotomy into the operating theatre, a positioning guide, similar in concept but larger than those described in Case study 7 in Section 5.10,

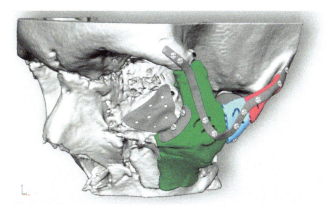

Figure 5.63 Orbital and zygomatic implant designs.

Chapter 5, was proposed. It was also agreed that this would also provide an appropriate mechanism to accurately locate custom implants that would remain once the guide had been removed, which dictated a complex design with locating features. Two implants were required: one to fix the zygomatic complex in the correct position and the second to reconstruct the orbital floor based on the new bone positions. The goal was to ensure the implants were as low profile as possible while providing the necessary fixation. FreeForm® was used to model a 0.7-mm-thick, 5-mm-wide plate that spanned from just anterior of the auditory meatus to the zygomaticofrontal suture. Orbital floor reconstruction and implant design was undertaken using techniques described and adapted from Case 2. Computer-aided design representations of 1.5-mm Synthes Cortex screws were located along the zygomatic implant and the locating tabs of the orbital floor implant. The thickness of the implants was increased locally around the screws to increase strength and maintain a smooth contour around the countersunk screw heads. The completed implant and guide designs are shown in Figure 5.63.

The positioning guide was agreed on as a crucial component in ensuring that the bone pieces were located correctly and that the implants were fixed correctly. It needed to provide a secure method of engaging the fixed portion of anatomy in two locations: at the posterior border of the zygomatic arch and on the supraorbital rim. The positioning guide was designed using the layer and multiple sculpting tools in FreeForm® Modelling Plus and the implants' profiles were used to Boolean subtract material from inside the guide shape (taking care to include clearance and remove undercuts that would prevent them slotting in). Computer-aided design representations of 1.5-mm Synthes Cortex screws were located on the positioning guide. These were required to hold the guide firmly in place while the osteotomy pieces were located and the implants fixed. The design is shown in Figure 5.64.

Stage 3: device fabrication

The implant designs were signed off by the prescribing surgeon for fabrication. The agreed-upon STL files were exported and sent to external ISO13485-accredited

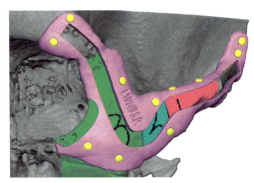

Figure 5.64 Positioning guide design.

Figure 5.65 Fabricated implants and guides.

manufacturers. The zygomatic implant was built using the Selective Laser Melting (SLM) process (LayerWise, Leuven, Belgium). A basic support-removal and bead-blasted finish was requested. Grade 23 Ti6Al4 ELI was specified as the build material on the basis of its biocompatibility and favourable mechanical properties. The orbital floor implant (0.5 mm nominal thickness) was built using the same process and with the same finish on the inferior surface. The superior surface was polished. Post-reaming of the fixation holes was undertaken as a precautionary measure in the hospital lab upon delivery. The guides were built using the Laser Sintering (LS) process (Renishaw, Wotton-under-Edge, UK). Dental-grade cobalt-chrome was specified as the build material owing to its compatibility for in vivo use as a transient device, high stiffness and lower cost relative to titanium. The implants and guides are shown in Figure 5.65.

Stage 4: surgery

The parts were checked and sterilised prior to surgery. In surgery the positioning guides and implants fitted as intended enabling the accurate reproduction of the surgical planning. One of the surgical guides and its associated implant is shown in Figure 5.66.

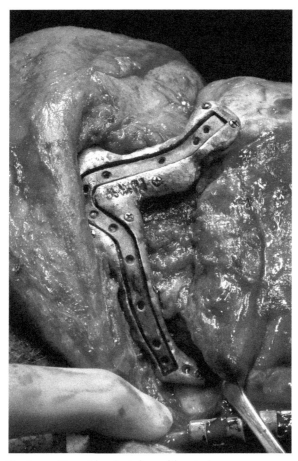

Figure 5.66 Positioning guide and implants in position during surgery.

Results and conclusions from case 4

The design process combined with the use of custom, AM-produced guides and implants resulted in excellent clinical outcome. Both of the guides fulfilled their functional requirements. The cutting guide fitted onto the existing anatomy precisely and securely. Each bone segment was osteotomised in the location planned virtually before the surgery. Extra in-theatre clarification and discussion was required to properly define the appropriate saw angle for the medial cut at the edge of the guide. A cutting ledge of greater thickness than the main body of the guide could mitigate this delay in future cases.

The repositioning guide fitted securely, accommodating each bone segment accurately in its preplanned position. During the process of fitting the smaller segments into the repositioning guide, however, the locations of the pieces anteriorly had to be cross-referenced with the planning imagery; they retained some freedom to move even with the guide in place. This could not be solved mechanically without endangering the guide's ability to interface with the pieces without undercut complications.

A solution for future cases may be to indicate the intended end point of each bone piece using embossed markings on the surface of the guide itself.

Each implant was positioned successfully and fixed without complication. However, despite the successful clinical outcome, the design process was highly involved (required numerous interactions between the clinical and design engineering team) and it required multiple concept iterations; therefore, it could be argued to be cost-inefficient. Although time was not a significant factor (because it was an elective procedure) and this was a particularly complex case, it helped to highlight the need for process efficiency improvements that would make the use of computer-aided planning, design and AM techniques more economically viable.

References

Bibb, R., Eggbeer, D., & Evans, P. (2010). Rapid prototyping technologies in soft tissue facial prosthetics: current state of the art. *Rapid Prototyping Journal, 16*(2), 130–137.

Bibb, R., Eggbeer, D., Evans, P., Bocca, A., & Sugar, A. W. (2009). Rapid manufacture of custom fitting surgical guides. *Rapid Prototyping Journal, 15*(5), 346–354.

Bibb, R., Eggbeer, D., & Williams, R. (2006). Rapid manufacture of removable partial denture frameworks. *Rapid Prototyping Journal, 12*(2), 95–99.

Chua, C. K., Leong, K. F., & Lim, C. S. (2010). *Rapid prototyping: principles and applications* (3rd ed.). WSPC. ISBN-13: 978–9812778987.

Eggbeer, D., Bibb, R., & Williams, R. (2005). The computer aided design and rapid prototyping of removable partial denture frameworks. *Proceedings of the Institute of Mechanical Engineers Part H: Journal of Engineering in Medicine, 219*(3), 195–202.

Eggbeer, D., Evans, P., & Bibb, R. (2006). A pilot study in the application of texture relief for digitally designed facial prostheses. *Proceedings of the Institute of Mechanical Engineers Part H: Journal of Engineering in Medicine, 220*(6), 705–714.

EOS Materials data sheet - EOS Titanium Ti64 and EOS Titanium Ti64 ELI, EOSINT M 270 Systems (Titanium Version), EOS GmbH - Electro Optical Systems, Robert-Stirling-Ring 1, D-82152 Krailling, München, Germany.

Gibson, I., Rosen, D. W., & Stucker, B. (2009). *Additive manufacturing technologies: rapid prototyping to direct digital manufacturing.* Springer. ISBN-13: 978–1441911193.

Herlin, C., Koppe, M., Béziat, J. L., & Gleizal, A. (2011). Rapid prototyping in craniofacial surgery: using a positioning guide after zygomatic osteotomy – a case report. *Journal of Cranio-Maxillofacial Surgery, 39*(5), 376–379.

Hopkinson, N., Hague, R., & Dickens, P. (Eds.). (2005). *Rapid manufacturing: an industrial revolution for the digital age.* Wiley Blackwell. ISBN-13: 978–0470016138.

Mazzoli, A., Germani, M., & Raffaeli, R. (2009). Direct fabrication through electron beam melting technology of custom cranial implants designed in a PHANToM-based haptic environment. *Materials & Design, 30*, 3186–3192.

Murr, L. E., Gaytan, S. M., Martinez, E., Medina, F., & Wicker, R. B. (2012). Next generation orthopaedic implants by additive manufacturing using electron beam melting. *International Journal of Biomaterials, 245727*, Available at http://www.hindawi.com/journals/ijbm/2012/245727/abs/.

Noorani, R. I. (2005). *Rapid prototyping: principles and applications.* John Wiley & Sons. ISBN-13: 978–0471730019.

Salmi, M., Tuomi, J., Paloheimo, K. S., Björkstrand, R., Paloheimo, M., Salo, J., et al. (2012). Patient-specific reconstruction with 3D modeling and DMLS additive manufacturing. *Rapid Prototyping Journal, 18*(3), 209–214.

Wehmöller, M., Warnke, P. H., Zilian, C., & Eufinger, H. (2005). Implant design and production: a new approach by selective laser melting. In H. U. Lemke, K. Inamura, K. Doi, M. W. Vannier, & A. G. Farman (Eds.), *CARS 2005: Computer assisted radiology and surgery* (vol. 1281) (pp. 690–695). ISBN: 0-444-51872-X.

Weihe, S., Wehmöller, M., Schliephake, H., Hassfeld, S., Tschakaloff, A., Raczkowsky, J., et al. (2000). Synthesis of CAD/CAM, robotics and biomaterial implant fabrication: single-step reconstruction in computer-aided frontotemporal bone resection. *International Journal of Oral Maxillofacial Surgery, 29*(5), 384–388.

Williams, R. J., Bibb, R., Eggbeer, D., & Collis, J. (2006). Use of CAD/CAM technology to fabricate a removable partial denture framework. *Journal of Prosthetic Dentistry, 96*(2), 96–99.

5.10   Surgical applications case study 7: computer-aided planning and additive manufacture for complex, mid-face osteotomies

Acknowledgments

The processes described in this case study were undertaken in collaboration with Mr Adrian Farrow and Steven Hutchison at Raigmore Hospital, Inverness, Scotland.

5.10.1   Introduction

The concept of using rapid prototyping (RP) to transfer a computer-aided surgical plan to the operating theatre has been discussed in the literature (Bibb, Bocca, Sugar, & Evans, 2003; Bibb & Eggbeer, 2009; Bibb, Eggbeer, Bocca, Evans, & Sugar, 2003; Goffin, Van Brussel, Vander Sloten, Van Aude & Smet, 2001; Goffin et al., 1999; Poukens, Verdonck, & de Cubber, 2005; Sarment, Al-Shammari, & Kazor, 2003; Sarment, Sukovic, & Clinthorne, 2003; SurgiGuides, 2014). Since these earlier case studies, the application of advanced computer-aided design (CAD) and fabrication technologies has grown to encompass more complex surgeries including the cutting and repositioning of bones (osteotomy) in multiple vectors. Accurate osteotomy of bone in multiple vectors presents a significant challenge, especially in the mid-face region, where small movements can have a large impact on the surgical result. The type of surgery described in this case study would not be possible to any degree of accuracy without the use of computer-aided techniques. The application of CAD techniques and RP to enable complex, multi-axis osteotomies of the mid-face has not been previously reported.

To date, three surgical guides have been designed and produced as described here and subsequently used in surgery. These were for similar cases, but each involved slightly different, case-specific challenges. All patients had blunt trauma to the face, which had caused the bones to fracture and become displaced posteriorly and laterally. The patients all had facial asymmetry and problems with vision, and therefore required surgery to restore symmetry and function.

The cases reported here describe the use of stereolithography (SL) as an AM method of producing a patient-specific, transient use guide devices. Since the studies reported in Chapter 5 Section 5.5, where metal guides were used, further work has been undertaken to demonstrate the ability of the chosen SL material to withstand autoclave sterilisation and measure the baseline bioburden of the material having gone through the cleaning and sterilisation process. The SL material chosen, ClearVue (3D Systems, USA), had also undergone prerequisite testing by the manufacturer to USP 23 Class 6, which demonstrated low toxicity and tissue reaction. The resin also exhibits a high degree of dimensional stability and accuracy in the presence of moisture. Stereolithography was also deemed appropriate because the guides produced spanned a relatively small area, were not used to drill or cut along, and could be made sufficiently robust through material thickness.

5.10.2   Methods

Over-the-Internet screen sharing was used to undertake the plan in real-time with the surgeon and prosthetist at the hospital available to guide the process, which was driven by a design engineer. The approach will be illustrated through one of the cases.

5.10.2.1   Step 1: three-dimensional computed tomography scanning and virtual model creation

The patients were scanned using computed tomography (CT) to capture the facial bony anatomy (see Chapter 2, Section 2.2). Techniques described in Chapter 3 were then used to produce a three-dimensional (3D) rendering of the bony anatomy and export the data as an STL file. The extent of the asymmetry and displacement of the zygomatic arch can be seen in Figure 5.67.

5.10.2.2   Step 2: computer-aided surgical planning

The STL files were then imported into CAD software, FreeForm® Modelling Plus (Geomagic, 3D Systems, USA) using the buck setting and duplicated. FreeForm® was selected for its capability in the design of complex, arbitrary, but well-defined shapes that are required when designing custom appliances and devices that must fit human anatomy.

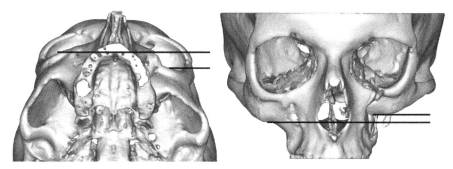

Figure 5.67 The 3D reconstruction of existing bony anatomy, highlighting the posterior and lateral displacement of the zygomatic bone.

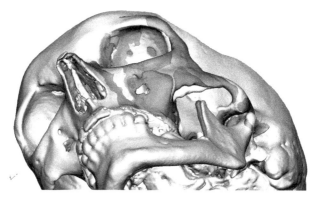

Figure 5.68 Mirrored bone overlaid on the defect area.

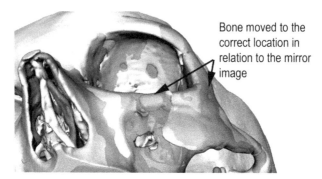

Bone moved to the correct location in relation to the mirror image

Figure 5.69 Sections of bone located in their osteotomised position.

FreeForm® is introduced in Chapter 3, Section 3.6.6. To create a symmetry-based target position for the osteotomised bone, a portion of the patient's right, unaffected side was selected using the Select Clay with Box function and then mirrored across the midsagittal (midline) plane to the left side. Manual repositioning of the mirrored piece was used to create the ideal target contour, to which the repositioned bones should match. The zygomaticofrontal suture was used as a key point of rotation in positioning the mirrored section because the fracture was hinged around this point. The mirror was agreed upon by the prescribing surgeon and is shown in Figure 5.68.

In each case, it was deemed unnecessary to use a cutting guide to free the zygomatic bone, but it was important to plan the approximate cutting locations to ensure the subsequent guide design would fit correctly. In the illustrated case, four primary cuts were required: two on the infra-orbital rim to free a small section of rim and the anterior portion of the zygomatic bone; a cut along the zygomaticofrontal suture; and a cut on the posterior portions of the zygomatic arch. These cuts were undertaken on the duplicate model to preserve a version of the original anatomy, and were created using the select with ball tool, cutting the selected areas and pasting them as new pieces of clay. The two freed sections of bone for osteotomising are shown in Figure 5.69. This also left a version of the bony anatomy with the cut sections

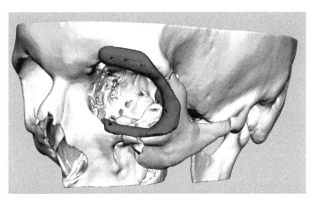

Figure 5.70 Completed guide design in CAD.

removed. The free sections of bone were then positioned by eye, using the superficial area of the zygomaticofrontal suture as a point of rotation and by making minor movements. The final bone positions that were agreed upon by the surgeon can be seen in Figure 5.69.

The digitally osteotomised bone pieces were then joined into the model of the cutaway anatomy, thereby creating the intended surgical result, onto which a guide could be designed. The model was selected and converted to buck to protect it from further modification by material removal.

A small section of the reconstructed left skull was then exported as an STL file for fabrication using SL. The SL model provided a physical jig around which mini-plates that would provide permanent fixation of the bone sections were bent in the prosthetics laboratory before being cleaned and sterilised before surgery.

5.10.2.3   Step 3: design of surgical positioning guide

The aim of the guide was to provide contours that would firmly locate using screws on the fixed areas of bone and enable the osteotomy pieces to locate and be fixed according to the planned position. It was agreed that the device needed to avoid being fixed where pre-bent plates would provide permanent fixation, low enough profile to minimise surgical incisions, but still provide adequate overlapping contours to locate on the fixed bone and engage the osteotomy pieces. The basic shape was created using the layer tool and then refined using various carving and shaping tools. Holes that would enable temporary fixation of the device using standard screws were also created. The completed design shown in Figure 5.70 was validated by the prescribing surgeon before being exported as an STL file ready for SL fabrication.

5.10.2.4   Step 4: guide additive manufacture

The guide STL file was imported into build preparation software (Magics Version 16, Materialise NV) and orientated to reduce the number of supports on the fitting surfaces and minimise the effects of stair stepping (described in Chapter 4, Section 4.1.3) on the hole features. Proprietary software (Lightyear V1.1, 3D Systems,

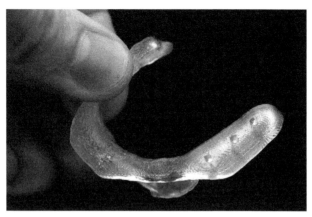

Figure 5.71 The SL-produced osteotomy guide.

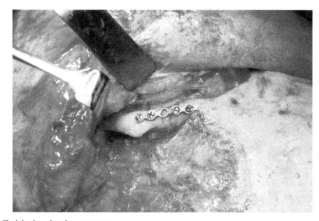

Figure 5.72 Guide in situ in surgery.

USA) was used to set SL build parameters and prepare the guide for fabrication in 0.1-mm layers using an SLA 250-50 machine (3D Systems, USA). ClearVue resin (3D Systems, USA) was used because it had undergone testing to demonstrate suitable levels of biocompatibility. In the illustrated case, two guides were produced: one as a lab reference and the other for use in surgery. Cleanup and post-processing involved removing support structures, wiping off excess resin from the guide surface, scrubbing and then soaking in fresh isopropanol, drying and post-curing in an ultraviolet oven. The SL-produced guide is shown in Figure 5.71.

5.10.3   Results

The surgical guide used to illustrate the process (and in all cases where they have been used) was assessed by the clinicians before going to surgery and were deemed satisfactory for use. Surgery took approximately 1 h (Figures 7.72 and 7.73).

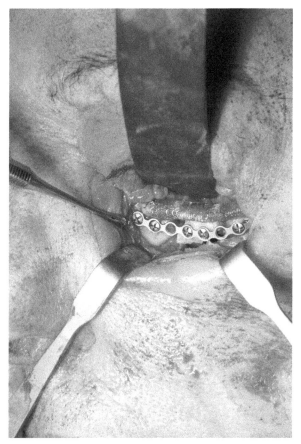

Figure 5.73 Guide being fitted to the patient during surgery.

5.10.4   Discussion

The application of surgical guides described here all aided in the successful transfer of computer-aided planning to surgery. Although improvements in accuracy and efficiency cannot be quantified, post-surgery analysis of each case provided the opportunity to evaluate the approach, understand the limitations and inform the development of revised design and manufacture guidelines. The first case (used to illustrate the process) demonstrated the value of a collaborative approach to osteotomy guide design. The profile was sufficiently small to enable easy insertion, yet it enabled the guide to be located securely to the fixed sections of anatomy and the osteotomy pieces to locate with minimal play (which would have been manifested as a rocking motion around the guide).

   The surgical team also noted potential improvements. The value of using a physical guide to accurately reproduce the planned vectors and locations of the bone cuts in surgery would have been useful to ensure accuracy of the subsequent osteotomy. This would be of increasing value in more complex cases. Directional arrows or other features that ensured the surgeon could easily identify anatomical landmarks in relation to the guide would also provide increased confidence that it was correctly located.

5.10.5   Conclusions

The approach illustrated represents a clinically viable and appropriate method of undertaking complex, mid-face osteotomies. Although the case illustrated did not require a cutting guide, techniques similar to those discussed in Section 5.5, case Study 2, could be incorporated if cut precision were more critical. The use of SL polymer was deemed appropriate in the cases undertaken to date, but the use of metal could assist where guides need to span a large area with sufficient stiffness, or provide a harder edge against which to cut.

References

Bibb, R., Bocca, A., Sugar, A., & Evans, P. (2003). Planning osseointegrated implant sites using computer aided design and rapid prototyping. *The Journal of Maxillofacial Prosthetics and Technology*, *6*(1), 1–4.

Bibb, R., & Eggbeer, D. (2009). Rapid manufacture of custom fitting surgical guides. *Rapid Prototyping Journal*, *15*(5), 346–354.

Bibb, R., Eggbeer, D., Bocca, A., Evans, P., & Sugar, A. (2003). Design and manufacture of drilling guides for osseointegrated implants using rapid prototyping techniques. In *Proceedings of the 4th national conference on rapid and virtual prototyping and applications* (pp. 3–12). London, UK: Professional Engineering Publishing, ISBN: 1-86058-411-X.

Goffin, J., Van Brussel, K., Vander Sloten, J., Van Audekercke, R., & Smet, M. H. (2001). Three-dimensional computed tomography-based, personalized drill guide for posterior cervical stabilization at C1-C2. *Spine*, *26*(12), 1343–1347.

Goffin, J., Van Brussel, K., Vander Sloten, J., Van Audekercke, R., Smet, M. H., Marchal, G., et al. (1999). 3D-CT based, personalized drill guide for posterior transarticular screw fixation at C1-C2: technical note. *Neuro-Orthopedics*, *25*(1–2), 47–56.

Poukens, J., Verdonck, H., & de Cubber, J. (2005). Stereolithographic surgical guides versus navigation assisted placement of extra-oral implants (Oral presentation). In *2nd international conference on advanced digital technology in head and neck reconstruction, Banff, Canada* (p. 61). Abstract.

Sarment, D. P., Al-Shammari, K., & Kazor, C. E. (2003). Stereolithographic surgical templates for placement of dental implants in complex cases. *International Journal of Periodontics and Restorative Dentistry*, *23*(3), 287–295.

Sarment, D. P., Sukovic, P., & Clinthorne, N. (2003). Accuracy of implant placement with a stereolithographic surgical guide. *International Journal of Oral and Maxillofacial Implants*, *18*(4), 571–577.

SurgiGuides. (2014). Company information, Materialise Dental NV. Technologielaan 15, 3001: Leuven, Belgium.

Maxillofacial rehabilitation

5.11 Maxillofacial rehabilitation case study 1: an investigation of the three-dimensional scanning of human body surfaces and its use in the design and manufacture of prostheses

Acknowledgments

The work described in this case study was first reported in the reference below and is reproduced here in part or in full with the permission of the Council of the Institute of Mechanical Engineers.

Bibb R, Freeman P, Brown R, Sugar A, Evans P, Bocca A, "An Investigation of three-dimensional scanning of human body parts and its use in the design and manufacture of prostheses" Proceeding of the Institute of Mechanical Engineers Part H: Journal of Engineering in Medicine, 2000, Volume 214, Number H6, pages 589–94. http://dx.doi.org/10.1243/0954411001535615

5.11.1   Introduction

Three-dimensional (3D) surface scanning, or reverse engineering, has been used in industry for many years as a method of integrating the surfaces of complex forms with computer-generated design data (Motavalli, 1998). Noncontact scanners operate by using light and camera technology to capture the exact position in space of points on the surface of objects. Computer software is then used to create surfaces from these points. These surfaces can then be analysed in their own right or integrated with computer-aided design (CAD) models. The general principles of noncontact surface scanning are described more fully in Chapter 2. This section includes a description of the potential difficulties that may be encountered when employing the technique and includes suggested methods to overcome them.

The scanner used in the work described in this section was a structured white light system that uses a projected fringe pattern of white light and digital camera technology to capture approximately 140,000 points on the surface of an object (Steinbichler USA, 40 000 Grand River, Suite 101, Novi, MI 48375). Scanners using this type of moiré fringe pattern have been used in the past in the assessment of spinal deformity (Wong, Balasubramaniam, Rajan, & Chng, 1997). In this case, the area to be scanned is distinguished from its surroundings by altering the contrast. For example, a white object may be placed on a dark background and vice versa.

Because of the high accuracy of this type of scanner, nominally accurate to within 0.05 mm, movement of the object was avoided during scanning. Even small movements result in noise and affect the quality of the data captured. In this case, the patient must remain motionless for approximately 40 s. Other systems that have been investigated for use in measuring and recording changes in a patient's topography employ multiple cameras and a fast capture time to eliminate the problems associated with motion. However, a smaller number of data points are captured at a slightly lower accuracy (Tricorder Technology plc, 1998) now known as 3dMD. Other systems based on scanning have been used to manufacture custom orthotics for podiatric patients (Bao, Soundar, & Yang, 1994; Bergman, 1993). At the time of this work, the application of captured surface data in the manufacture of facial prostheses had not been fully investigated.

Although prosthetic rehabilitation of the human face offers many potential applications that could exploit this technology, scanning human faces presents particular problems. A primary difficulty is presented by the presence of hair. Hair does not form a coherent surface and the scanner will not pick up data from areas such as the eyebrows and lashes. This problem can be overcome to a certain degree by dusting a fine white powder over the hair. When considering the scanning of faces, the area around the eyes may also be particularly difficult. As described previously, movement leads to the capture of inaccurate data, to minimise problems caused by blinking during the scan it is more comfortable for the subject to keep their eyes closed. This also alleviates discomfort caused by the bright light emitted by the scanner. If the eye is held open during the scan, watering of the eye may cause problems. In addition, the surface of the eyeball is highly reflective, making data capture difficult.

Line-of-sight issues are also encountered when scanning faces. For example, a single scan will not acquire data where the nose casts a shadow. However, this is overcome by taking several overlapping scans (as illustrated in Chapter 2).

This case study describes how these issues were approached by an investigation into the scanning of human faces and describes the application of these techniques in the manufacture of a facial prosthesis to restore the appearance of a patient recovering from the excision of a rare form of tumour called an olfactory neuroblastoma. Removal of this tumour had necessitated the surgical removal of the patient's left eye.

5.11.2 Methods

5.11.2.1 Preliminary trial of facial scanning

As a preliminary investigation of the practicality of scanning human faces, a male subject was scanned using the system described previously. Initial attempts at scanning the face of the seated subject were poor because of slight involuntary movement of the head despite being seated in a comfortable position. Therefore, additional support was fashioned from a block of polystyrene foam to locate the back of the head and minimise movement. With the subject thus supported in a semireclining position, a series of three scans were taken, each from a different viewpoint. Scans were taken from the patient's left side, from directly in front and from the right side. A fourth scan was taken from a central position below the second scan to allow the acquisition of data from the area below the eyebrow ridge. Each scan took approximately 40 s, during which the subject remained motionless. The whole process of arranging the subject and taking these four scans took approximately 10 min.

Once completed, the scans were aligned using the proprietary scanner software. To achieve this alignment, four notable points or landmarks were manually selected in an overlapping area in each of two separate scans. The software then aligns the landmarks and calculates the best fit between the two data sets. Subsequent scans were aligned in a similar fashion.

In this experiment, four scans resulted in the capture of accurate data describing the whole face with the exception of areas obscured by hair, such as eyebrows, eyelashes and facial hair. Other areas lacking data not immediately obvious in the figure include areas beyond the line of sight, such as the nostrils. It was therefore concluded that the approach could be applied to the scanning of patients provided they could be kept still during the scan.

5.11.2.2   Scanning a surgical subject

In this case, the subject was a patient recovering from reconstructive craniofacial surgery. The surgery to excise a tumour necessitated the removal of bone and soft tissue including the left eye. After successful operations to replace the orbital rim, an osseointegrated (bone-anchored) prosthetic was planned for the missing eye and surrounding tissue (Branemark and De Oliveira, 1997). To aid in the construction of this prosthesis, the right (unaffected) side of the patient was scanned and the data used to create a laterally inverted ('mirrored') model that would be used as a guide when creating the prosthesis.

Four scans were taken of the patient's face with the chin supported on a polystyrene block to minimise movement. Because the data were intended to aid in the construction of a prosthetic of an open eye, a scan of the open eye was attempted. The scans were taken from angles similar to those described previously; however, care was taken to ensure that the bright light from the scanner did not shine directly into the patient's eye. As before, data were not captured from areas obscured by the eyebrows and lashes. Small unavoidable movements of the eye and eyelids and the reflective nature of the surface of the eye itself affected the accuracy of the captured data. However, as this inaccuracy was extremely small it did not affect the overall quality of the data. The resulting scan data are shown in Figure 5.74. The whole process of scanning the patient took approximately 10 min.

The next step required the building of a model of the area around the unaffected eye. To further aid the creation of the prosthesis, the data would be laterally inverted ('mirrored' left to right) before building the model. This model could then be used to guide the production of a prosthetic with good size, fit and aesthetic symmetry.

To create a model from the scan data it was translated into an STL file format (Manners, 1993). This is a triangular facetted surface normally used in rapid prototyping systems. The STL file format is more fully described in Chapter 3. However, before the data could be used to produce a model, gaps in the data, such as the area at the eyebrows, were filled. This was achieved by using surface creation software to create a patch that continues the shape of the captured data surface. The patch was created to follow the natural curves of the surrounding data and replicate the surface as best as possible. This required a certain amount of judgement on the part of the operator. However, this case did not present great difficulty in this respect, as the missing areas were relatively small.

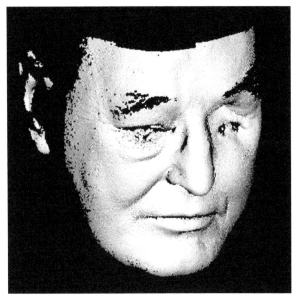

Figure 5.74 The aligned scan data.

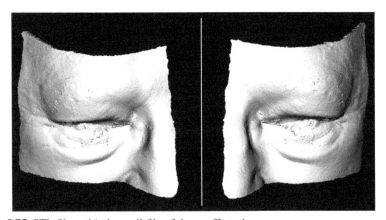

Figure 5.75 STL file and 'mirrored' file of the unaffected eye.

The file size was reduced at this stage by removing unnecessary points. This was achieved without sacrificing accuracy because there were a vast number of points in the captured data coupled with the fact that the accuracy of the scan data is greater than can be achieved by subsequent rapid prototyping (RP) processes. An offset surface was created from the captured data and the gap between them closed to create a finite bound volume. To minimise file size and model cost, only the specific area of interest was selected. The resulting data were stored as an STL file. The STL file size was reduced to 3.6 MB. This was then laterally inverted ('mirrored'), as shown in Figure 5.75. In this case, the process of creating a valid STL file from the scan data took approximately 3 h, but this will vary from case to case.

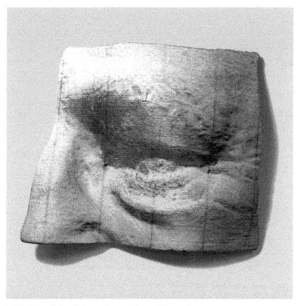

Figure 5.76 An LOM model of the mirrored unaffected eye.

The model in this case was produced using laminated object manufacturing (LOM) (Jacobs, 1996). The model is shown in Figure 5.76. The scan data and STL file were archived, in case it was necessary to reproduce the model or for future reference.

5.11.2.3   Prosthesis manufacture

This case involved the manufacture of an osseointegrated implant retained silicone prosthetic of the left eye (see the explanatory note in Section 7.2.1 for an explanation of osseointegrated implants). Previously, two titanium fixtures had been attached to the zygomatic bone (cheekbone) and allowed to integrate to the bone. Six months later, they were exposed and percutaneous titanium abutments attached to them (this means that the abutments passed through the skin to form an anchor for the subsequent prosthesis). The eventual prosthesis would attach to these abutments with magnets. To aid the construction of the prosthesis, the LOM model was used to cast a wax replica. The prosthetist then removed excess material from the wax until it approximated the required shape. The traditional procedure would have required the prosthetist to carve this piece from wax. This would have taken the prosthetist approximately half a day in this particular case, during which the patient would have been required to sit with the prosthetist for visual reference. The use of the model therefore not only saved approximately half a day of work for the prosthetist (from a total of three) but importantly also reduced the time required for the patient to attend the clinic.

The aperture for the eye was opened to allow the positioning of the artificial eye. This is considered crucial to the overall success of the prosthesis (Branemark and De Oliveira, 1997). It was noted that, compared with the traditional methods, the mirrored nature of the model allowed far greater accuracy when locating the artificial eye, especially concerning anterior-posterior positioning. Once the eye position was fixed, the fine details were built up in wax. The areas immediately around the eye were dealt with in particular, as this is where the original scan data, and therefore the LOM model, had lost some detail.

An impression was taken from the patient and used to shape the rear surface of the prosthesis. A small acrylic base plate that would hold the magnets used to locate the prosthesis was also made. When the prosthetist was satisfied with the visual appearance of the fine details and the fit of the prosthesis, it was cast in colour-matched silicone in the usual manner.

5.11.3   Results

5.11.3.1   Accuracy

The accuracy of the scan data is nominally within 0.05 mm. From the data, the theoretical height of the model was 76.76 mm. LOM models are nominally accurate to within 0.2 mm. However, when measured, the height of the completed LOM model was found to measure 76.7 mm. Therefore, the accuracy of the model can be estimated in the order of ±0.1 mm. As all human faces are somewhat asymmetric and the surface of the skin is pliable, the wax replica was manually manipulated and adjusted to fit the desired area. Therefore, an accuracy of around 0.1 mm is more than adequate for facial prosthesis manufacture. The model also proved to be a good match in terms of reproducing a realistic visual appearance for the prosthesis.

5.11.3.2   Outcome analysis

The success of this experiment proves the feasibility of 3D scanning of human body surfaces. The ease and relative speed of the scanning allow the complex forms of human features to be permanently captured without hindrance or discomfort to the patient. The accuracy of the data were found more than adequate for prosthesis construction. This method would compare favourably with the current practice of taking impressions, proving to be quicker, more accurate and aiding the reproduction of a realistic visual appearance. In particular, the use of 'mirrored' medical models was felt to be of great help to the prosthetist when positioning artificial eyes in orbital prostheses.

The cost of the scanning equipment is considerable and it may be difficult for hospitals to justify the initial investment. For this reason, it may prove more feasible for hospitals to use external service providers for the cases where the approach is expected to produce superior results. The cost of the scanning described in this paper would probably amount to several hundred pounds with the LOM model costing approximately £120. The costs incurred by this approach should be balanced against the improved results and crucially the time saved over traditional methods. The reduction in time taken allows more patients to be treated, reducing waiting lists (a major goal of the British National Health Service).

The noncontact nature of the scanning means there is less discomfort for the patient and no distortion of soft tissues caused by the pressure applied when taking impressions. This advantage in combination with the ability to 'mirror' data may have many applications in rehabilitation. It is difficult, for example, to take a satisfactory impression of a breast; therefore, a similar technique may be used in the creation of symmetrical prostheses for mastectomy patients. From the results of this case study, it can be concluded that 3D scanning and medical modelling can save a significant amount of time for both the patient and the prosthetist. Lateral inversion and high accuracy can be a significant aid in prosthesis manufacture, especially for large or complex cases. These techniques may be a valuable aid to shaping and positioning the prosthesis but the skill and knowledge of the clinicians will determine the best method of creating, colour matching and attaching the prosthesis to the patient.

5.11.4 Update

Scanning and processing software has developed considerably since this paper was first published greatly speeding up the process. However, the fundamental approach and advantages identified remain true and many of the issues encountered with line of sight, hair and involuntary movement are still present and need to be carefully considered when scanning patients.

References

Bao, H. P., Soundar, P., & Yang, T. (1994). Integrated approach to design and manufacture of shoe lasts for orthopaedic use: reverse engineering in industry: research issues and applications. *Computers in Industrial Engineering*, *26*(2), 411–421.

Bergman, J. N. (1993). The Bergman foot scanner for automated orthotic fabrication. *Clinics in Podiatry Medicine and Surgery*, *10*(3), 363–375. Treatment biomechanical assessment using computers.

Branemark, P., & De Oliveira, M. F. (Eds.). (1997). *Craniofacial prostheses, anaplastology and osseointegration* (pp. 101–110). Carol Stream, IL: Quintessence Publishing.

Jacobs, P. F. (1996). *Stereolithography and other RP&M Technologies*. Dearborn, MI: Society of Manufacturing Engineering.

Manners, C. R. (1993). *STL file format*. Valencia, CA: 3D Systems Inc.

Motavalli, S. (1998). Review of reverse engineering approaches. *Computers in Industrial Engineering*, *35*(1–2), 25–28.

Tricorder Technology plc. (1998). *Tricorder ships new measurement software*. December 8th Press Release. The Long Room, Coppermill Lock, Summerhouse Lane, Harefield, Middlesex, UB9 6JA, UK: Tricorder Technology Ltd.

Wong, H. K., Balasubramaniam, P., Rajan, U., & Chng, S. Y. (1997). Direct spinal curvature digitization in scoliosis screening: a comparative study with Moiré contourography. *Journal of Spinal Disorders*, *10*(3), 185–192.

5.12   Maxillofacial rehabilitation case study 2: producing burns therapy conformers using noncontact scanning and rapid prototyping

Acknowledgements

The work described in this case study was first reported in the reference below and is reproduced here in part or in full with the permission of First Numerics Ltd.

Bibb R, Bocca A, Hartles F, "Producing Burns Therapy Conformers Using Non-Contact Scanning and Rapid Prototyping", Proceedings of the 6th International Symposium on Computer Methods in Biomechanics & Biomedical Engineering, Madrid, Spain, February 2004, ISBN: 0-9549670-0-3 (Published on CD-ROM by First Numerics Ltd. Cardiff, UK)

5.12.1   Introduction

This case study describes the use of 3D noncontact scanning, CAD software and RP techniques in the production of burns therapy masks, also known as conformers. Such masks are used in the management of hypertrophic scars on the face resulting from burns injuries (see 'Glossary and explanatory notes').

Two case studies were undertaken where noncontact laser scanning techniques were used to capture accurate data of burns patients' faces. The surface data were then manipulated using two different CAD techniques to achieve a reduction in prominence of the scarring. This reduction in height of the scarring on the vacuum forming mould results in a conforming facemask that fits the face precisely whilst applying localised pressure to the scars. This pressure on the scars produces the beneficial effect from such masks. Once manipulated to achieve this effect, the data were then used to create vacuum forming moulds via a selection of RP methods.

The effectiveness of the CAD techniques and RP processes for this application are evaluated. The case studies illustrate the benefits of the approach in comparison to traditional practices whilst indicating operational and technical difficulties that may be encountered. Finally, the cost effectiveness, patient benefits and opportunities for further research are discussed.

Closely fitting masks have been shown to provide a beneficial effect on the reduction of scarring resulting from burns, particularly to the face and neck (Leach, 2002; Powell, Haylock, & Clarke, 1985; Rivers, Strate, & Solem, 1979; Shons, Rivers, & Solem, 1981). These masks are typically vacuum-formed from the strong clear plastic material, polyethyleneterephthalate glycol (PETG). Traditionally, the vacuum-forming mould is made from a plaster cast of the patient, which itself is made from an alginate impression. Taking a facial impression is uncomfortable, time-consuming for the patient, and may be particularly disturbing following the physical and psychological trauma of burns.

Published work has indicated that optical scanning and computer-aided manufacturing techniques can be used for various clinical applications (Bibb & Brown, 2000; Bibb et al., 2000; Sanghera, Amis, & McGurk, 2002) including the fabrication of burns masks (Lin & Nagler, 2003; Rogers et al., 2003; Whitestone, Richard, & Slemker, 1995). The potential benefit of this approach is the noncontact nature of the data capture, which has been shown to be more accurate, quicker, more comfortable and less distressing for burns patients compared to the traditional impression. The aim of this research is to explore the practical implications of employing such an approach to the treatment of facial burns and to assess various methods of adapting and physically reproducing the data to create a vacuum-forming mould.

5.12.2   Methods

3D surface scanning has been used in industry for many years to integrate surfaces of objects with computer-generated designs. Noncontact scanners operate by using structured light or lasers and digital camera technology to capture the exact position in space of a large number of points on the surface of objects. Computer software is then used to create surfaces based on these points. These surfaces can then be analysed or integrated with CAD models. The general principles of non-contact surface scanning are described more fully in Chapter 2.

The optical scanner used in this work uses a laser and digital camera technology to capture the surface of an object (Vivid 900, Konica Minolta Photo Imaging UK Ltd., Milton Keynes, UK). This scanner was selected because the specifications suggested that the accuracy, resolution and range of capture were more than adequate for capturing the human face. It also benefited from ready availability, manufacturer after sales support, comparatively low price and compact size compared with other systems that have been reported, which have been specialised and expensive or locally made prototypes (Lin & Nagler, 2003; Rogers et al., 2003).

Although the acquisition time for this type of scanner is only a fraction of a second, movement would still lead to inaccuracy in the captured data. Therefore, the patients remained motionless in a comfortable position during the acquisition. All light-based scanners are limited by line of sight during each acquisition and this is typically overcome by taking several overlapping scans (as illustrated in Chapter 2). However, in these cases, a pair of scanners was used to capture both sides of the patient's face. The scanners were positioned low to ensure data were captured from the areas under the chin and eyebrow ridge (Rogers et al., 2003).

Once the data points are acquired, software is used to create surfaces based on them. The simplest method of creating a surface from point data is polygonisation. Neighbouring data points are joined together to form triangular facets, which form the computerised surface model.

5.12.2.1   Case 1

This patient was recovering from burns to the head, neck and arms. Scans were taken as described previously, taking approximately 5 min. Eyebrows, eyelashes and blinking affected the accuracy of the data around the eyes but as the mask is intended to avoid the eyes, this did not present problems.

The point data were then polygonised to create a triangular faceted surface model of the patient's face. This was created in the STL file format (the STL file format is more fully described in Chapter 3), which is commonly used in RP. However, RP requires STL files that represent a single fully enclosed volume. Therefore, before the data can be used to produce a physical model, gaps in the data have to be filled and the surface needs to be given a thickness to produce a finite bound volume. This was achieved by using CAD software that extruded the perimeter of the captured data surface towards an arbitrary plane to create a solid model as shown in Figure 5.77 (FreeForm, Geomagic, 3D Systems, USA). The resulting STL file was thus created in less than 5 min.

In traditional practice, the plaster replica of the patient's face would be ground back in areas of scarring to produce localised pressure on scar sites whilst conforming comfortably to the rest of the face. For this research, the reduction in the height of the scarring was undertaken on the computer before producing the physical RP model. Although software has been reported that has been designed for this purpose, it is not widely available (Rogers et al., 2003). Therefore, this research specifically applied readily available software.

To achieve the desired effect, software that is commonly used to prepare STL files for RP was used (Magics, Materialise NV, Technologielaan 15, 3001 Leuven, Belgium). A 'smoothing' function in this software averages out the STL surface. The effect is to reduce the height of raised features on the surface and produce a simpler, smoother surface. However, the affect is applied to the whole surface of the object.

The second approach utilised CAD software (FreeForm®) that enables the user to conduct virtual sculpture on 3D computer models using a touch-feedback stylus. The software tools mimic those of traditional handcrafting. This makes the software

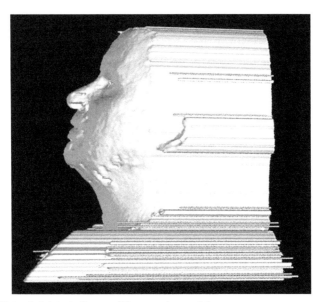

Figure 5.77 Extruded data to form solid computer model.

easy to learn and intuitive to use, particularly for prosthetists and it has been success-fully used by the authors in other maxillofacial laboratory applications.

Initially, a smoothing operation was carried out, which produced an effect similar to that obtained using the RP software. Second, the smoothing function was used only over areas selected by the user. Finally, a carving tool was used to carve away small areas of scarring locally. The effect is illustrated in an exaggerated manner in Figure 5.78. In practice, a combination of these functions enabled the prosthetist to produce a surface that met the needs of the individual case rapidly.

The data were then cropped so the physical model would form a good vacuum-forming mould. Then LOM was used to produce the vacuum-forming mould shown in Figure 5.79.

5.12.2.2 Case 2

The scan for the second case was carried out as before. In the interests of compari-son, a mould was manufactured using stereolithography (SL) (3D Systems Inc.). SL materials and process time are more expensive than LOM; therefore, to reduce cost, the data were reduced to a thin shell shown in Figure 5.80. As the glass transition temperature of SL materials may be exceeded during vacuum forming, the shell was

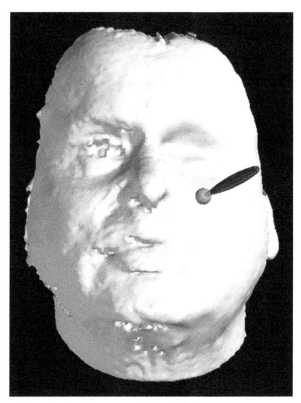

Figure 5.78 Smoothing the data (exaggerated for clarity).

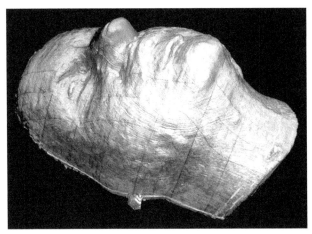

Figure 5.79 LOM vacuum-forming mould.

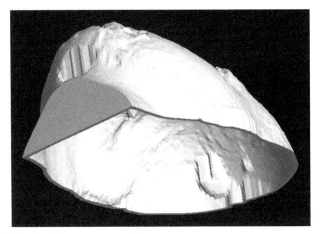

Figure 5.80 Shelled solid computer model.

filled with plaster as shown in Figure 5.81. This prevented the SL shell from distorting under the load and heat encountered in vacuum forming.

5.12.2.3 Experimental moulds

The ThermoJet® (obsolete, 3D Systems, USA) RP process prints 3D models using a wax material in a layered manner (the ThermoJet process is no longer available). The layers are very thin, leading to models with excellent surface finish. However, the low melting point of the wax precluded its use as a vacuum-forming mould. Instead, reversing the shelling operation used previously resulted in a negative pattern. The intention was to cast plaster a vacuum-forming mould from the wax pattern. However, in practice the pattern was so fragile that it was destroyed during transport to the laboratory.

Other centres have successfully employed computer numerically controlled (CNC) machining in this and similar applications (Rogers et al., 2003; Whitestone et al., 1995).

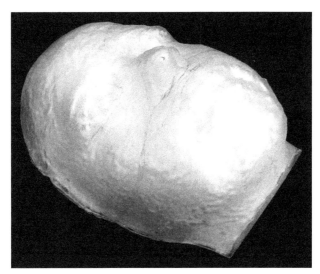

Figure 5.81 Plaster-filled SL mould.

In comparison to RP processes, CNC is a viable option for this application as it is unlikely to encounter undercuts or re-entrant features. To investigate the approach, a trial mould (case 1) was machined from a medium-density board, typically used in industrial model making. Unlike reported techniques that use soft foams for ease and speed of machining, this mould proved to be perfectly adequate for direct use as a vacuum forming mould requiring no surface treatment or modification (Rogers et al., 2003).

5.12.2.4 Mask manufacture

The masks were manufactured in the usual manner by vacuum forming sheet PETG over the RP mould. It is common practice to drill holes through moulds to provide even spread of the vacuum. However, holes were not drilled in these cases and the masks were formed perfectly well without them.

5.12.3 Results

5.12.3.1 Treatment outcome

Both cases responded extremely well to treatment and showed considerable reduction in scarring. The patients' masks fitted well, performed as intended, and proved equally as effective as those produced by traditional methods.

5.12.3.2 Vacuum moulding

The LOM model proved an excellent mould and was unaffected by vacuum forming. The plaster-filled SL mould also proved a satisfactory mould. Both types of mould can be worked on with grinding tools should they need to be altered to produce subsequent masks as treatment progresses. However, it was noted that the transparent nature of the SL material combined with the white plaster infill made the surface features difficult to see clearly. The layered surface finish of neither the LOM nor SL mould was transferred to the vacuum-formed PETG mask.

5.12.3.3 Accuracy

The accuracy of the scan data is nominally 0.1 mm. As the skin is somewhat mobile and pliable, the accuracy of the captured data proved to be more than adequate. SL models are typically accurate to within 0.1 mm of the data from which they are built and LOM models are nominally accurate to within 0.2 mm. These accuracies proved to be well within the requirements of this application. For this application, relatively inexpensive CNC machines are capable of producing satisfactory moulds.

5.12.4   Discussion

The success of this study illustrates the efficacy of 3D scanning, CAD and RP for this application. The accuracy of the scan data were found more than adequate. The noncontact nature of the scanning imposes less discomfort on patients and eliminates the potential risk of distortion of soft tissues, which may occur when taking impressions. The method therefore compares favourably with current practice. In addition, the reduction in anxiety or claustrophobia when taking facial impressions should be seen as an advantage.

Although this research deliberately utilised readily available, manufacturer-supported equipment, software, the cost of the scanners, associated software and RP equipment remains considerable, and it may be difficult for hospitals to justify the investment. As with all treatments, the costs must be balanced against the benefits.

5.12.5   Conclusions

As reported by other researchers, the principal benefits observed in this research are increased comfort and speed of the process compared with the traditional impression (Lin & Nagler, 2003; Rogers et al., 2003). The scans took no more than a few minutes of the patient's time and presented no discomfort or distress. The time saving also benefits the prosthetist.

The vacuum-forming moulds produced by LOM and CNC require no special treatment. The production of the moulds from the scan data presented no inconvenience to the prosthetist and they could be delivered in a matter of days in most circumstances.

References

Bibb, R., & Brown, R. (2000). The application of computer aided product development techniques in medical modelling. *Biomedical Sciences Instrumentation, 36*, 319–324.

Bibb, R., Freeman, P., Brown, R., Sugar, A., Evans, P., & Bocca, A. (2000). An investigation of three-dimensional scanning of human body surfaces and its use in the design and manufacture of prostheses. *Proceedings of the Institute of Mechanical Engineers Part H, Journal of Engineering in Medicine, 214*(6), 589–594.

Leach, V. S. E. (2002). A method of producing a tissue compressive impression for use in fabricating conformers in hypertrophic scar management. *The Journal of Maxillofacial Prosthetics and Technology, 5*, 21–23.

Lin, J. T., & Nagler, W. (2003). Use of surface scanning for creation of transparent facial orthoses: a report of two cases. *Burns, 29*, 599–602.

Powell, E. M., Haylock, C., & Clarke, J. A. (1985). A semi-rigid transparent facemask in the treatment of post burn hypertrophic scars. *British Journal of Plastic Surgery, 38*, 561–566.

Rivers, E. A., Strate, R. G., & Solem, L. D. (1979). The transparent facemask. *American Journal of Occupational Therapy, 33*, 108.

Rogers, B., Chapman, T., Rettele, J., Gatica, J., Darm, T., Beebe, M., et al. (2003). Computerized manufacturing of transparent facemasks for the treatment of facial scarring. *Journal of Burn Care and Rehabilitation, 24*, 91–96.

Sanghera, B., Amis, A., & McGurk, M. (2002). Preliminary study of potential for rapid prototype and surface scanned radiotherapy facemask production technique. *Journal of Medical Engineering and Technology, 26*, 16–21.

Shons, A. R., Rivers, E. A., & Solem, L. D. (1981). A rigid transparent face mask for control of scar hypertrophy. *Annals of Plastic Surgery, 6*, 245–248.

Whitestone, J. J., Richard, R. L., Slemker, T. C., & Ause-Ellias, K. L. (1995). Fabrication of total-contact burn masks by use of human body topography and computer-aided design and manufacturing. *Journal of Burn Care and Rehabilitation, 16*, 543.

5.13   Maxillofacial rehabilitation case study 3: an appropriate approach to computer-aided design and manufacture of cranioplasty plates

Acknowledgements

Some of the work described in this case study was first reported in the reference below and is reproduced here in part or in full with the permission of the Institute of Maxillofacial Prosthetics and Technologists.

Bibb R, Bocca A, Evans P, "An Appropriate Approach to Computer Aided Design and Manufacture of Cranioplasty Plates", The Journal of Maxillofacial Prosthetics & Technology, 2002, Volume 5, Issue Number 1, pages 28–31.

The authors would like to gratefully acknowledge Sarah Orlamuender, Greta Green and James Mason from (at the time of original case study development) SensAble Technologies Inc. for their assistance in this project. The authors would also like to thank Brendan McPhillips, Principal Maxillofacial Prosthetist & Technologist/Laboratory Manager, Maxillofacial Laboratory, Royal Preston Hospital, for the images of cranioplasty plate pressing in case study 3.

5.13.1   Introduction

It has long been recognised in product design and engineering that computer-aided design and rapid prototyping (also called additive manufacturing, CAD/RP or AM) can have significant advantages over traditional techniques, particularly in terms of speed and accuracy. These advantages can be realised at all stages from concept through to mass production.

As these processes have become more widespread in industry, attempts have been made to transfer the technology to medical procedures. For example, computer-aided production methods such as RP have been used to build highly accurate anatomical models from medical scan data. These models have proved to be a valuable aid in the production of reconstructive implants such as cranioplasty plates (a cranioplasty plate is an artificial plate that is fitted to the skull to restore the shape of the head and protect the brain). Typically, RP is used to create accurate models of internal skeletal structures, such as skull defects (Klein, Schneider, Alzen, Voy, & Gunther, 1992). The cranioplasty plate is then handcrafted in wax on the model by the prosthetist (Bibb & Brown, 2000; Bibb et al., 2000; D'Urso & Redmond, 2000). Alternatively, the anatomical model can be used to create moulds or formers (Joffe et al., 1999). Although this process has dramatically improved the accuracy of cranioplasty plate manufacture, it incurs significant time and cost to produce the anatomical model. This current route does not fully exploit the potential advantages of an integrated and optimised CAD/RP process. A major impediment to the application of CAD in cranioplasty design is the fact that it requires the creation of complex, naturally occurring free form shapes that are necessary to accurately reconstruct the defect. Although efforts have been made to investigate the use of CAD in cranioplasty plate design, they have proved time-consuming and only served to highlight these limitations

(Taha et al., 2001). Advances in voxel-based CAD software described in Chapter 3, Section 3.6.6, such as FreeForm®, has enabled these challenging reconstructions to be undertaken more efficiently (Massie & Salisbury, 1994; FreeForm Modelling Plus Software).

In addition to advances in CAD, metal AM has enabled implants, such as cranioplasty plates, to be fabricated directly from CAD models (Lethaus et al., 2011). This section will describe the development of CAD techniques from an initial concept through to current best practice that employs AM. Although the tools available within FreeForm® have developed since early applications, the essential functions described in the initial study remain relevant to current versions.

5.13.2   Initial case

To investigate the application of voxel CAD technologies for reconstruction, a clinical case requiring a cranioplasty plate was attempted using FreeForm®. The case began with importing a 3D model of the cranial defect. This data were derived from a 3D computed tomography (CT) scan and imported into the software as a 'buck'. This means that the user can feel the surface of the skull but not alter it with any subsequent tools. The next step was to create a piece of 'virtual clay' that could then be worked into the correct shape. In prosthetics terms, it may be more appropriate to refer to this as virtual or digital wax up.

First, a sketch plane is positioned by eye over the defect and a two-dimensional perimeter drawn around it. The planes are then offset either side of the defect and the perimeters joined to form a 3D piece of digital wax approximately the right size, as shown in Figure 5.82. Material removal tools were then used to 'grind' away material. The tools used are similes of physical sculpting tools such as scrapers, grinding wheels, etc. However, the size and shape of the tools can be arbitrarily altered to suit the job in hand. A wide 'grinding' tool was used to work the surface back as shown in Figure 5.83. In addition, the simulated physical properties of the virtual clay can be altered to represent differing hardness and tack strength. When using the system the user can feel tactile resistance when removing material and when they reach the surface of the skull resistance is total. The approach is therefore a digital equivalent of waxing up on a medical model.

However, operating in the digital domain enables some useful techniques to be exploited. In this case, control curves were created on the surface of the plate. These curves can then be moved using control points on the surface or tangents to the surface. A single control curve applied to the sagittal plane is shown in Figure 5.84. Many control curves can be added in various planes to achieve a very smooth and well-defined natural curvature that matches the surrounding tissue. In addition, whole areas can be selected by painting on the model with the stylus. Then various operations can be performed on the entire selected area in a single operation. For example, the area could be pulled out to create a bulge or pushed in to create and depression.

The material removal and smoothing process was then repeated for the inner surface illustrated in Figure 5.85. Then the buck (skull) is subtracted from the wax to leave a free form shape that would repair the defect as shown in Figure 5.86. To allow the implant to be inserted from the outside of the skull the inner edges of the plate were worked back as shown in Figure 5.87. The final shape can then be accurately

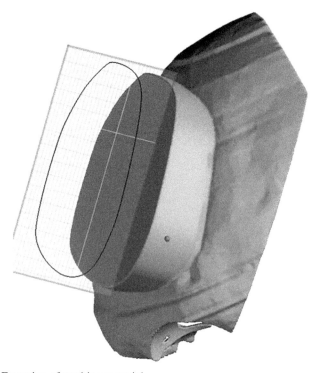

Figure 5.82 Extrusion of working material.

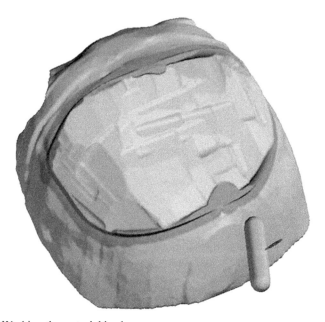

Figure 5.83 Working the material back.

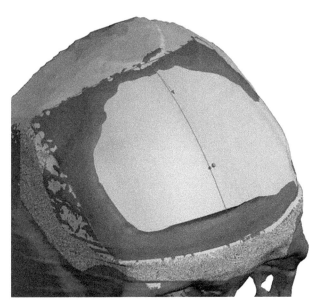

Figure 5.84 Controlling curvature.

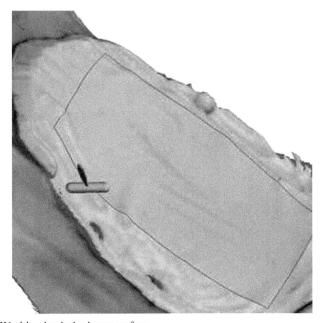

Figure 5.85 Working back the inner surface.

fabricated using RP techniques. This can then be used as a pattern from which an impl
ant can be made using an appropriate material.

   In this case, a highly accurate stereolithography model was made of both the implant
and the defect to test the design for fit and accuracy (see Figures 5.88 and 5.89). In this case,
the fit was excellent and comparable to the results achieved when using stereolithography

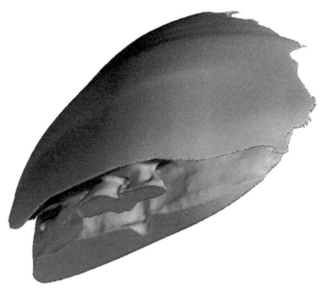

Figure 5.86 The result of the Boolean subtraction.

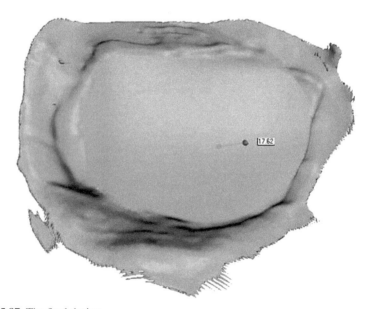

Figure 5.87 The final design.

models and traditional wax up methods. The stereolithography resin used is approved for patient contact under theatre conditions, which will allow the plate to be test fitted to the actual defect at the time of surgery. However, the material is not approved for implantation and therefore the actual implant used will be cast from the stereolithography master into an approved acrylate.

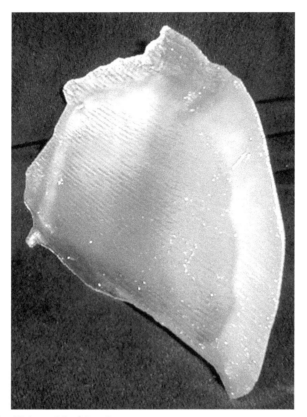

Figure 5.88 The SLA model of the plate.

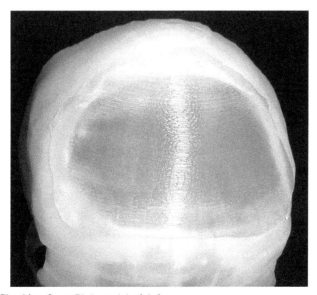

Figure 5.89 Checking fit on SLA model of defect.

5.13.3   Second case

This case followed similar techniques to the initial case yet proved more challenging because of the complex nature of the implant required. The added complexity further illustrated the potential benefits of utilising this approach. The unaffected side of the patient's skull was copied and laterally inverted to provide a base part from which a well-matched implant could be designed (Figure 5.90). As in all cases attempted in this manner, thus far the laterally inverted copy does not precisely fit the defect, however, the FreeForm® software enables the form to modified and adapted to produce a

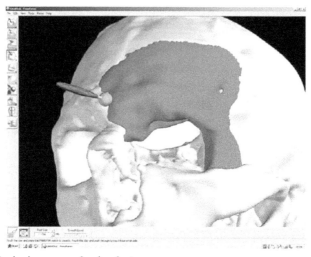

Figure 5.90 Designing a complex implant.

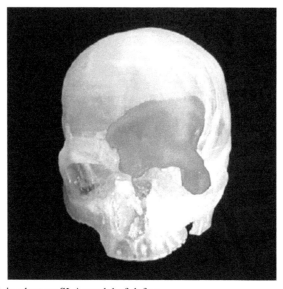

Figure 5.91 SLA implant on SLA model of defect.

prosthetic design that shows good aesthetic appearance and a precise fit at the margins. The software also enabled the implant design to be produced in a very smooth manner, which will prove suitable for a double processed and polished acrylic implant. As in the previous case, the implant design was manufactured using stereolithography and test fitted to an existing model of the patient defect, as shown in Figure 5.91.

5.13.4   Third case: press tool design

The first case studies demonstrated the feasibility of a CAD approach. With minor refinement of the techniques, a clinically viable method of producing mould tools that enabled pressed titanium plates was developed. This method has subsequently become widely adopted and used by the UK National Health Service and can be illustrated through a case study.

Techniques described in the first two studies were used to reconstruct the defect. At the point of completed reconstruction, the model consisting of buck anatomy and clay reconstruction was duplicated. The 'remove buck' option was used to leave just the clay reconstruction, which showed jagged edges around the area where it had previously blended into the anatomy (Figure 5.92). The clay coarseness was refined by 0.1 mm, which had the effect of smoothing the jagged edges and creating a small gap between the original anatomy and reconstruction (Figure 5.93). This gap defined the boundary between anatomy and CAD reconstruction, thus creating a margin, beyond that a pressed plate should extend. The smoothed reconstruction was then combined with a copy of the original anatomy, creating a reconstructed CAD model, with a physical groove representing the defect margin. This was cropped according to a reasonable margin around the defect that would allow the plate to be pressed.

The model was then exported as an STL file and it was fabricated using SL. The SL model was then back filled with plaster and used as the male section of a steel-bolstered press tool (Figure 5.94). Conventional, laboratory-based methods of pressing sheet titanium similar to those described in research literature (Bartlett, Carter, & Russell, 2009) were then used to complete the implant fabrication process (Figure 5.95).

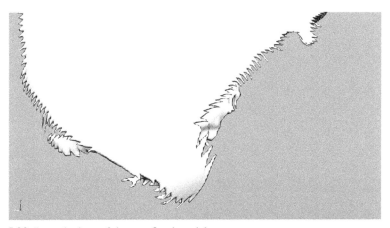

Figure 5.92 Jagged edges of the unrefined model.

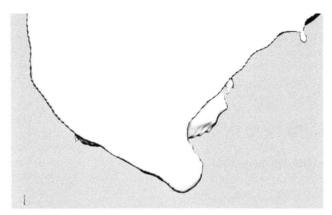

Figure 5.93 Refined edges.

Figure 5.94 Plaster-back, steel-bolstered press tool.

Figure 5.95 Completed pressed titanium plate.

5.13.5   Fourth case: implant design for AM fabrication

With further CAD modelling, it is also possible to use these techniques to design the definitive implant that incorporates a wide range of design features that can be produced using metal AM technologies. This has the potential to improve functional and aesthetic outcomes. This can be illustrated through the same case study as previously, but going a stage further to design an implant that sits just inside the defect.

At the stage where reconstruction has been completed and a duplication of the model has been made, the process is different. A boundary line enclosing the defect was drawn just outside the defect margin on the buck anatomy. The 'emboss clay' function was used to lower the clay of the reconstruction (note that the buck remains un-modified) by the desired thickness of the plate, in this case, 1 mm. The entire model was then converted to buck. In this case, a hexagonal pattern was desired to create a perforated plate, which would be lighter than a comparable solid structure. A boundary line, inside which the pattern would be created, was drawn on the buck model just inside the defect margin. A repeating, uniform, hexagonal mesh pattern was created using Photoshop (Adobe Systems, USA) and the 'emboss with wrapped image' tool used to create a 1-mm raised thickness of clay in the shape of the pattern. The 'layer' tool was then used to manually draw in 1-mm-thick plate around the periphery of the plate to the point where it met the defect margin, also filling in incomplete perforations. The same tool was used to design 0.6-mm-thick tabs that extended over the defect margin to provide fixation points for the plate and prevent it recessing too far into the defect during surgery. The 'remove buck' tool was then used to remove the reference buck model, leaving just the implant design. 'Select lump of clay' was used to select the implant design, with 'invert selection' and delete used to remove any unwanted, unattached pieces of clay. CAD designs of the intended 1.5-mm screws that incorporated countersink head features were then imported and positioned according to the implant tab locations, ensuring they did not protrude beneath the fitting surface of the implant. Boolean subtraction was undertaken to remove the screws from the plate, leaving the holes with countersink. The completed plate design clay coarseness was then refined by a factor of 0.1 mm to smooth jagged features and create a tolerance to the fixation holes. 'Select lump of clay' was used again to select the implant design, with 'invert selection' and delete used to remove any unwanted, unattached pieces of clay. The implant design, shown in Figure 5.96 was then exported as an STL file.

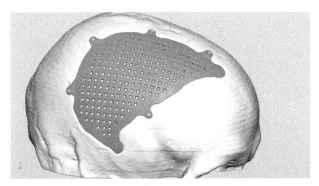

Figure 5.96 CAD cranioplasty implant design.

For subsequent fabrication, it is also necessary to undertake further checks to validate both the design and integrity of the STL file.

For the illustrated case study, electron beam melting (Q10, Arcam AB) (described in Chapter 4) was the intended fabrication method. The plate was not fabricated due to cost limitation.

5.13.6   Future development and benefits

This incremental development has proved that it is feasible to use CAD techniques to design and fabricate implants such as cranioplasty plates, but there is further work required to validate the economic and design process impacts of introducing these techniques. Some benefits are already apparent. The CAD model of the implant can be altered and refabricated any number of times if it is required. Data can be archived easily. There is reduced dependence on laboratory-based materials during the design process (the process as a whole is clean and requires minimal floor space, equipment and consumables) and there is potential for increased precision over pressing sheet titanium. This combination of advantages could lead to significant improvements in treatment times and cost effectiveness. An estimate of potential cost- and time saving is indicated in Table 5.1.

Notwithstanding the benefits, further work is still required to better understand the design process implications of using a CAD/AM approach. Introducing these

Table 5.1 **Comparison of current technique and potential future technique**

Current technique			Future technique		
3D CT scan			3D CT scan		
Send data to RP service provider	1 day	£5	Derive 3D CAD data from CT data	5 min	Minimal
Derive 3D CAD data from CT data	5 min	Minimal	Import into FreeForm	5 min	Minimal
Derive RP files from CAD data	1 h	£22	Design implant	1 h	£22
Build RP model of anatomy	1 day	£1000	Direct manufacture of implant	2 h	£200
Post model to prosthetist	1 day	£18	Sterilise for surgery		
Prosthetist wax up on model	1 day	£250			
Make plaster mould					
Cast implant					
Sterilise for surgery					
Total	4 days	£1295	Total	3 h	£222

techniques impacts on when design decisions need to be made, resources committed and costs are borne. Current laboratory-based techniques used in the production of sheet titanium plates have developed over many years and they are clinically proven. Although there are design limitations, sheet materials can be modified easily even at the point of surgery and are low cost. CAD/AM must offer a significant functional or economic advantage over current methods before they become more widely adopted.

Since the early studies into applications of CAD in cranioplasty design, parallel work has developed in other medical applications, such as external (wearable) facial prosthetics. Noninvasive, light-based optical scanning of faces combined with voxel-based CAD modelling has shown promise in the area but represents an even greater challenge due to the difficulties in handling point cloud data and converting it quickly and simply into a useable format (Bibb et al., 2000).

References

Bartlett, P., Carter, L. M., & Russell, J. L. (2009). The Leeds method for titanium cranioplasty construction. *British Journal of Oral and Maxillofacial Surgery*, *47*, 238–240.

Bibb, R., & Brown, R. (2000). The application of computer aided product development techniques in medical modelling. *Biomedical Sciences Instrumentation*, *36*, 319–324.

Bibb, R., Brown, R., Williamson, T., Sugar, A., Evans, P., & Bocca, A. (2000). The application of product development technologies in craniofacial reconstruction. In *Proceedings of the 9th european conference on rapid prototyping and manufacturing* (pp. 113–122). Athens, Greece.

Bibb, R., Freeman, P., Brown, R., Sugar, A., Evans, P., & Bocca, A. (2000). An investigation of three-dimensional scanning of human body surfaces and its use in the design and manufacture of prostheses. *Proceedings of the Institute of Mechanical Engineers Part H, Journal of Engineering in Medicine*, *214*(6), 589–594.

D'Urso, P. S., & Redmond, M. J. (2000). A method for the resection of cranial tumours and skull reconstruction. *British Journal of Neurosurgery*, *14*(6), 555–559.

FreeForm Modelling Plus Software, Geomagic, 3D Systems, 333 three D systems Circle, Rock Hill, SC 29730, USA.

Joffe, J., Harris, M., Kahugu, F., Nicoll, S., Linney, A., & Richards, R. (1999). A prospective study of computer-aided design and manufacture of titanium plate for cranioplasty and its clinical outcome. *British Journal of Neurosurgery*, *13*(6), 576–580.

Klein, H. M., Schneider, W., Alzen, G., Voy, E. D., & Gunther, R. W. (1992). Pediatric craniofacial surgery: comparison of milling and stereolithography for 3D-model manufacturing. *Pediatric Radiology*, *22*, 458–460.

Lethaus, B., Poort ter Laak, M., Laeven, P., Beerens, M., Koper, B., Poukens, J., et al. (2011). A treatment algorithm for patients with large skull bone defects and first results. *Journal of Cranio-Maxillo-Facial Surgery*, *39*(6), 435–440.

Massie, T. H., & Salisbury, K. J. (1994). PHANToM haptic interface: a device for probing virtual objects. In *Proceedings of the 1994 international mechanical engineering congress and exposition (code 42353)* (Vol. 55–1, pp. 295–299). American Society of Mechanical Engineers, Dynamic Systems and Control Division (Publication) DSC.

Taha, F., Lengele, B., Boscherini, D., & Testelin, S. (June 2001). "Phidias report" No. 6. In N. Moos (Ed.), *Danish technological institute teknologiparken*, 8000 Aarhus C, Denmark.

5.14   Maxillofacial rehabilitation case study 4: evaluation of advanced technologies in the design and manufacture of an implant retained facial prosthesis

Acknowledgements

This paper was written by Dominic Eggbeer and it is based on PhD research he conducted at PDR under the supervision of Richard Bibb and in collaboration with Morriston Hospital, Swansea. The authors would like to thank Frank Hartles, Head of the Dental Illustration Unit, Media Resources Centre, Wales College of Medicine, Biology, Life and Health Sciences, Cardiff University for his help in using the Konica-Minolta scanners.

5.14.1   Introduction

Despite the widespread application of CAD/RP technologies in the production of medical models to assist maxillofacial surgery, advanced technologies remain underdeveloped in the design and fabrication of facial prosthetics. Research studies to date have achieved some limited success in the application CAD/RP technologies but very few have addressed the whole design and manufacture process or incorporated all of the necessary components (Bibb et al., 2000; Cheah, Chua, & Tan, 2003; Cheah, Chua, Tan, & Teo, 2003; Coward, Watson, & Wilkinson, 1999; Eggbeer, Bibb, & Evans, 2004; Evans, Eggbeer, & Bibb, 2004; Reitemeier et al., 2004; Sykes, Parrott, Owen, & Snaddon, 2004; Tsuji et al., 2004; Verdonck, Poukens, Overveld, & Riediger, 2003; Wolfaardt, Sugar, & Wilkes, 2003). In particular, the components concerned with retention have been neglected. Given that implant retention is now widely considered state of the art, the incorporation of this into digital facial prosthesis techniques must be addressed. Implant-retained prostheses are described in an explanatory in 'Glossary and explanatory notes'.

Being able to undertake all of the prosthesis design and construction stages without the patient present has the potential to dramatically reduce clinic time and make the entire process more flexible. Reducing the number of clinic visits and the time involved in them would help to reduce patient inconvenience and improve efficiency and flexibility by allowing the prosthetist to work on any given design at any period.

This paper reports on doctoral research that ultimately aimed to identify the target specification requirements for advanced digital technologies that may be used to drive development of advanced technologies such that they will be suitable for the design and manufacture of complex, soft-tissue facial prostheses. The study reported here tested the capability of currently available technologies in the design and manufacture of an implant retained prosthesis. Evaluation of the results will clarify the current position and, where they fall short, direct further research that will identify the direction and magnitude of the developments required.

5.14.2   Existing facial prosthetics technique

Facial prosthesis design and construction techniques have changed little in 40 years and they are described well in textbooks (McKinstry, 1995; Thomas, 1994) and papers (Postema, van Waas, & van Lokven, 1994; Seals, Cortes, & Parel, 1989). By their nature, prostheses are one-off, patient-specific devices that cannot benefit from batch or mass manufacture. Hand crafting techniques are therefore used to fabricate the prosthesis form and retentive components and in some cases join them to prefabricated components that enable the prosthesis to be attached to the implants.

Various retention methods may be used to secure a facial prosthesis such as magnets, bar and clip, adhesives or engaging anatomical undercuts. However, in many cases implant retained prostheses are now considered the optimum solution. In implant-retained cases, the prosthesis typically consists of three components; the soft tissue prosthesis itself, a rigid sub-structure incorporating the retention parts and the corresponding retention parts that remain attached to the patient. The attachment between two retention components can be by bar and clip or by magnets. Bar and clip gives the highest retention force and the strength may be altered by crimping the metal clips. Magnets can provide a range or retentive forces (around 500–1000 g) depending on the number and type used. Magnets may either be screwed directly on to the abutments, or located on a framework. The prosthesis-mounted components may be bonded directly into the silicone if the prosthesis is small or a sub-structure is not necessary.

Prosthesis design is typically undertaken by shaping wax on a plaster replica of the patient's anatomy. Realism is predominantly achieved though the prosthetist's ability to interpret the correct location and physically recreate the anatomical shape and detail. Colour matching of the silicone also helps to complete a good blend into the surrounding anatomy.

Although these existing techniques are time-consuming, they can be applied to a wide range of situations. Previous studies have shown that to be effective digital technologies must be sympathetically integrated into these existing techniques so that the skills and flexibility of the prosthetist are not hampered (Eggbeer et al., 2004; Evans et al., 2004; Sykes et al., 2004).

5.14.3   Review of advanced technologies in facial prosthetics

A review of previous research highlights a range of advanced technologies that may be used to design and manufacture a facial prosthesis (Bibb et al., 2000; Cheah, Chua, & Tan, 2003; Cheah, Chua, Tan, & Teo, 2003; Coward et al., 1999; Eggbeer et al., 2004; Evans et al., 2004; Reitemeier et al., 2004; Sykes et al., 2004; Tsuji et al., 2004; Verdonck et al., 2003; Wolfaardt et al., 2003).

- **Data capture**   Noncontact surface scanning to digitise the surface of the affected
                     anatomy
                     Various structured white light scanners, laser scanners, computerised
                     photogrammetry
- **Design**   Flexible CAD software
                     FreeForm (FreeForm Modelling Plus, Geomagic, 3D Systems, USA,
                     http://geomagic.com/en/products/freeform-plus/overview); Magics
                     (Materialise NV, Technologielaan 15, 3001 Leuven, Belgium, www.
                     materialise.com); Rhino (Robert McNeel & Associates, 3670 Woodland
                     Park Avenue North, Seattle, WA 98103, USA, www.rhino3d.com)
- **Manufacture**   RP processes
                     ThermoJet wax printing (obsolete, 3D Systems, USA. The ThermoJet
                     process is now obsolete with the most analogous replacement process
                     being the 3D Systems ProJet 3510CP and CPX wax printers); Selective
                     Laser melting (at time of original publication – MCP, now Renishaw,
                     UK. The technology has now been acquired by Renishaw Plc, New
                     Mills, UK); stereolithography (3D Systems), selective laser sintering
                     (3D Systems) and various CNC machining processes

The review of previous work has shown that these advanced techniques and technologies demonstrate a number of limitations and identifies a range of technical challenges. Specifically, there are three notable areas: the capture of data that describes the anatomy and implant abutment features, the design and alignment of the prosthesis components and the manufacture of components in appropriate materials.

5.14.3.1   Data capture

Although noncontact surface scanning technologies has been used to capture anatomical forms limitations have also been identified. Areas of hair, undercut surfaces, highly reflective surfaces patient movement give poor results (Bibb et al., 2000; Cheah, Chua, & Tan, 2003; Cheah, Chua, Tan, & Teo, 2003; Reitemeier et al., 2004). Insufficient data resolution and errors in the form of 'noise' also limit the ability of scanning technologies to capture sharp edges and small geometrical features at the scale required (Chen and Lin, 1997). The ideal scanning technology must therefore be capable of capturing both anatomical surfaces and implant components with sufficient accuracy, resolution and speed to overcome these limitations.

5.14.3.2   Design

To design the various components of the prosthesis such that they accurately fit together using CAD the operator must be able to import, manipulate, create and align both anatomical and geometric forms. Engineering CAD software packages typically work with geometric shapes and provide methods of aligning components. However, engineering software is poorly suited to handing complex and individual anatomical forms. CAD software such as FreeForm (Geomagic, 3D Systems, USA) provides a more intuitive solution to handling anatomical forms (more akin to a digital sculpting package) yet does not provide suitable tools for aligning the various components.

A suitable CAD software package must provide tools for precisely aligning geometric shapes as well as the manipulation of complex anatomical forms.

5.14.3.3 Manufacture

Material requirements for maxillofacial prostheses are varied according to the separate components. For the soft tissue elements that are currently made from colour-matched silicone, no technology exists that is able to build the final prosthesis form directly from CAD data. Therefore, a pattern must be produced instead. The review of previous research and experiments carried out at PDR and Morriston Hospital have shown that producing the pattern in a material compatible with conventional sculpting techniques is highly desirable. Building the pattern in wax allows the prosthetist to easily adjust the pattern using their existing techniques and skills, particularly during test fitting on the patient (Eggbeer et al., 2004; Evans et al., 2004; Sykes et al., 2004). The substructures need to be accurate and rigid enough to contain the retentive elements and the forces experienced during attachment and removal. The retentive components that remain attached to the patient via the implants must be noncorrosive and unreactive (similar to jewellery) and rigid enough to withstand the retentive forces. The materials for all of the components must also not react with each other and resist the effects of being included in the manufacture of the final prosthesis from the pattern, such as mould heating. Finally, they must also provide adequate wear resistance, resist permanent distortion and provide adequate retention for the service life of the prosthesis.

5.14.4   Case 1

To assess the capabilities of current advanced technologies in the design and manufacture of an entire implant-retained prosthesis accurately an exploratory study was undertaken. The study would not only evaluate the ability of current technologies but measurements and observations made would inform future research. A bar and clip, implant-retained auricular prosthesis case was selected. A 3D CT scan had already been undertaken and the data used to plan the placement of two implants in a single-stage operation. A healing period of 6 weeks was allowed before prosthesis construction.

5.14.4.1 Data capture

An impression and dental stone replica that recorded the implant abutment locations and the surrounding anatomy was made using conventional methods. In addition, the patient was digitally scanned using a pair of laser scanners (Konica-Minolta Vivid 900 laser scanners, Osaka, Japan) to allow for subsequent digital prosthesis design. Previous work has shown that these scanners had a relatively fast capture time and an accuracy level appropriate to the scanning of faces (Kau, Knox, & Richmond, 2004; Kau, Zhurov, Scheer, Bouwman, & Richmond, 2004). The actual number of points captured per mm^2 is determined by scanner's field of view. At a distance of 1.35 m, the scanners each captured and area of 445×333 mm, resulting in a point density of one point per 0.69 mm^2. A paired setup was used to capture a wider field of view

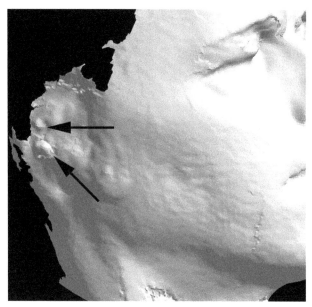

Figure 5.97 Scan data of the defect site (arrows indicating the implant abutments).

without having to move the patient. In this configuration, the scanners are triggered consecutively (simultaneous capture would lead to the scanners interfering with each other). The patient was seated with their head positioned 1.35 m away from the scanners and a 14 mm focal length lens used. Although the specified capture time is 0.6 s for each camera, a short pause between scans meant that the patient had to remain motionless and with the same facial expression for approximately 8 s. The point-cloud scan data were aligned and converted to an STL file using Rapidform software (INUS Technology Inc., Seoul, South Korea). Shadow areas behind the ears were not captured. However, the 3D CT data that had been acquired for the implant planning was also available to be used as the basis for the prosthesis design.

5.14.4.2 Design

The scan data were imported into the sculpting CAD package, FreeForm using the 'thickness' option to make a solid model. Whilst the scanners have been shown to be able to capture anatomical detail well, this study demonstrated that the data resolution was insufficient to describe the implant abutments accurately (see Figure 5.97). However, the data were good enough to allow the abutment locations to be identified, which allowed the overall prosthesis form to be designed around them with sufficient accuracy. The patient's opposite healthy ear was obtained from the CT data, imported into FreeForm and mirrored to the defect site (see Figure 5.98). The tools in FreeForm were then used to blend this ear into the surrounding anatomy and then subtracted from it to leave an accurate fitting surface using a technique that has also been reported in the digital design of other prostheses (Eggbeer et al., 2004; Evans et al., 2004; Sykes et al., 2004).

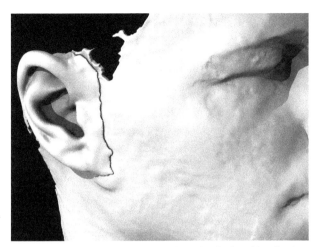

Figure 5.98 The mirrored ear positioned at the defect site (before blending).

5.14.4.3   Manufacture

The final design of the prosthesis pattern was physically manufactured using the Thermo-Jet printing process in a wax material. The use of wax allowed modifications to include the retentive components to be made using conventional methods. A colour-matched, silicone prosthetic ear was then fabricated for the patient using conventional methods.

5.14.4.4   Initial findings from case 1

This initial trial highlighted the limitations of noncontact scanning to capture anatomy and finely detailed abutments with sufficient resolution.

5.14.5   Case 2

5.14.5.1   Data capture

The first case showed that the data captured was insufficient to be used in the design of the retentive components. Therefore, the same case was repeated using a higher-resolution structured white light scanner (Steinbichler Optotechnik GmbH, Germany). This type of scanner is typically used for engineering and has a much longer capture time. Therefore, it was used to digitise the dental stone replica of the patient produced using conventional impression methods. This scanner captures approximately nine points per mm^2 (three per mm in the x, y plane) and around 140,000 points per scan over an area of approximately 250×250 mm and a working range of approximately 180 mm. To make the abutment locations easier to scan, magnetic keepers (Technovent Ltd., Leeds, UK) were screwed on to the abutments to provide a flat surface and the model was coated in a fine matt white powder to reduce reflectivity. Six overlapping scans covering the entire model were taken and the data aligned using Polyworks software (InnovMetric Software

Inc., Quebec, Canada). STL file data were created from the point cloud information using Spider (Alias-Wavefront, Toronto, Canada) and the data imported into Magics. Magics provides alignment and modification tools for STL file data and was used to digitally remove the abutment caps. The flat surfaces of each abutment cap in turn were aligned by selecting a triangle on the top surface and using a Magics function to make it down facing to the x-y plane. The 'sectioning' and 'cut' tools were then used to remove the exact depth of the cap. This effectively left a perfectly flat surface representing the top surface of the abutments. The modified STL data were then imported into FreeForm.

5.14.5.2   Design

As in the previous study, FreeForm was used to manually align the ear taken from the CT data to the digital cast based upon the estimated aesthetic requirements and possible sub-structure location. However, unlike the first case the data quality in this case study enabled the design of all of the components to be attempted. Digital versions of the screws used to attach frameworks to the abutments and cylinder components were designed using the drawing and rotation tools in FreeForm. A circular-section framework linking the two cylinders was created using the 'add clay' tool. As in conventional methods, this followed the thickest section of the ear to ensure all of the components would fit within the ear profile.

Clip designs were created using the two-dimensional drawing and 'extrude' tools. These were copied three times and manually located along the bar structure at key points of maximum prosthesis thickness. To secure the clips into the prosthesis body and to assist application, a sub-structure shell that would be bonded to the silicone was required. This had to provide enough clearance for the clips to spring open and closed, but provide firm anchorage for bonding to the silicone. Digital 'clay' that enclosed the framework and clips, leaving just their top features as a point of attachment, was built up from the cast model. The 'paint on selection' tool was used to select and copy then paste the raised section surrounding the framework. The 'create offset piece' tool was then used to create a shell 1.5 mm thick surrounding the clips and bar. Boolean subtraction operations and hand carving were used to finalise the shell before joining it to the clip components. The bar, clip and shell design are shown in Figures 5.99 and 5.100.

The prosthesis profile was thickened around the clip areas to accommodate the shell component and a Boolean subtraction operation used to create a fitting recess. The bar design was finalised with Boolean subtraction operations to provide location for the screws and smoothing operations were applied around the joints with the cylinder features. Hemispherical dimples were also created where cylinders located on the abutments. Figure 5.101 shows an exploded view of the components in the FreeForm environment. Each of the components was then exported as high quality STL files ready for RP fabrication (see Chapter 3 for more information about the STL file format).

5.14.5.3   Manufacture

A range of RP technologies were selected to produce the components in the most suitable materials available. Selective laser melting (SLM) was selected to produce

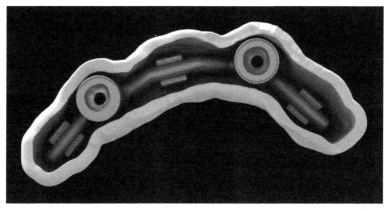

Figure 5.99 The bar located in the clips inside the sub-structure shell (in FreeForm®).

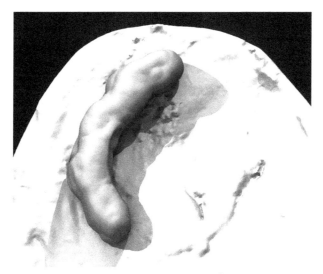

Figure 5.100 The located sub-structure shell (in FreeForm®).

the bar component due to its ability to produce parts in corrosion resistant and rigid metals and alloys (see Chapter 4 for a description of the SLM process). The bar was created using SLM in 0.05-mm-thick layers using 316L Stainless Steel. Grit blasting and light polishing was used to remove the rough finish produced by the process.

Stereolithography was used to produce the shell component in DSM Somos 10110-epoxy resin (see Chapter 4 for more information on stereolithography). ThermoJet wax printing was used to produce the prosthesis pattern. The physical components are shown in Figure 5.102.

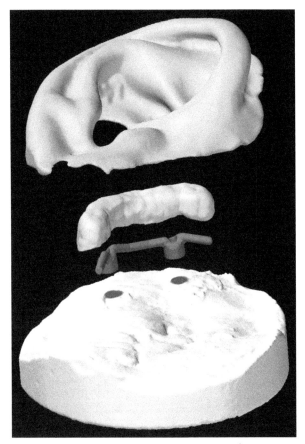

Figure 5.101 An exploded view of the components in FreeForm®.

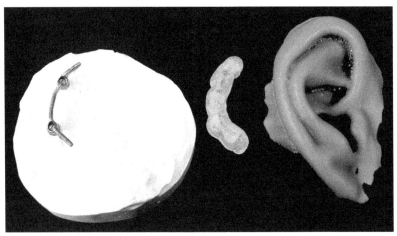

Figure 5.102 The manufactured components: SLM™ bar (left), stereolithography sub-structure (middle) and ThermoJet® pattern (right).

5.14.6 Results

The parts designed and produced in the second case study did produce a whole and complete prosthesis. However, the fit between the components required a small amount of adjustment. Therefore, although the finished components would not have been suitable for use in an actual prosthesis, it was possible to evaluate them when fitted to the dental stone model.

The fit between the implant abutments and the bar. A passive fit as described by Henry (Henry, 1987) was not achieved and although the bar did screw on to the stone model securely, visible gaps remained. In addition, the surface finish of the SLM bar was slightly pitted when compared to soldered gold bars.

Fit between bar and clips/sub-structure. Clip retention was initially good, but repeated application caused the clips to wear. This suggested that the relatively soft epoxy resin is not suitable for use in a clip that undergoes repeated applications.

Fit between the sub-structure and prosthesis pattern. The fit was tight due to the rough finish left after removing supports from the ThermoJet part's down facing surfaces. This was easily corrected using a heated scalpel enabling a secure location and close fit.

Detail – the fragile nature of the ThermoJet wax prevented the pattern's edges from being made any thinner. Thin edges allow the prosthesis margin to blend naturally into the skin when made in silicone. These thin edges were added by the prosthetist using sculpting wax. Heated metal sculpting tools were also used to blend the join between the different waxes and add further anatomical details that helped to achieve a more realistic appearance.

5.14.7 Discussion

This study has highlighted the potential of digital technologies to assist facial prosthesis design, but also demonstrated that there are many limitations that must be addressed to improve their effectiveness. The limitations encountered are discussed in three categories: data capture, design and manufacture.

5.14.7.1 Data capture

Scanning small detailed abutments proved particularly difficult. The distributed nature of point cloud data captured by optical scanning technologies and the effects of noise meant that even with a high point density, small features were subject to loss of edge definition. Inadvertent patient movement during data capture exasperates this problem. With current technologies, a compromise must be made between detail and speed of capture. However, as digital camera technology and computer processor power increases, it can be foreseen that the desired capability to capture rich, high-quality data sufficiently quickly to enable the scanning of patients may be achieved in the near future.

5.14.7.2 Design

This study has shown that digital techniques can be used to design all of the components of a prosthesis. However, as the major aim of embracing digital technologies is the gain in efficiency it is clear that more work is needed to address software capability. This study has shown that although digital design is possible it requires the use of multiple software packages to achieve specific tasks. This reduces efficiency, increases costs and introduces more opportunities for error as data is translated or transferred from one source to another.

Future studies will explore alternative software solutions and identify practical methods of overcoming the issues identified in this research. The authors intend to evaluate other potentially suitable technologies in future studies once an initial specification has been developed.

5.14.7.3 Manufacture

The technologies used in this study have shown that they are capable of producing a complete prosthesis. However, the processes all require improvement to match or improve upon existing techniques. The ThermoJet wax process produced a good quality pattern in an appropriate and useful material that integrated with existing skills and techniques. However, it requires the ability to generate thinner edges if it is to produce a complete pattern without modification. The ThermoJet process is now also obsolete. The closest analogous product currently available that prints in wax is the ProJet CP range of 3D printers (3D Systems, USA).

The SLM bar was sufficiently strong and rigid enough for the application and the material should prove corrosion resistant enough for most patients. The bar however did not fit as precisely as would be expected of a bar made by existing techniques and the surface was slightly pitted. As the overall shape and accuracy appeared adequate, finer control of the process may yield parts with better detail and surface finish.

The stereolithography shell component was accurate and rigid enough for the application. The retention strength of the clips was not very high, although it may have been high enough for the purpose. However, the clips did not withstand repeated use and quickly wore down, severely degrading retention strength. The process therefore could prove adequate for the purpose if a harder wearing material was available.

5.14.8 Conclusions

Literature to date and the findings of this study have demonstrated that whilst advanced technologies enable the digital design and RP fabrication of complete facial prostheses further work is needed before they produce results comparable to existing techniques. Without a specification against which potential technologies may be measured and towards which they may be developed, quantifying success is based on subjective assessment and expert opinion. The authors intend to use the findings of this study to direct further research that will develop a specification that will provide quantifiable and objective measures against which advanced technologies may be assessed.

References

Bibb, R., Freeman, P., Brown, R., Sugar, A., Evans, P., & Bocca, A. (2000). An investigation of three-dimensional scanning of human body surfaces and its use in the design and manufacture of prostheses. *Proceedings of the Institute of Mechanical Engineers Part H, Journal of Engineering in Medicine, 214*(6), 589–594.

Cheah, C. M., Chua, C. K., & Tan, K. H. (2003). Integration of laser surface digitizing with CAD/CAM techniques for developing facial prostheses Part 2: development of molding techniques for casting prosthetic parts. *International Journal of Prosthodontics, 16*(5), 543–548.

Cheah, C. M., Chua, C. K., Tan, K. H., & Teo, C. K. (2003). Integration of laser surface digitizing with CAD/CAM techniques for developing facial prostheses Part 1: design and fabrication of prosthesis replicas. *International Journal of Prosthodontics, 16*(4), 435–441.

Chen, L. C., & Lin, G. C. (1997). An integrated reverse engineering approach to reconstructing free-form surface. *Computer Integrated manufacturing systems, 10*, 49–60.

Coward, T. J., Watson, R. M., & Wilkinson, I. C. (1999). Fabrication of a wax ear by rapid-process modelling using stereolithography. *International Journal of Prosthodontics, 12*(1), 20–27.

Eggbeer, D., Bibb, R., & Evans, P. (2004). The appropriate application of computer aided design and manufacture techniques in silicone facial prosthetics. In *Proceedings of the 5th national conference on rapid design, prototyping, and manufacture* (pp. 45–52). London: Professional Engineering Publishing.

Evans, P., Eggbeer, D., & Bibb, R. (2004). Orbital prosthesis wax pattern production using computer aided design and rapid prototyping techniques. *The Journal of Maxillofacial Prosthetics and Technology, 7*, 11–15.

Henry, P. J. (1987). An alternative method for the production of accurate casts and occlusal records in the osseointegrated implant rehabilitation. *Journal of Prosthetic Dentistry, 58*(6), pp. 694–677.

Kau, C. H., Knox, J., & Richmond, S. (2004). Validity and reliability of a portable 3D optical scanning device for field studies. In *Proceedings of the 7th european craniofacial congress*. Bologna: Monduzzi Editore-International Proceedings Division.

Kau, C. H., Zhurov, A., Scheer, R., Bouwman, S., & Richmond, S. (2004). The feasibility of measuring three-dimensional facial morphology in children. *Orthodontic Craniofacial Research, 7*, 198–204.

McKinstry, R. L. (1995). *Fundamentals of facial prosthetics*. Arlington: ABI Professional, ISBN: 1-886236-00-3.

Postema, N., van Waas, M. A., & van Lokven, J. (1994). Procedure for fabrication of an implant-supported auricular prosthesis. *Journal of Investigative Surgery, 7*(4), 305–320.

Reitemeier, B., Notni, G., Heinze, M., Schöne, C., Schmidt, A., & Fichtner, D. (2004). Optical modeling of extraoral defects. *Journal of Prosthetic Dentistry, 91*(1), 80–84.

Seals, R. R., Cortes, A. L., & Parel, S. (1989). Fabrication of facial prostheses by applying the osseointegration concept for retention. *Journal of Prosthetic Dentistry, 61*(6), 712–716.

Sykes, L. M., Parrott, A. M., Owen, C. P., & Snaddon, D. R. (2004). Application of rapid prototyping technology in maxillofacial prosthetics. *International Journal of Prosthodontics*,
17(4), 454–459.

Thomas, K. F. (1994). *Prosthetic rehabilitation*. Chicago: Quintessence Publishing, ISBN:
1-85097-032-7.

Tsuji, M., Noguchi, N., Ihara, K., Yamashita, Y., Shikimori, M., & Goto, M. (2004). Fabrication
of a maxillofacial prosthesis using a computer-aided design and manufacturing system.
Journal of Prosthodontics, *13*(3), 179–183.

Verdonck, H. W. D., Poukens, J., Overveld, H. V., & Riediger, D. (2003). Computer-assisted
maxillofacial prosthodontics: a new treatment protocol. *International Journal of Prosthodontics*, *16*(3), 326–328.

Wolfaardt, J., Sugar, A., & Wilkes, G. (2003). Advanced technology and the future of facial
prosthetics in head and neck reconstruction. *International Journal of Oral and Maxillofacial Surgery*, *32*(2), 121–123.

5.15   Maxillofacial rehabilitation case study 5: rapid prototyping technologies in soft-tissue facial prosthetics – current state of the art

Acknowledgements

The work described in this case study was first reported in the reference below and is reproduced here with the permission of Emerald Publishing Ltd.

Bibb R, Eggbeer D, Evans P, "Rapid prototyping technologies in soft tissue facial prosthetics: current state of the art", Rapid Prototyping Journal 2010; 16(2): 130-137, ISSN: 1355–2546, http://dx.doi.org/10.1108/13552541011025852

5.15.1   Introduction

Patients who suffer from facial deformity, either, congenital, traumatic or from ablative surgery are treated by maxillofacial units using a variety of surgical and prosthetic techniques. Maxillofacial prosthetics and technology covers the treatment and rehabilitation of these patients by producing facial prostheses using artificial materials. Improvements in medicine, surgical techniques and in particular cancer survival rates are resulting in ever increasing patient numbers. However, these same drivers are also leading to higher costs, which is putting pressure on health care providers to improve efficiency during a period when the number of newly qualified maxillofacial prosthetists and technologists is only matching the number retiring (Wolfaardt, Sugar, & Wilkes, 2003).

These pressures have led researchers to explore whether the cost- and time savings associated with advanced design and product development technologies can be realised in maxillofacial prosthetics. Technologies such as 3D surface capture (3D scanning), 3D CAD and layer additive manufacturing processes (or RP and manufacturing [RP&M]) have been investigated in maxillofacial prosthetic applications. However, the literature mostly comprises reports of single case studies that describe a given technology or application. Much early work involved making an anatomical form using RP processes such as stereolithography or laminated object manufacture (Bibb et al., 2000; Chen, Tsutsumi, & Iizuka, 1997; Chua, Chou, Lin, Lee, & Saw, 2000; Coward, Watson, & Wilkinson, 1999). This research however did not attempt to integrate the RP technologies into existing prosthetic practice or was limited to the production of an anatomical form that was used as a pattern for replication into more appropriate materials via secondary processes such as silicone moulds or vacuum casting. Later research attempted to use RP methods to produce moulds from which prosthesis forms could be moulded but these required extensive time and CAD facilities and did not attempt to exploit the advantages of RP processes that could be better integrated with existing prosthetic practice (Cheah, Chua, & Tan, 2003; Cheah,

Chua, Tan, & Teo, 2003). Other researchers attempted to exploit the ability of the ThermoJet (obsolete, 3D Systems, USA) process to produce prosthesis forms in a wax material that was comparable to the waxes used in the typical prosthetic laboratory (Chandra et al., 2005; Reitemeier, Notni, & Heinze, 2004; Sykes, Parrott, Owen, & Snaddon, 2004; Verdonck, Poukens, Overveld, & Riediger, 2003). This enabled a more integrated approach that incorporated the advantages of RP into the existing workflow. However, these studies lacked critical evaluation of the physical properties of the prostheses produced or the technical capabilities and appropriateness of the RP technologies used for developing an integrated and efficient digital prosthetic process.

A wider investigation was therefore required to explore how these technologies may be applied in combination to a variety of applications. The research described here was part of a wider project to investigate whether available design technologies could be successfully applied to produce gains in efficacy and efficiency in a busy maxillofacial unit. The work resulted in a PhD thesis which covers all of the work in greater detail than is possible in this paper (Eggbeer, 2008). Four case studies (with three previously reported) were explored over a 4-year period in collaboration with a regional maxillofacial unit and complemented by two technical experiments (with one previously reported). This article focuses on the aspects of the research that addressed the current capabilities and limitations of layer additive manufacturing technologies more commonly referred to as RP&M technologies in maxillofacial prosthetics. The paper summarizes 4 years of study and draws conclusions on the current state of the art of RP&M in maxillofacial prosthetics and includes recommendations and target technical specifications towards which future RP&M developments should be made to meet the needs of patients and clinicians in this field.

5.15.2   Methodology

The case studies were planned and carried out using an action research (AR) approach. AR methods utilise an iterative process of research design, research implementation and evaluation. AR approaches and case studies are typically applied in the social sciences for studying 'real-life' situations where the researcher cannot control all of the variables or the research environment (Yin, 2003). Typically, these involve complex, changing situations and small samples. The nature of maxillofacial treatment means that each case is unique and therefore repetitive, or series studies are not possible in a clinical setting, making the AR approach appropriate for this research. Therefore, the research was undertaken through iterative case studies, the findings from each case study informing the design and implementation of the subsequent case study or experiment. The practical methods used in each case study or experiment varied according to the needs of the individual case study. A number of the cases required flexibility and adaptation to unforeseen challenges. Throughout the research, a small number of complementary experiments were carried out to establish technical capabilities that informed subsequent case studies and did not require clinical investigation.

5.15.3   Summary of case studies

This section summarises the methods and results from the cases that directly employed RP&M technologies. Additional, complementary case studies and experiments were carried out into related digital technologies such as 3D scanning but are not reported here.

5.15.3.1   Case 1: orbital prosthesis

This study was intended to explore the use of CAD and RP&M methods that had been identified in the literature. The case used CT, FreeForm Modelling Plus Computer-Aided Design software (Geomagic, 3D Systems, USA) and ThermoJet rapid prototyping (obsolete, 3D Systems, USA) combined with conventional fitting and finishing techniques to produce the final prosthesis.

This case explored the idea that clinical 3D CT data acquired for maxillofacial surgery purposes could also be subsequently applied to aid in prosthetic design and manufacture. As it would be inappropriate to undertake unnecessary CT scans without full clinical justification, this work was based on exploiting existing CT scans that the patient had previously undergone for diagnostic or surgical planning reasons. This case demonstrated that such CT data were sufficient to capture the gross facial anatomy, but it could not capture fine details such as skin texture and wrinkles, for example at the corners of the eyes. This is because a typical CT scan taken for maxillofacial surgery will require a field of view wide enough to acquire images of the whole head; this would be typically 25 cm. A typical CT scanner in clinical use will have a pixel array of 512×512 leading to a pixel size of 0.488 mm. As skin textures typically have depths in the range 0.1–0.8 mm, it is clear that the pixel size is too great to be able to adequately describe skin texture (Lemperle, Holmes, Cohen, & Lemperle, 2001). Whilst it is technically feasible that a CT scan could be optimised to enable the capture of skin texture the change in CT protocol required would increase scan time and X-ray dosage for the patient that would be difficult to justify.

This case also indicated that FreeForm was an appropriate CAD tool for the design of the basic shape of a prosthesis based on the surface data captured from the patient. The ThermoJet process successfully manufactured the prosthesis form. The CAD design and final ThermoJet model are shown in Figure 5.103. It was noted by the prosthetist undertaking the case that the wax ThermoJet material was compatible with his conventional

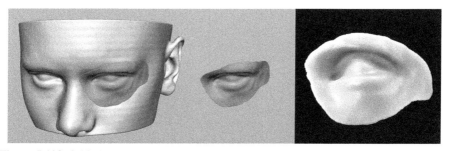

Figure 5.103 Orbital prosthesis designs in CAD and ThermoJet wax pattern.

maxillofacial laboratory techniques. The wax prosthesis form was therefore finished and fitted to the patient in the normal manner. It was then used as a sacrificial pattern in the production of the final silicone rubber prosthesis. This case proved that the CAD and RP&M technologies described in the literature were appropriate for further development in facial prosthetics and was subsequently published (Evans, Eggbeer, & Bibb, 2004).

5.15.3.2   Case 2: texture experiment

The first case highlighted the importance of capturing and reproducing fine skin features such as wrinkles and texture. The experiment involved developing 3D skin texture relief on an anatomical shape using FreeForm CAD and reproducing the relief at a series of depths identified from dermatology literature ranging from approximately 0.1–0.8 mm in depth (Lemperle et al., 2001). The parts were produced using the ThermoJet process. The ThermoJet process was found to be capable of reproducing a convincing skin texture on a prosthesis and the study was published in 2006 (Eggbeer, Evans, & Bibb, 2006). One of the CAD models and the resulting ThermoJet model are shown in Figure 5.104.

5.15.3.3   Case 3: auricular prosthesis A

A review of the literature revealed that although RP&M had been used in maxillofacial prosthetics the cases had all addressed only the creation of the overall anatomical form of the prosthesis. The accepted 'gold standard' for maxillofacial prostheses is the osseointegrated implant retained silicone prosthesis. Implant retained maxillofacial prostheses typically consist of multiple components, each having different physical requirements. A typical maxillofacial prosthesis will consist of the following items.

- Main body of the prosthesis to provide the anatomical form and appearance in a soft silicone material
- Rigid sub-structure to support the main body and provide a firm location for retention mechanisms

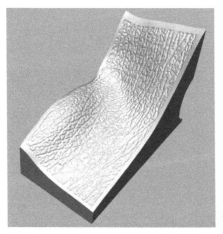

Figure 5.104 Skin texture sample in CAD and ThermoJet wax pattern.

- Retention mechanisms in the prosthesis (clips or magnets)
- Retention mechanisms attached to the patient (bar for clips or magnets)

This case investigated the use of 3D scanning, CAD and RP&M techniques in the design and manufacture of a magnet retained ear prosthesis. 3D scanning was used to capture a plaster replica of the patient's defect site and contralateral (unaffected) ear and then FreeForm CAD and ThermoJet technologies were used combined with standard fitting and finishing techniques. The resulting prosthesis showed that advanced technologies and RP&M could be used to develop a magnet retained ear prosthesis design and the case reported in 2006 (Eggbeer, Bibb, & Evans, 2006).

5.15.3.4   Case 4: auricular prosthesis B

This built on the experience gained in the previous case but addresses the bar and clip retention method. As previously the defect site and contralateral ear were scanned and all of the prosthesis components were designed using FreeForm CAD. This case explored the application of a number of RP&M processes in the manufacture of the various prosthesis components. As it had been shown to be successful in previous cases the ThermoJet process was used to produce a wax pattern of the main body of the prosthesis. The rigid sub-structure of the prosthesis was attempted using stereolithography (SLA, 3D Systems, USA). The design of the sub-structure incorporated the retention clips. The bar (which remains attached to the patient by the implants) is usually made from a suitable metal such as Titanium or Gold. In this case, Selective Laser Melting (SLM – then MTT Technologies Ltd., now Renishaw, UK) was utilised to produce the bar in 316L stainless steel. This study identified significant limitations in capturing sufficient detail of the implants abutments to design a precisely fitting bar. The stereolithography sub-structure proved adequate in design, rigidity and strength and was found to be very comparable to the physical properties of the light-cure acrylic materials that are typically used in current maxillofacial laboratories. However, it was readily apparent upon attaching and removing the sub-structure to the bar that the integral clips provided insufficient retention strength as they could be easily removed using very slight finger pressure. As the retention strength was so apparently low it was decided that accurate measurement of the retention force was redundant. In addition, it was also readily apparent that the service life would be insufficient for clinical use. Visible wear was apparent on the clips after fewer than 100 cycles of attachment and removal at which point the retention strength reduced to the point where the clips no longer functioned. The SLM bar demonstrated that the process had the potential to produce a clinically acceptable bar if the quality of the detail and finish could be enhanced. The SLM bar, SLA sub-structure and ThermoJet ear are shown in Figure 5.105. These findings were reported in 2006 (Eggbeer, Bibb, & Evans, 2006).

5.15.3.5   Case 5: nasal prosthesis

This case was undertaken to apply the findings of the previous cases and compare them to entirely conventional methods. The case involved the design and manufacture of a nasal prosthesis incorporating magnetic retention. To overcome the limitations of 3D scanning the implant abutments described in previous cases, a 3D scan of a plaster replica of the patient's defect site was used rather than scanning the patient directly. The design was undertaken using FreeForm and stereolithography (was used to produce a rigid sub-structure and the ThermoJet process was used to produce the main prosthesis body in wax as shown in Figure 5.106. This case demonstrated that digital methods (combined with conventional fitting and finishing) were capable of producing a clinically acceptable facial prosthesis. It was shown that there were no significant differences in the positional accuracy and reproduction of anatomical shape achieved by both techniques. The only significant difference was in the quality of the margin of the prosthesis. The margins of facial prostheses are made to be as thin as possible. This causes the silicone to become transparent and extremely flexible which allows the edge of the prosthesis to be closely adapted to the patients' skin and follow their facial contours without leaving a conspicuous gap. The digitally produced designed and manufactured prosthesis was not able to reproduce a sufficiently thin margin which led to a noticeable gap.

In addition to investigating the technical capabilities of advanced technologies, this case was also used to compare time- and cost-effectiveness. This case suggested that reductions in the overall design and construction time were possible when utilising digital techniques. The application of digital technologies in this case gave a 1 h

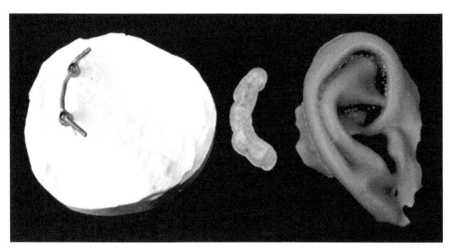

Figure 5.105 SLM retention bar, SLA sub-structure and ThermoJet wax pattern.

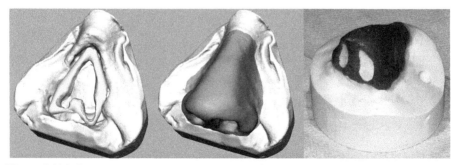

Figure 5.106 CAD designs for nasal prosthesis and ThermoJet wax pattern.

reduction in the time spent by the prosthetist and reduced the patient time spent in the clinic by 2 h and 35 min. Whilst these reductions in time do not appear hugely significant, the ability to go straight to colour matching effectively removes a much longer period where the patient must be either in clinic or waiting nearby. The waiting and in-clinic period may effectively be reduced from a day to a single morning of work.

Reduction in consultation and laboratory time was however offset by the additional stages of scanning and RP fabrication. Whilst this extra time does not require an operator, clinician or patient input, it does add to the overall delivery time. RP fabrication is typically an overnight process with an additional delay for postage if built by a service provider. The 40-min period required to set up scans and process the data were represented by four block periods: three, 5-min sessions to set the individual scans up and a final 20-min period to process the data. This effectively meant that the operator had to attend to the scanner despite being able to carry on with other work. The same situation is reflected for the patient and prosthetist during periods between construction stages, such as material curing and boiling out of moulds.

5.15.3.6   Case 6: direct manufacture of retention bar experiment

This experiment addressed only the design and direct manufacture of a prosthesis retention bar. To overcome the accuracy issues when capturing the implant abutments this experiment utilised a touch-probe scan of a plaster cast of a patient's defect site (Roland Pix-30 – Roland ASD, 25,691 Atlantic Ocean Drive, B-7, Lake Forest, CA 92630, USA). The design of the bar was undertaken using FreeForm and the bar produced directly using SLM. Unlike the previous attempts to produce a bar using SLM this bar was produced on an SLM machine specifically designed to produce small, accurate parts (SLM-100 – then MTT Technologies Ltd., now Renishaw). The resulting bar proved to be a good fit when it was screwed on to the abutments located in the replica plaster cast as shown in Figure 5.107. Accuracy and surface finish of the bar were deemed acceptable for clinical use by the prosthetists.

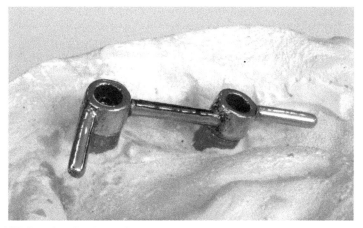

Figure 5.107 SLM bar fitted to patient cast.

5.15.4   Discussion

The case studies were analysed in three areas: quality, economic impact and clinical implications. Quality refers to characteristics such as fit (marginal integrity of the prosthesis against the skin and fit between components), accuracy (anatomical shape and position), resolution (reproduction of folds, wrinkles, texture, and sub-structure components), and materials (mechanical and chemical properties).

5.15.4.1   Fit

The margins produced using traditional techniques were measured to range from 40 to 130 μm using a dial test indicator. Whilst some RP technologies are capable of producing very thin layers, most could not produce parts in a material that could be readily incorporated into the workflow of the maxillofacial prosthetist. The vast majority of maxillofacial prosthetic sculpting is carried out using wax. Currently available RP processes that are capable of sufficiently thin layers include Objet 3D printing (Objet Geometries Ltd.), Perfactory (EnvisionTEC), Solidscape (Solidscape Inc.), ThermoJet (obsolete, 3D Systems, USA) and ProJet (3D Systems, USA). Of these, only the Solidscape, ThermoJet and ProJet systems are capable of building in a wax-based material that could be incorporated into conventional lab techniques. The Solidscape machine was deemed too slow for the size and mass of the typical facial prosthesis. Although the ThermoJet process proved capable of producing patterns in an appropriate wax material and produced parts in 40-μm thin layers, limitations with the CAD process and fragility of the wax meant that patterns were fragile and very thin edges were prone to breaking when support structures were removed. To produce sufficiently thin margins, the edges of the wax RP pattern were heated and blended into the lower section of the dental stone mould using flexible metal sculpting tools.

Despite some reports in the literature there is no recognised standard method of assessing the fit between maxillofacial prosthesis retention components, e.g. between bar and implant abutments (Brånemark, 1983; Jemt, 1991; Kan, Rungcharassaeng, Bohsali, Goodacre, & Lang, 1999). Clinical methods often rely on identifying gaps by visual inspection and finger pressure tests for movement. The cases reported here suggested that the SLM process was capable of producing a satisfactory bar but that the optical scanning methods were not able to capture data of sufficient quality directly from the patient to enable the design of the bar.

5.15.4.2   Accuracy

The accuracy of a maxillofacial prosthesis is essentially a subjective visual assessment. The prosthesis should be convincing in restoring the appearance of the particular individual. Therefore it is very difficult to assess the accuracy using a quantifiable system. The approach taken in these studies has been to assess the accuracy of the prostheses created using RP&M techniques to those produced using entirely conventional techniques. In case 5, a five-point scale was used to rate the digital and conventionally produced prostheses in terms of quality of edge, positional accuracy and overall shape.

Thirteen clinical staff members from the maxillofacial unit at Morriston Hospital were asked to rate each of the prostheses based upon photographs provided of each prosthesis. All were blinded to the production methods used (none were involved in providing these prostheses) and the images were provided sequentially. Table 5.4 shows the results. The results were analysed using a paired, Student t-test ($p = 0.05$) to identify the significance between the results.

In terms of edge quality, there was statistical significance in opinions between the two prostheses ($p = 0.003695$) in favour of the conventional prosthesis. The average score for the conventional prosthesis was 3 (standard deviation [SD] = 1.35). The average score for the digitally produced prosthesis was 1.8 (SD = 0.93). For positional accuracy, there was no statistical significance in opinions between the two prostheses ($p = 0.179533$). The average score for the conventionally produced prosthesis was 4.1 (SD = 0.76). The average score for the digitally produced prosthesis was 3.5 (SD = 1.2). For shape, there was no statistical significance in opinions between the two prostheses ($p = 0.064649$). The average score for the conventionally produced prosthesis was 4 (SD = 1.15). The average score for the digitally produced prosthesis was 3.2 (SD = 1.1). In this respect the RP methods used were deemed capable of producing an accurately formed and located maxillofacial prosthesis.

5.15.4.3   Resolution and texture

Whilst the 3D scanners investigated in these cases proved able to accurately capture overall anatomical form very well the case studies demonstrated that 3D scanning technologies are not yet able to capture data of a sufficiently detailed nature to enable the reproduction of delicate skin folds, wrinkles and texture. This limitation was also apparent when attempting to capture the precise shape and location of implant abutments. The use of plaster cast replicas of patient anatomy showed that some touch-probe scanners are capable of scanning to a high enough resolution but these involve unwanted extra process steps resulting in higher costs and longer delivery times.

Case 2, the texture experiment, demonstrated that CAD packages are able to create 3D relief at an appropriate scale to produce convincing skin textures. It also demonstrated that the ThermoJet process was able to physically reproduce these textures over an anatomically shaped surface. However, whilst the experiment was successful on small samples, the associated computer file sizes were large (in the order of 100 MB). Applying texture to the surface of a large facial prosthesis with this technique may present difficulties due to the large data file sizes it produces. A large contributor to this problem is the fact that most RP&M technologies rely on the STL file format, which is very inefficient for describing highly detailed surface relief.

5.15.4.4   Physical properties

Attempts to incorporate retention clips into substructures manufactured using RP&M techniques proved unsuccessful. Whilst RP&M technologies were shown to be capable of producing a functioning retention incorporated into the rigid sub-structure to a sufficient accuracy the retention strength was poor and they wore rapidly reducing retention strength still further. As a prosthesis are likely to be applied and removed twice a day over a typical 12-month service life retentive components should be expected to provide strong retention over 1460 cycles. The RP&M retention clips proved to suffer an unacceptable loss of retention strength after fewer than 100 cycles.

Retention bars need to be small, stiff and have good surface finish to enable easy cleaning. Due to their proximity to the skin nonreactive metals are required, such as gold, cobalt-chrome, stainless steel, tantalum, nitinol or titanium. The bars produced in this study using 316L stainless steel and cobalt-chrome showed great potential and indicated that with optimised build parameters and minimal finishing a clinically acceptable bar can be produced using RP&M techniques.

To date silicone elastomer rubber has the most suitable physical properties for maxillo-facial prostheses and can be colour matched to a produce a highly convincing appearance (Aziz, Waters, & Jagger, 2003). Currently no RP technology is capable of producing a prosthesis in silicone rubber or with physical properties similar to those required. This remains a great challenge to the RP&M industry. In the cases reported here the optimum RP&M process available proved to be the ThermoJet process. The wax parts produced were successfully incorporated into the conventional prosthetic production process with adaptation to produce finer edges, add fine details or make minor adjustments to the shape.

5.15.4.5 Economics

The economic impact of changes in maxillofacial prosthetics is difficult to quantify because each case is unique and costs vary regionally. In the United Kingdom, the situation is further complicated as some costs are direct, such as materials, whilst many of the greater costs are hidden in overheads and fixed costs, such as clinical staff salaries.

Digital techniques can potentially have a significant impact on the costs effective-ness of maxillofacial prosthesis delivery. The research undertaken, and illustrated in the time savings measured in the nasal prosthesis case study, explored the impact of the digital workflow compared to the traditional practice. As such, it is difficult to identify the individual impact that the RP&M technologies can have unless they are incorporated as part of a well resolved digital technology based workflow.

The time savings indicated in Tables 5.2 and 5.3 identified that there was poten-tial for savings in direct costs and opportunity costs. Whilst the productivity of rapid prototyping technologies cannot be objectively assessed from a single case and there would be a learning curve associated with moving to a new procedure, the fundamen-tal differences between a digital workflow and traditional techniques enables a more flexible approach to workflow management and reduces the necessity of patient atten-dance at clinics. Data acquisition can be undertaken rapidly and efficiently in a clinical setting. This enables the design stages to be scheduled and undertaken at the discretion of the prosthetist without requiring the patient to be present. The reduction in both the duration of patient attendance and the number of occasions they are required to attend would result in significant savings in travel, accommodation and missed appointments (referred to as 'did not attends'). In addition, the saving to the patient in terms of time away from home or work is also beneficial.

Table 5.2 **The time taken to construct the nasal prosthesis using conventional methods**

Stage	Prosthetist time (min)	Patient time (min)	Setting/curing time (min)
Initial consultation and impression taking	50	50	Included
Production of stone replica	50	0	15
Base-plate design	40	15	0
Pattern design	195	140	0
Mould production	95	0	40
Colour match	60	60	0
Curing	0	0	75
Finishing	80	60	0
Total	9 h 30 min	5 h 25 min	2 h 10 min

Table 5.3 **Target specifications for RP&M technologies**

Stage	Prosthetist/ operator time (min)	Patient time (min)	Setting/curing/ fabrication time (min)
Initial consultation and impression taking	50	50	Included
Production of stone replica	55	0	15
Scanning and conversion to STL	40	0	120
Base-plate design	15	0	0
Pattern design	76	0	0
Pattern fabrication	15	0	180
Sub-structure fabrication	20	0	90
Mould production (assume same as conventional)	95	0	40
Colour match (assume same as conventional)	60	60	0
Curing (assume same as conventional)	0	0	75
Finishing (assume same as conventional)	80	60	0
Total	8 h, 30 min	2 h, 50 min	8 h, 40 min

In a digital workflow, the physical production of prosthetic components can be achieved in a small batch basis which would enhance the cost effectiveness of running RP&M machines in a clinical setting. However, current technologies would require a level of investment that would be difficult to justify for all but the largest and busiest prosthetic units (Table 5.4).

Table 5.4 **Responses to aspects of prosthesis quality**

	Poor		Fair		Average		Good		Excellent	
	A	B	A	B	A	B	A	B	A	B
Feature										
Edge quality	3	5	1	5	3	2	5	1	1	0
Positional	0	1	0	1	3	4	5	4	5	3
accuracy										
Shape	0	1	2	1	2	6	3	3	6	2

A, conventionally produced prosthesis; B, digitally produced prosthesis.

Clearly, future developments in this area depend on the identification of a market opportunity. Maxillofacial prosthetics is relatively small when compared with other medical sectors, which combined with the varying models of health care delivery across different nations makes estimating the size of the facial prosthetics market extremely difficult. However, the demand for facial prosthetics is increasing with the improved detection and surgical intervention of head and neck cancer. A 2006 survey conducted by Watson et al. (Watson, Cannavina, Stokes, & Kent, 2006) found that maxillofacial prosthetists in 50 hospitals in the United Kingdom produced 4259 prostheses annually. This includes other work typically undertaken by maxillofacial prosthetists such as breast, nipple, hand and finger prostheses.

Whilst the United Kingdom enjoys a comprehensive of maxillofacial prosthetics service through the National Health Service other nations that rely on health care insurance have lower levels of provision. However even taking that into account it would be reasonable to anticipate similar levels of activity throughout countries in Western Europe, North America, Japan, Australia and New Zealand. Using UK figures, a crude estimate based on population sizes would suggest that more than 64,000 facial prostheses are made each year in the wealthiest nations. However, there is potentially enormous demand for facial prostheses throughout the developing world that is currently unmet. The developing world could benefit greatly from rapid, low cost methods of providing many thousands of facial prostheses. It is therefore reasonable to suggest that a potentially valuable global market could be developed for RP processes dedicated to the manufacture of facial prostheses.

5.15.5   RP&M Specification

This research analysed current best practice in prosthetic design and manufacture to identify key characteristics of prostheses. These characteristics were used to identify target specifications for digital technologies that would meet the particular needs of maxillofacial prosthetics. It is anticipated that dissemination of these target specifications will enable the RP&M industry to adapt or develop processes and machines specifically for the rapid and cost-effective production of maxillofacial prosthetics. Table 5.5 contains the target specifications for RP&M technologies.

Table 5.5 **Target specifications for RP&M technologies**

Bar structure	
Material	Stiffness approximately equal to or greater than 18 carat (75% gold) >75 GPa
	Suitable to polish
	Biocompatible
Resolution	Sufficient to build 1.3-mm-diameter holes with sharp detail
Pattern	
Wax material	Softening temperature in the order of 35–43 °C; melt point approximately 60–63 °C; 0% flow at 23 °C as per ISO standard 15854:2005; on the order of 25–30% flow at 37 °C (specification of Anutex wax by Kemdent)
Resolution	Equal to or better than a ThermoJet printer 300 × 400 × 600 dpi at 40-μm layer thickness
Sub-structure/clips	
Resolution	Equal to or better than an Objet or per factory; Objet = 600 × 300 × 1600 dpi, per factory = 90-μm minimum pixel size; 15-μm minimum layer thickness
Material	Wear/fatigue resistant (approximately 1460 cycles to represent 1 year of use).
	Clips approximately 150 kg/mm^2 hardness equivalent 18 carat gold
	Able to bond to the prosthesis body
	Resist hot water at 90 °C during mould release
	Hydrophobic (will not soften in the sustained presence of moisture/body fluids) water sorption equal to or less than 0.6 mg/cm^2 (that of heat-processed acrylic)
Other	Clip strength should be adjustable
Prosthesis body	
Colour	Production process capable of creating millions of colours from digital colour matches of a patient's skin wrapped around a CAD model
Resolution	Equal to or better than a ThermoJet printer 300 × 400 × 600 dpi at 40-μm layer thickness
Material	A20-30 shore hardness, >500% elongation at break, >16 kN/m tear strength, 4.8 N/mm^2 tensile strength
Environment	Degradation resistant to ultraviolet light, dirt and body secretions

5.15.6 Conclusions

This research has highlighted the fact that RP&M technologies have not been developed specifically towards the needs of maxillofacial prosthetics. However, whilst much research has demonstrated the potential effectiveness of RP&M technologies has in maxillofacial prosthetics the research described here has indicated that this potential cannot be fully exploited by currently available technologies. The full benefits of digital technologies will only be achieved through the adoption of an appropriately

<mark>254   Medical Modelling</mark>

devised, implemented and evaluated workflow. In addition, RP&M technologies need to be developed to address the specific materials and process requirements of the field of maxillofacial prosthetics.

References

Aziz, T., Waters, M., & Jagger, R. (2003). Analysis of the properties of silicone rubber maxillofacial prosthetic materials. *Journal of Dentistry, 31*, 67–74.

Bibb, R., Freeman, P., Brown, R., Sugar, A., Evans, P., & Bocca, A. (2000c). An investigation of three-dimensional scanning of human body surfaces and its use in the design and manufacture of prostheses. *Proceedings of the Institution of Mechanical Engineers. Part H, Journal of Engineering in Medicine, 214*(6), 589–594.

Brånemark, P. I. (1983). Osseointegration and its experimental background. *Journal of Prosthetic Dentistry, 50*, 399–410.

Chandra, A., Watson, J., Rowson, J. E., Holland, J., Harris, R. A., & Williams, D. J. (2005). Application of rapid manufacturing techniques in support of maxillofacial treatment: evidence of the requirements of clinical application. *Proceedings of the Institution of Mechanical Engineers, Part B: Journal of Engineering Manufacture, 219*(6), 469–476.

Cheah, C. M., Chua, C. K., & Tan, K. H. (2003). Integration of laser surface digitizing with CAD/CAM techniques for developing facial prostheses. Part 2: development of molding techniques for casting prosthetic parts. *International Journal of Prosthodontics, 16*(5), 543–548.

Cheah, C. M., Chua, C. K., Tan, K. H., & Teo, C. K. (2003). Integration of laser surface digitizing with CAD/CAM techniques for developing facial prostheses. Part 1: design and fabrication of prosthesis replicas. *International Journal of Prosthodontics, 16*(4), 435–441.

Chen, L. H., Tsutsumi, S., & Iizuka, T. (1997). A CAD/CAM technique for fabricating facial prostheses: a preliminary report. *International Journal of Prosthodontics, 10*(5), 467–472.

Chua, C. K., Chou, S. M., Lin, S. C., Lee, S. T., & Saw, C. A. (2000). Facial prosthetic model fabrication using rapid prototyping tools. *Integrated Manufacturing Systems, 11*(1), 42–53.

Coward, T. J., Watson, R. M., & Wilkinson, I. C. (1999). Fabrication of a wax ear by rapid-process modelling using stereolithography. *International Journal of prosthodontics, 12*(1), 20–27.

Eggbeer, D. (2008). *The computer aided design and fabrication of facial prostheses (PhD thesis)*. Cardiff, UK: University of Wales Institute Cardiff.

Eggbeer, D., Bibb, R., & Evans, P. (2006). Assessment of digital technologies in the design of a magnetic retained auricular prosthesis. *Journal of Institute of Maxillofacial Prosthetists and Technologists, 9*, 1–4.

Eggbeer, D., Bibb, R., & Evans, P. (2006). Towards identifying specification requirements for digital bone anchored prosthesis design incorporating substructure fabrication: a pilot study. *International Journal of Prosthodontics, 19*(3), 258–263.

Eggbeer, D., Evans, P., & Bibb, R. (2006). A pilot study in the application of texture relief for digitally designed facial prostheses. *Proceedings of the Institution of Mechanical Engineers. Part H, Journal of Engineering in Medicine, 220*(6), 705–714.

Evans, P., Eggbeer, D., & Bibb, R. (2004). Orbital prosthesis wax pattern production using computer Aided design and rapid prototyping techniques. *Journal of Maxillofacial Prosthetics & Technology, 7*, 11–15.

Jemt, T. (1991). Failures and complications in 391 consecutively inserted fixed prostheses supported by Brånemark implant in the edentulous jaw: a study of treatment from the time
of prosthesis placement to the first annual check-up. *International Journal of Oral and
Maxillofacial Implants*, *6*(3), 270–276.

Kan, J. Y., Rungcharassaeng, K., Bohsali, K., Goodacre, C. J., & Lang, B. R. (1999). Clinical methods for evaluating implant framework fit. *Journal of Prosthetic Dentistry*, *81*(1),
7–13.

Lemperle, G., Holmes, R. E., Cohen, S. R., & Lemperle, S. M. (2001). A classification of facial
wrinkles. *Plastic Reconstructive Surgery*, *108*(6), 1735–1750.

Reitemeier, B., Notni, G., & Heinze, M. (2004). Optical modeling of extraoral defects. *Journal
of Prosthetic Dentistry*, *91*(1), 80–84.

Sykes, L. M., Parrott, A. M., Owen, P., & Snaddon, R. (2004). Applications of rapid prototyping
technology in maxillofacial prosthetics. *International Journal of Prosthodontics*, *17*(4),
454–459.

Verdonck, H. W. D., Poukens, J., Overveld, H. V., & Riediger, D. (2003). Computer-assisted
Maxillofacial Prosthodontics: a new treatment protocol. *International Journal of Prosthodontics*, *16*(3), 326–328.

Watson, J., Cannavina, G., Stokes, C. W., & Kent, G. (2006). A survey of the UK maxillofacial
laboratory service: Profiles of staff and work. *British Journal of Oral & Maxillofacial
Surgery*, *44*(5), 406–410.

Wolfaardt, J., Sugar, A., & Wilkes, G. (2003). Advanced technology and the future of facial
prosthetics in head and neck reconstruction. *International Journal of Oral Maxillofacial
Surgery*, *32*(2), 121–123.

Yin, R. K. (2003). *Case study research: Design and methods* (3rd ed.). London, UK: Sage
Publishing.

5.16   Maxillofacial rehabilitation case study 6: evaluation of direct and indirect additive manufacture of maxillofacial prostheses using 3D printing technologies

Acknowledgements

This work was first reported in the reference below and is reproduced here with kind permission of Sage Publishing.

Evaluation of direct and indirect additive manufacture of maxillofacial prostheses, Proceedings of the Institution of Mechanical Engineers, Part H, Journal of Engineering in Medicine, **2012;** 226(9): **718–728,** ISSN: 0954-4119, http://dx.doi.org/10.1177/0954411912451826.

5.16.1   Introduction

Increasing patient numbers, the need to improve process efficiency, the desire to add value to the profession and the lack of access to facial prostheses provision in some areas of the world has led researchers to investigate the potential benefits of computer-aided technologies. Technologies such as 3D surface scanning, CAD and RP/AM have been applied in a number of research cases yet they are not in widespread clinical application. Within the published literature, computer-aided technologies have been employed in different ways, the most common method being to digitise pattern design and incorporate this into conventional mould and final prosthesis production (Chandra et al., 2005; Cheah, Chua, Tan, & Teo, 2003; Chen, Tsutsumi, & Iizuka, 1997; Chua, Chou, Lin, Lee, & Saw, 2000; Coward, Watson, & Wilkinson, 1999; Eggbeer, Bibb, & Evans, 2006; Eggbeer, Evans, & Bibb, 2006; Evans, Eggbeer, & Bibb, 2004; Reitemeier, Notni, & Heinze, 2004; Runte et al., 2002; Sykes, Parrott, Owen, & Snaddon, 2004; Verdonck, Poukens, Overveld, & Riediger, 2003). These methods rely on laboratory-based methods to produce moulds or time-consuming techniques such as vacuum casting. Computer-aided mould tool production has also been attempted (Cheah, Chua, & Tan, 2003); however, the techniques presented were not sympathetic to the skills of maxillofacial prosthetics, prosthodontics or anaplastologist professions and were not able to address the subtlety of design that makes a facial prosthesis realistic.

Previous research has predominantly been reports of individual cases and has not attempted critical evaluation. Consequently, they do not provide robust evidence to support or dismiss either the clinical efficacy or cost effectiveness of computer-aided technologies. Such evidence is recognised as critical to the adoption of a high value, technology-based approach (Wolfaardt, Sugar, & Wilkes, 2003). A more recent, comprehensive review concluded that 'the full benefits of digital technologies will only be achieved through the adoption of an appropriately devised, implemented and evaluated work flow' (Bibb, Eggbeer, & Evans, 2010).

Furthermore, silicone elastomer is proven in clinical application, material characteristics have been identified (Aziz, Waters, & Jagger, 2003), but how this compares to currently available RP/AM materials has not been explored.

This paper addresses the limitations of previous research and considers three aspects of evaluating the effectiveness of computer-aided technologies in facial prosthesis production: workflow, aesthetic outcome and material characteristics.

Computer-aided methods were evaluated through a case study and controlled experiments to ISO standards. Through consultation with the prosthetist undertaking the case, criteria for aesthetic evaluation were established and barriers encountered in previous research were considered. This helped to ensure an appropriate and intuitive process with outcomes that could be evaluated. This paper also compares the physical properties of an AM material that could be used in the fabrication of facial prosthetics with those of a benchmark silicone commonly used in facial prosthesis production.

5.16.2　Methods

A magnet-retained nasal prosthesis case was chosen as a case study to compare computer-aided with conventional methods. The patient had undergone a rhinectomy (total nose removal) following cancer and had been wearing a prosthesis for 2 years before this revision. Informed patient consent was obtained and the study was undertaken as part of routine treatment to provide a new prosthesis. This ensured that minimal additional procedures were required. The processes are illustrated in Figure 5.108.

On reviewing available AM technologies, it was apparent that no existing technology is capable of producing a realistic, detailed, coloured facial prosthesis directly. Therefore, an analysis of existing workflows revealed two possible applications of AM that could potentially improve the efficiency of prosthesis construction, whilst maintaining a viable outcome:

1. Direct AM prosthesis production of the body of the prosthesis from a digital design, which could then be wrapped in a very thin layer of colour-matched, detailed silicone
2. Indirect production of the prosthesis body in a colour matched silicone by moulding in a mould produced from a digital design and made using AM

Both approaches relied on 3D photogrammetry to capture patient anatomy data and FreeForm CAD for the initial design of the prosthesis form. A benchmark prosthesis was also fabricated by a highly skilled, chief maxillofacial prosthetist with 22 years' experience using conventional methods based on recognised best practice and published literature (Thomas, 2006). This is shown in Figure 5.109.

5.16.2.1　Common stage patient scanning and design

A base plate was fabricated using light cure acrylic material (TranSheet, Dentsply, York, USA) on the original replica cast. This was also coated in a grey paint to aid 3D topography capture.

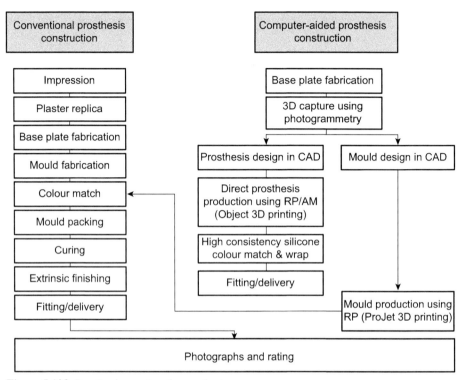

Figure 5.108 Prosthesis construction methods.

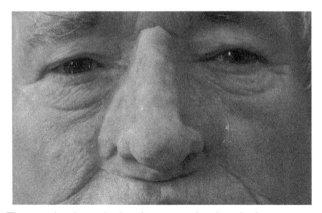

Figure 5.109 The completed prosthesis using conventional methods.

Photogrammetry (3DMD, Face Capture system, Atlanta, GA) was used to capture the facial topography with the base plate in position and with the patient's original prosthesis (since they were happy with the shape). The proprietary 3D Patient (3DMD) software used to create a STereoLithography (STL) file of the two data sets. Figure 5.110 shows the resulting mesh structure around the nasal area.

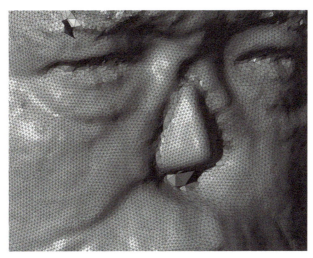

Figure 5.110 The STL file polygon mesh data of the baseplate and surrounding anatomy.

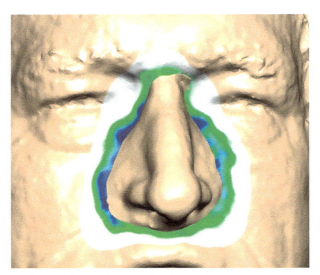

Figure 5.111 Creating relief around the prosthesis margin.

The STL data were imported into the CAD software, FreeForm Modelling Plus (Geomagic, 3D Systems, USA) as 'digital clay' models with 0.2-mm edge sharpness using the 'hole filling' option. FreeForm has previously been shown to be an appropriate software application for the CAD of prostheses (Eggbeer, Bibb, & Evans, 2007; Sykes et al., 2004; Verdonck et al., 2003).

Areas of the face where the prosthesis margins required positive pressure to form a blended seal with the skin were adjusted using the 'smudge' tool to press and reduce the thickness of skin by 1–2 mm. The 'tolerance map' tool was used to gauge the depth of the modification (Figure 5.111).

The digital face was then converted to a 'buck' model, which protects it from further modification. Areas of 'clay' representing air voids were then built up and the model turned into a 'buck' which prevents further unwanted modification (Figure 5.112).

The original prosthesis form was used as the basis for the new version. This was modified to reintroduce nostrils, blend in to the surrounding anatomy and had texture details added using techniques previously described (Eggbeer et al., 2006). The final result is shown in Figure 5.113.

5.16.2.2 Indirect approach: mould design

A copy of the completed design was made to assist in mould design. The 'buck' model of the face was subtracted in a Boolean operation, leaving the prosthesis form. The main volume of the form was selected, then 'inverse selection' and

Figure 5.112 Area of cavity inside the prosthesis.

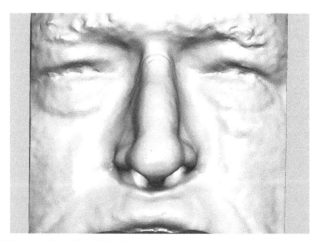

Figure 5.113 The complete digital design with texture.

'delete' were used to remove any stray unattached pieces of 'clay'. The edge sharpness was refined and smoothed to 0.18 mm, which improved the smoothness of the margins. The digital prosthesis form is shown in Figure 5.114. The 'reduce for export' option was used to reduce the STL file to 50 MB to make it manageable by a modern desktop computer.

Mould design was then undertaken in FreeForm. The copy of the prosthesis joined to the face was used to create the outer mould surface. The nostrils were blocked off with 'clay' to avoid major undercuts. A copy of the completed design was then made. A line representing the mould edge was drawn on the surface of the model around 15 mm offset from the prosthesis margin. The 'emboss with curve' tool was used to create an overlaying shell with a thickness of 2.5 mm on the model copy. The original version of the design was then used to perform a Boolean subtraction, leaving just the outer shell representing the top section of the mould. A flat section around the centre of the mould was created to allow the mould to be clamped together.

The lower mould section was created on the face model by building up 'clay' material to represent the internal cavity and leaving a cavity into which the silicone prosthesis would be formed. The design was 'shut off' with the outer mould at the nostril openings and the area around the base plate was kept clear to allow silicone to encase them. Care was taken to avoid large undercuts that would prevent the prosthesis from being released from the mould. An area of the face around the nose was selected to

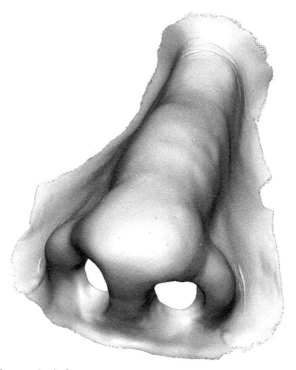

Figure 5.114 The prosthesis form.

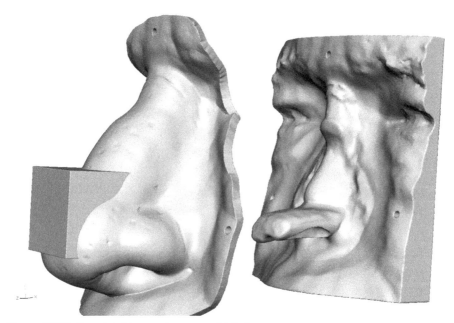

Figure 5.115 Completed two-part nose mould design.

form the rear section of the mould tool. This was shelled to 2.5-mm thick and the rear section removed to reduce material and therefore cost. A flat area corresponding to that on the outer shell was also created to allow the mould to be clamped together. The final mould design is shown in Figure 5.115.

5.16.2.3   Indirect approach: additive manufacture

The two mould sections were fabricated using 3D printing (ProJet HD 3000 Plus, 3D Systems, USA) in Xtreme High Definition mode (the resolution of the machine in this mode is given as $750 \times 750 \times 1600$ DPI (x–y–z) and the layer thickness is 16 μm). This provided sufficient detail to reproduce the textures created in the computer model and a smooth surface finish. The build took 16 h, 40 min. Once complete, the wax supporting material was removed from the mould halves (90 min at 80 °C in a temperature-controlled oven) and they were cleaned and grit blasted to create a smooth, matt surface finish (approximately 30 min of manual labour).

The 3D printed mould was used to complete the prosthesis body by moulding silicone in a manner similar to conventional methods. A base shade colour-matched silicone (Reality Series, Spectromatch Ltd., UK) was first mixed and used as the basis for creating different tones and shades that matched the surrounding skin. The inner surface of the front mould was painted with a mixture of base shade and variations to match the local surrounding anatomy colours and flocking to mimic capillaries before being packed out with the base shade. The mould was then closed and clamped ready for curing (Figure 5.116). Figure 5.117 shows the final result.

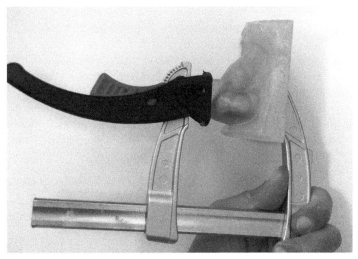

Figure 5.116 Mould clamped together.

Figure 5.117 Completed prosthesis from the RP-fabricated mould.

5.16.2.4 Direct approach

The direct prosthesis body production method used 'Polyjet modelling' 3D printing process (Objet Connex 500, Objet Geometries, Rehevot, Israel) in a soft, transparent, acrylate-based material (TangoPlus – an Objet trade name) with a specified Shore hardness of 26-A. This was the only available AM process capable of producing objects in a soft material with similar physical properties to the silicone rubber typically used in prosthesis production. However, at the time of the study the material was not approved for clinical application or undergone skin sensitivity trials.

The directly manufactured prosthesis body resulted in a clear transparent form. Therefore a novel method was required to produce a realistic colour matched surface. High Consistency HC20 silicone (Technovent Ltd, Newport, UK) was mixed with base shade and flocking. This was mill rolled to create a thin pliable sheet approximately 0.4 mm thick. This was wrapped around the prosthesis pattern which had been pre-coated with G604 Primer (Technovent Ltd) used to form a strong adhesive bond. Another layer of silicone sheet was then wrapped over to create a stronger colour and the edges blended out by pressing against a hard surface with a metal sculpting tool. This was cured at 60 °C for 3 h. The final result is shown in Figure 5.118.

5.16.2.5   Qualitative rating of aesthetic outcome

Of the three production methods, only two were judged clinically viable and worth rating in terms of aesthetic quality: the conventionally produced prosthesis and CAD/AM mould version.

Photos looking straight on, at a 45° angle and side on were taken to record the aesthetic result of each prosthesis. The set of three images for each prosthesis was printed on separate sheets for the reviewers to rate the results. Nineteen people who worked within the hospital unit in other dental specialties, but were blinded to the design and construction methods used were asked to evaluate the prostheses. A Likert five-point rating scale was used to evaluate four aspects of prosthesis appearance: positional accuracy, shape, colour and quality of edge. A Student t-test (two tails, type 2, $p = 0.05$) was also used to identify the significance between the results.

5.16.2.6   Material testing

Since no specific standard yet exists for evaluating the performance of AM-produced samples, mechanical testing was undertaken as a pilot study to provide a benchmark. Three

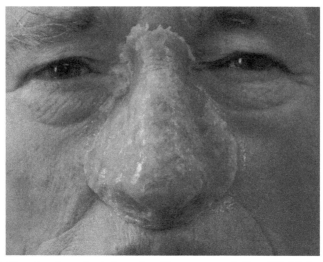

Figure 5.118 Completed prosthesis from the silicone wrapped direct RP-fabricated pattern.

aspects of mechanical performance were tested: Tensile testing, elongation at break (ISO 37:2005) and tear strength (ISO 34-1:2010). Each test was repeated five times for each material at room temperature (approximately 21 °C). Three-dimensional computer models were made of the test specimens using computer-aided design software (ProEngineer Wildfire 4, PTC, Needham, MA, USA) according to the dimensions specified in the ISO standards and shown in Tables 5.6 and 5.7. The CAD files were exported as STL files (Figures 5.119 and 5.120) suitable for AM. All of the samples were produced in a single build using an Objet Connex 500 in TangoPlus material (Objet Ltd., Rehovot, Israel).

Table 5.6 **Dimensions of the tensile test bar**

Overall length A/mm	Width of ends B/mm	Length of narrow portion C/mm	Width of narrow portion D/mm	Transition radius outside E/mm	Transition radius inside F/mm
115	25.0 ± 1	33 ± 2	6 ± 0.4	14 ± 1	25 ± 2

Table 5.7 **Dimensions of the tear test strip**

Overall length A/mm	Width of ends B/mm	Length of cut C/mm
≥100	15 ± 1	40 ± 5

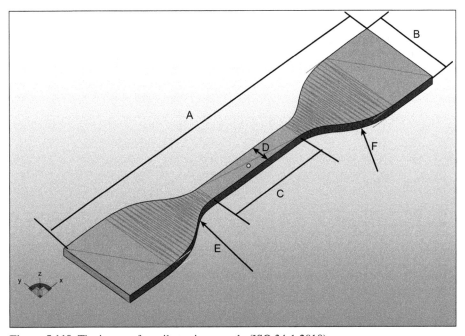

Figure 5.119 The image of tensile testing sample (ISO 34-1:2010).

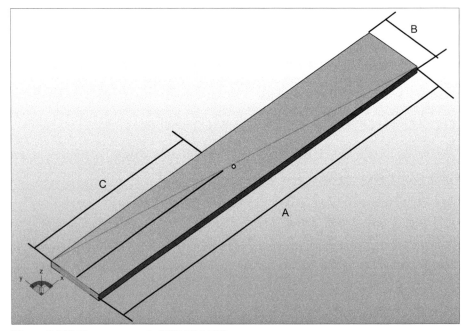

Figure 5.120 Tear test sample (ISO 37:2005).

For the benchmark comparison, test specimens were produced from a noncoloured silicone rubber (M511, Technovent Ltd., Newport, UK). This is the same chemistry of silicone that was used for the traditionally manufactured silicone prosthesis, but sold by a different company and was not precoloured. Test specimens were cut from a cast sheet of the silicone by mechanical die cutter to the same ISO specifications.

Tensile testing and elongation at break were ascertained using a Lloyds LR50KPlus testing machine set with a 1-kN load cell and an elongation speed of 500 mm/min. The same Lloyds testing machine was used to establish tear strengths; a 1-kN load cell was used and the specimens were fixed into two crossheads and torn at the speed of 100 mm/min.

5.16.3   Results

5.16.3.1   Aesthetic outcomes

The results of the Likert-scale ratings are shown in Tables 5.8 and 5.9.

The student t-test results are shown in Table 5.10.

5.16.3.2   Mechanical properties

The calculation of tensile strength is:

$$T_s = \frac{F_m}{Wt}$$

where T_s—tensile strength (MPa), F_m—maximum force (N), W—width of the gauge section (mm), t—thickness of the test length (mm). The average dimensions, maximum force and tensile strength values of selected materials are shown in Table 5.11.

Elongation is defined as the increase of the length of narrow portion (ΔL) subjected to a tension force, divided by the original length of the test sample (L). It also can be named as strain (ε). The calculation of strain is shown as:

$$\varepsilon = \frac{\Delta L}{L} \times 100\%$$

The average values of elongation at break for TangoPlus and M511 silicone rubber material are shown in Table 5.12.

Table 5.8 Mean average ratings and standard deviations of aesthetic quality for the conventionally produced prosthesis

Conventional	Mean	SD
Position	3.158	1.068
Shape	2.474	1.219
Colour	3.000	1.202
Edge	1.947	1.129

Table 5.9 Mean average ratings and standard deviations of aesthetic quality for the RP mould produced prosthesis

RP mould	Mean	SD
Position	3.842	0.834
Shape	4.000	0.745
Colour	3.842	0.688
Edge	3.526	0.772

Table 5.10 Student *t*-test results identifying significance between the rated aesthetic outcomes of each prosthesis

Aesthetic factor	Significance
Position	0.034241
Shape	0.000043
Colour	0.011873
Edge	0.000014

Table 5.11 **Average dimensions, maximum force and tensile strength values of TangoPlus and M511 silicone**

Testing sample	Average maximum force (Fm)/N	Average width (W)/mm	Average thickness (t)/mm	Average tensile strength (Ts)/MPa
TangoPlus	19.05	6.02	2.99	1.1
M511 maxillofacial silicone rubber	61.46	6.01	3.00	3.4

Table 5.12 **Average values of elongation at break for TangoPlus and M511 silicone rubber**

Test sample	Elongation at break (%)
TangoPlus	365.36
M511 maxillofacial silicone rubber	1181.87

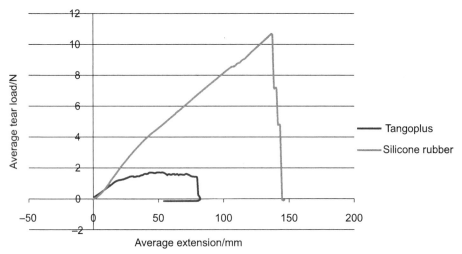

Figure 5.121 Tear load versus extension for M511 silicone and TangoPlus.

The calculation of tear strength is shown as follows:

$$T_s = \frac{F}{d}$$

where F—the maximum force/N, d—the median thickness of the test piece/mm. The curves of tear load versus extension are shown in Figure 5.121. Table 5.13 shows the average tear strength of M511 silicone and TangoPlus material.

Table 5.13 **Average tear strength of M511 silicone and TangoPlus material**

Testing sample	Average tear strength N.mm^{-1}
TangoPlus	0.897
M511 maxillofacial silicone rubber	5.471

5.16.4 Discussion

The qualitative rating results indicated a high degree of confidence that the digital mould design and production process created an aesthetically acceptable prosthesis. The results demonstrate a significant difference in favour of the AM mould prosthesis over the conventional version in all aspects, but especially in edge quality. Given that the final prosthesis was also produced in a suitably biocompatible material, with demonstrated clinical acceptance, it provided a prosthesis that was fit for purpose and viable for use. The shape was also rated significantly better on the digital version, perhaps due to the ability to analyse and adjust it from viewing angles that are more difficult to achieve when observing a patient sat in a chair (Figure 5.122).

From a process perspective, computer-aided technologies improved flexibility of working for the prosthetist. Design work was undertaken independently of the patient and mould production was semiautomated. The computer-aided mould technique described also has the potential to remove a full day of clinic and waiting between processes for the patient. Conventional stages of taking a physical impression and hand carving were condensed into two short consultation periods, the first to 3D capture the facial topography and the second to colour match the silicone and final fit the prosthesis.

The research presented also highlights specific limitations of using a computer-aided approach. These can be classified as process, cost, material and technical limitations. Despite demonstrated improvements in process flexibility, further, in-depth measurement of the resources and time taken to fabricate prostheses is required to accurately determine the actual cost and overall viability of computer-aided technologies. This is challenging when dealing with low case numbers and with every case being unique, but essential if a computer-based approach is to be widely adopted in clinical practice. Even with process efficiency savings, there is a significant offset in the cost of technology investment and machine time that should be considered if critically comparing the economics of each method.

Although the concept of direct AM body production was demonstrated, the mechanical testing results highlight the limitations that prevent it being used in the manufacture of a definitive prosthesis. Whilst the TangoPlus material currently represents the closest match to the physical properties of the benchmark silicone, it is not sufficiently robust. TangoPlus has a tensile strength of just 1.06 MPa and a tear strength of less than 1 N/mm, which when subjected to daily wear and tear would likely result in premature breakdown of thin wall sections. Since the AM material used

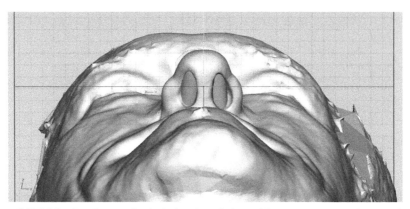

Figure 5.122 Viewing the prosthesis design from below with a measurement plane used to evaluate positioning.

is an acrylate-based, ultraviolet-curing photopolymer, prolonged exposure to ultraviolet light and other weathering may cause further degradation of mechanical properties and this should be investigated in future work.

To make the direct AM fabrication process viable, it would be necessary to additively manufacture prostheses in a suitably biocompatible material with mechanical properties closer to the benchmark silicones currently used. The ability to selectively print multiple materials and transparent materials means that, in principle, the Objet process could be expanded to also selectively print colour to produce a colour matched prosthesis body in a soft, pliable material. However, this would require a great deal of materials and process research and development.

Another aspect that requires further technical refinement is digital base plate design. Due to technology limitations of photogrammetry or any other currently available 3D surface scanning technique, it is not possible to capture the relatively large volume of the gross facial topography and the very small abutment or magnet details in the same scan with sufficient resolution and accuracy. Refinement of computer-aided techniques in base plate and retention mechanism design would enable an entirely digital prosthesis creation route. Experiments to refine suitable techniques are ongoing.

5.16.5   Conclusions

Two alternative methods of using computer-aided technologies were used to produce a facial prosthesis. The method utilising the AM-produced mould resulted in a prosthesis that was judged by experts to be clinically acceptable and rated superior to a benchmark prosthesis produced using conventional methods. Despite the potential for using AM to produce a prosthesis body directly, poor mechanical properties and untested biological responses of the chosen material currently prevent it from being used in clinical application. Further research is required to test the biological response of AM materials, improve the mechanical properties and optimise the digital design process around direct fabrication. Further work is also necessary to incorporate base plate design to create an entirely digital process.

Both computer-aided methods enabled the prosthetist to work in a more flexible manner without relying on long patient consultations. They also reduced the length of consultation time for the patient, who only had to attend for a surface scan, then a colour match on a separate day.

This research contributes towards the understanding of how computer-aided technologies may most effectively be used in clinical extra-oral prosthesis cases and provides direction to future research efforts.

References

Aziz, T., Waters, M., & Jagger, R. (2003). Analysis of the properties of silicone rubber maxillofacial prosthetic materials. *Journal of Dentistry*, *31*, 67–74.

Bibb, R., Eggbeer, D., & Evans, P. (2010). Rapid prototyping technologies in soft tissue facial prosthetics: current state of the art. *Rapid Prototyping Journal*, *16*(2), 130–137.

Chandra, A., Watson, J., Rowson, J. E., Holland, J., Harris, R. A., & Williams, D. J. (2005). Application of rapid manufacturing techniques in support of maxillofacial treatment: evidence of the requirements of clinical application. *Proceedings of the Institution of Mechanical Engineers, Part B: Journal of Engineering Manufacture*, *219*(6), 469–476.

Cheah, C. M., Chua, C. K., & Tan, K. H. (2003). Integration of laser surface digitizing with CAD/CAM techniques for developing facial prostheses. Part 2: development of molding techniques for casting prosthetic parts. *International Journal of prosthodontics*, *16*(5), 543–548.

Cheah, C. M., Chua, C. K., Tan, K. H., & Teo, C. K. (2003). Integration of laser surface digitizing with CAD/CAM techniques for developing facial prostheses. Part 1: design and fabrication of prosthesis replicas. *International Journal of prosthodontics*, *16*(4), 435–441.

Chen, L. H., Tsutsumi, S., & Iizuka, T. (1997). A CAD/CAM technique for fabricating facial prostheses: a preliminary report. *International Journal of prosthodontics*, *10*(5), 467–472.

Chua, C. K., Chou, S. M., Lin, S. C., Lee, S. T., & Saw, C. A. (2000). "Facial prosthetic model fabrication using rapid prototyping tools. *Integrated Manufacturing Systems*, *11*(1), 42–53.

Coward, T. J., Watson, R. M., & Wilkinson, I. C. (1999). Fabrication of a wax ear by rapid-process modelling using stereolithography. *International Journal of prosthodontics*, *12*(1), 20–27.

Eggbeer, D., Evans, P., & Bibb, R. (2006). A pilot study in the application of texture relief for digitally designed facial prostheses. *Proceedings of the Institution of Mechanical Engineering – Journal of Engineering Medical*, *220*(6), 705–714.

Eggbeer, D., Bibb, R., & Evans, P. (2006a). Assessment of digital technologies in the design of a magnetic retained auricular prosthesis. *Journal of Institute of Maxillofacial Prosthetists and Technologists*, *9*, 1–4.

Eggbeer, D., Bibb, R., & Evans, P. (2006b). Specifications for non-contact scanning, computer aided design and rapid prototyping technologies in the production of soft tissue, facial prostheses. In *Proceedings of the 7th national conference on rapid design, prototyping & manufacturing* (pp. 67–77). Beaconsfield: MJA Print.

Eggbeer, D., Bibb, R., & Evans, P. (2007). Digital technologies in extra-oral, soft tissue facial prosthetics: current state of the art. *Journal of Institute of Maxillofacial Prosthetists and Technologists*, *10*, 9–18.

Evans, P., Eggbeer, D., & Bibb, R. (2004). Orbital prosthesis wax pattern production using computer aided design and rapid prototyping techniques. *The Journal of Maxillofacial Prosthetics & Technology, 7*, 11–15.

Reitemeier, B., Notni, G., & Heinze, M. (2004). Optical modeling of extraoral defects. *Journal of Prosthetic Dentistry, 91*(1), 80–84.

Runte, C., Dirksen, D., Delere, H., Thomas, C., Runte, B., Meyer, U., et al. (2002). Optical data acquisition for computer-assisted design of facial prostheses. *International Journal of prosthodontics, 15*(2), 129–132.

Sykes, L. M., Parrott, A. M., Owen, P., & Snaddon, D. R. (2004). Applications of rapid prototyping technology in maxillofacial prosthetics. *International Journal of Prosthodontics, 17*(4), 454–459.

Thomas, K. (2006). The art of clinical Anaplastology. In *Chapter 4: Published by S Thomas.* UK: Printed by 4Edge Ltd.

Verdonck, H. W. D., Poukens, J., Overveld, H. V., & Riediger, D. (2003). Computer-assisted maxillofacial prosthodontics: a new treatment protocol. *International Journal of prosthodontics, 16*(3), 326–328.

Wolfaardt, J., Sugar, A., & Wilkes, G. (2003). Advanced technology and the future of facial prosthetics in head and neck reconstruction. *International Journal of Oral Maxillofacial Surgery, 32*(2), 121–123.

5.17 Maxillofacial rehabilitation case study 7: computer-aided methods in bespoke breast prosthesis design and fabrication

Acknowledgements

The work described in this case study was first reported in the references below and is reproduced here in part or in full with the permission of Sage Publishing.

Eggbeer D, Evans P, "Computer-aided methods in bespoke breast prosthesis design and fabrication", Proceedings of the Institution of Mechanical Engineers, Part H, Journal of Engineering in Medicine 2011; 225(1): 94–99, http://dx.doi.org/10.1243/09544119JEIM755.

5.17.1 Introduction

Around 38,000 women were diagnosed with breast cancer in 2006 in England and Scotland (UK National Statistics). In the 2007–2008 period in England, this resulted in 19,334 breast removals (NHS Hospital Episode Statistics; Health Solutions Wales). In England, there were 4209 breast reconstructions and 7786 breast prostheses in 2007–2008 (NHS Hospital Episode Statistics). There are various options available to reconstruct the breast including autologous reconstruction and prosthetic implant (Rozen, Rajkomer, Anavekar, & Ashton, 2009). These may require additional surgery, which may not be suitable or desirable. In the majority of these cases, brassiere-retained, external prostheses are provided.

The goal of an external prosthesis is to restore the aesthetic contours of the chest region and help to maintain good posture. The prosthesis should be comfortable to wear for long periods during the day and help to maintain good posture by counteracting the weight of the remaining breast (Glaus & Carlson, 2009). Studies have shown that satisfaction is significantly associated with how well the prosthesis fits, the weight and movement (Gallagher, Buckmaster, O'Carroll, Kiernan, & Geraghty, 2009; Livingston et al., 2005). Traditionally, prostheses are available commercially as 'off the shelf' in a range of sizes or are custom made, typically by maxillofacial prosthetists. Both are usually fabricated in a soft, skin-like silicone. The bespoke route offers an improved fit, better contour, improved colour match and therefore a more lifelike appearance. However, techniques used in the production of bespoke prostheses are time-consuming, complicated and material intensive. There is also limited literature describing the typical, laboratory-based stages involved with producing bespoke breast prostheses (Thomas, 1994).

Laboratory-based techniques also do not account for the shape of the breast when it is supported by a brassiere. There remains a need to develop an efficient method of producing well-fitted, patient-specific breast prostheses.

The introduction of computer-aided technologies such as 3D topographic scanning, photogrammetry, CAD and RP&M into patient-specific medical applications has provided new opportunities to improve the delivery of prostheses and other patient-specific medical devises (Eggbeer, Bibb, & Williams, 2005; Eggbeer, Bibb, & Evans, 2006; Evans, Eggbeer, & Bibb, 2004; Sykes, Parrott, Owen, & Snaddon, 2004; Verdonck, Poukens, Overveld, & Riediger, 2003). However, an efficient production chain for the delivery of bespoke prostheses that match the contralateral, brassiere-supported breast has not been reported. This paper introduces a technique that uses 3D photogrammetry, haptic computer sculpting and RP&M manufacturing methods to produce bespoke breast prostheses.

5.17.2   Methods

Four patients to date have had a prosthesis fabricated using the technique described, however the methods will be illustrated through a single patient case study. An overview of the digital procedure is show in Figure 5.123.

A 3dMDtorso, four-pod photogrammetry system (3dMD, USA) was used to capture the chest topography of the patient without a brassiere and with a plain, unpatterned white brassiere. The anatomy was captured pointing slightly upwards and towards the breast fold (Figure 5.124). Data acquisition took ~1.5 ms and was undertaken as part of a multipatient clinic.

Data processing took approximately 3 min and created a mesh of 363,346 and 190,441 triangles for the brassiere and non-brassiere wearing anatomy, respectively. The meshes represented surface topography of the chest area and also included a colour map, which provided sufficient detail to create the final prosthesis. Figure 5.125 shows the surface topography captured of the patient's chest with the defect.

The mesh data were exported as the de facto industry standard STL file (STereoLithography) and imported into CAD software for prosthesis design (FreeForm Modelling Plus, Geomagic, 3D Systems, USA). FreeForm CAD was chosen due to its suitability when working with anatomical data and tools analogous to conventional laboratory sculpting methods (Evans et al., 2004). The fill holes option was used to create a solid model in the CAD software and an edge sharpness of 0.4 mm chosen to provide sufficient detail. The unaffected side of the chest with the brassiere on was copied, pasted and mirrored to the defect side on the second scan (Figure 5.126).

Software carving and shaping tools were used to further shape the digital prosthesis design to fit the defect side with the brassiere off and remove the brassiere strap. Particular attention was given to the fit of the prosthesis under the arm, which is an area of particular discomfort when wearing a stock sized prosthesis. Once satisfied with the contours, a Boolean subtraction operation was undertaken to form the fitting contours and leave just the digital prosthesis design. This could be checked for correct size and position using the with brassiere data (Figure 5.127).

At this point, there were two options: (1) fabricate the prosthesis pattern using RP&M tools, then use laboratory techniques to create a mould or (2) design a tool using the CAD software and produce that using RP&M technologies. Option 2 was chosen because this further reduced the dependence on lengthy lab-based techniques.

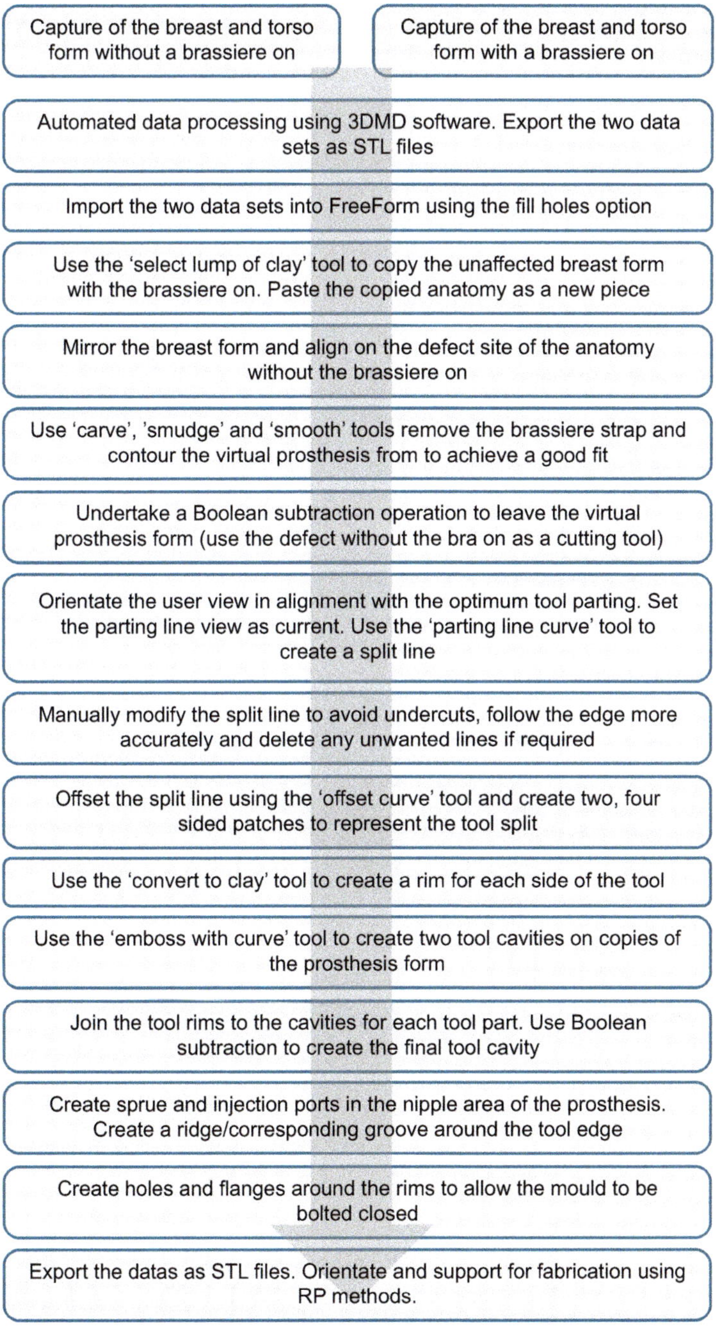

Capture of the breast and torso form without a brassiere on	Capture of the breast and torso form with a brassiere on

Automated data processing using 3DMD software. Export the two data sets as STL files

Import the two data sets into FreeForm using the fill holes option

Use the 'select lump of clay' tool to copy the unaffected breast form with the brassiere on. Paste the copied anatomy as a new piece

Mirror the breast form and align on the defect site of the anatomy without the brassiere on

Use 'carve', 'smudge' and 'smooth' tools remove the brassiere strap and contour the virtual prosthesis from to achieve a good fit

Undertake a Boolean subtraction operation to leave the virtual prosthesis form (use the defect without the bra on as a cutting tool)

Orientate the user view in alignment with the optimum tool parting. Set the parting line view as current. Use the 'parting line curve' tool to create a split line

Manually modify the split line to avoid undercuts, follow the edge more accurately and delete any unwanted lines if required

Offset the split line using the 'offset curve' tool and create two, four sided patches to represent the tool split

Use the 'convert to clay' tool to create a rim for each side of the tool

Use the 'emboss with curve' tool to create two tool cavities on copies of the prosthesis form

Join the tool rims to the cavities for each tool part. Use Boolean subtraction to create the final tool cavity

Create sprue and injection ports in the nipple area of the prosthesis. Create a ridge/corresponding groove around the tool edge

Create holes and flanges around the rims to allow the mould to be bolted closed

Export the datas as STL files. Orientate and support for fabrication using RP methods.

Figure 5.123 Illustration of the digital process tools.

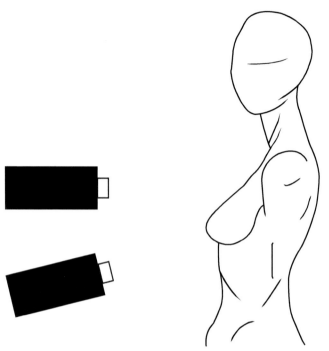

Figure 5.124 Illustration of the angle used to capture the breast contour and defect.

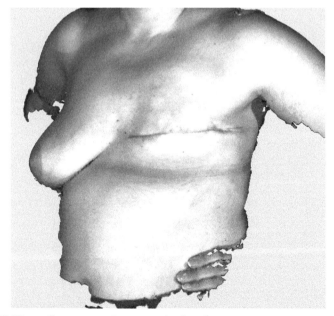

Figure 5.125 The surface topography captured using photogrammetry.

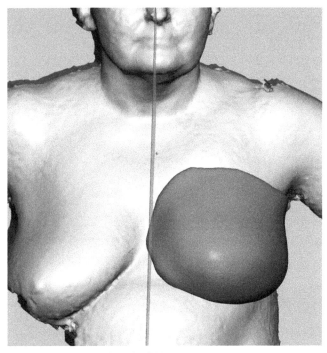

Figure 5.126 The mirrored breast form in CAD.

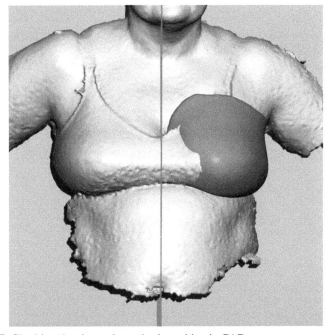

Figure 5.127 Checking the size and prosthesis position in CAD.

The parting line for the mould was first chosen and highlighted in FreeForm. This represented the most bulbous line around the rim of the prosthesis where the tool would split. This line was offset by 15 mm using the 'offset curve' tool. Two, four sided patches were created using the split line and offset split line. These represented the splitting surface of the tool. A new empty piece of clay with an edge sharpness of 0.4 mm was created for each tool part. The patches were given a thickness of 3 mm using the 'convert to clay' tool to form the rim of each tool side. The 'emboss with curve tool' was then used to create the cavities of each tool side with a thickness of 1.8 mm on a copy of the prosthesis form. The rim sections were joined to the cavities for each tool side and holes with flanges created around the rim to allow the tool to be bolted closed. A Boolean subtraction operation was then undertaken with the prosthesis form used to remove material from the tool. A groove and corresponding ridge were also created around the rim to prevent silicone from leaking around the split line. Sprue injection and vent holes were then created on the top section of the mould to allow the final prosthesis material to be injected and air to escape. Optional holes and flanges can then be created, which allow the tool to be bolted closed. The completed mould design is shown in Figure 5.128. The sections were exported as STL files.

Magics (Materialise, Belgium) was used to support the two mould sections. The sections were fabricated in 0.15-mm layers using Stereolithography (SLA 250-50, 3D Systems, USA), in WaterShed XC resin (DSM Somos, USA). This resin was chosen due to its ease of hand finishing and translucency, which would make it easier to see when the mould was fully filled with silicone. The tool sections were built on edge to allow both to fit within the build volume. The build took overnight to complete. Once completed, the sections were cleaned and hand finished with glass paper to remove the layer steps and achieve a smooth matt surface.

The mould was then passed to the maxillofacial prosthetics laboratory.

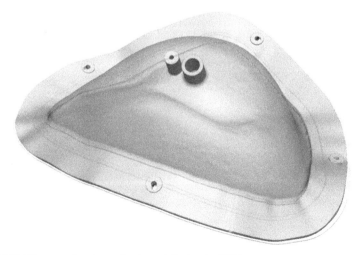

Figure 5.128 The completed prosthesis tool design in CAD.

Research has suggested that the weight of the prosthesis should match the remaining breast to prevent back and neck problems for the patient (Kovacs et al., 2007). However, all previous patients treated complained about the prosthesis weight. It was therefore decided to reduce the weight of the prosthesis from a theoretical 1100 g (if the prosthesis was entirely gel silicone–based) to a final weight of 845 g using a low-density, open-cell foam polyurethane core. This was fabricated by part filling and contouring dental plaster within the mould to create an insert. This reduced the volume of the mould and formed the desired contour for the insert. A two-part, foaming polyurethane material (Technovent Ltd, UK) was injected into the mould and left to cure.

The SLA mould was first coated with a 1-mm layer of silicone elastomer (M511, Technovent Ltd, UK), which was applied incorporating Cosmesil base colour pigments to achieve a 'skin like' colouration (Technovent Ltd, UK). The polyurethane

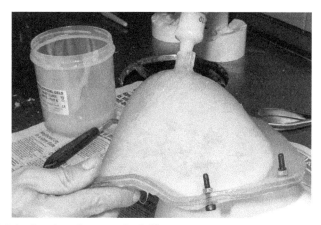

Figure 5.129 Injecting the colour-matched silicone.

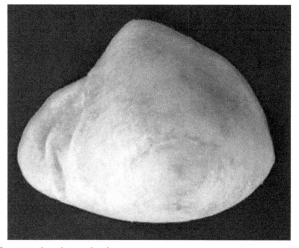

Figure 5.130 The completed prosthesis.

Table 5.14 A comparison of traditional methods and computer-aided highlighting the resource reduction and opportunities identified

Conventional	Computer-aided method	Resource reduction/ opportunity
Take an impression Create a replica model of the anatomy	Photogrammetry with bra and without bra Import data into FreeForm	Time saving Reduced material usage Improved accuracy Captures the defect and breast form in a brassiere
Manually carve wax prosthesis form	Design prosthesis form using CAD	Time saving More flexible working No need for patient to be present Reduced material usage
Mould the prosthesis in a two-part plaster mould Boil out wax residue Apply separator	Identify and mark the tool split line Design the tool sections Export as STL data Orientate and support for RP&M process	Automated split line identification Reduced material usage Time saving Material saving Mistakes easy to correct
	Fabrication	Can be duplicated easily Automated fabrication, but additional cost over conventional methods
Create foam mould insert Apply colour matched silicone to the mould cavity Inject with skin tone colour silicone Cure Trim and deliver	Hand finishing Same as conventional	

core was placed on the lower half of the mould and the mould closed and bolted together. The catalysation of the elastomer was accelerated by heating the coated mould at 80 °C. The silicone elastomer gel (M512, Technovent Ltd, UK) was pigmented to match the patient's skin colour and then injected through the injection port with a syringe (Figure 5.129).

The mould was left for 24 h to allow the gel to set. Once set, the prosthesis was removed from the mould and trimmed to remove any flash from the moulding process. The injection port area was sealed over with a silicone nipple to complete the prosthesis. The completed prosthesis is shown in Figure 5.130.

From data capture to final prosthesis delivery took approximately 5.5 prosthetist hours, plus additional automated fabrication time. Table 5.14 compares conventional, laboratory-based methods with the computer-aided method.

5.17.3   Discussion

The techniques described provided the patient with a satisfactory cosmetic result, which established symmetry with the healthy, brassier supported breast and that fitted well against the chest wall. Follow-up reports by the breast care nurse team suggested that this helped to ensure stability and comfort of the prosthesis when the patient was mobile. The application of computer-aided technologies also provided other process benefits. Using photogrammetry removed the need for impression taking, which is labour and material intensive, time-consuming, intrusive for the patient and messy. The advantages and limitations of photogrammetry and other 3D topographic scanning methods compared to direct impression techniques have been discussed and in this case study, provided an ideal solution (Kovacs et al., 2007; Weinberg et al., 2006). The design stages did not require the patient to be present and FreeForm provided the ideal tool for shaping the prosthesis. Mould designing in FreeForm could be more intuitive and efficient, however it is estimated that it was still faster than conventional methods.

Alternative RP&M methods such as Selective Laser Sintering (EOS, GmbH, Germany), Z-Corp printing (Z Corporation, USA) and Fused Deposition Modelling (Stratasys Inc., USA) could also be viable for mould production, but would be more difficult to achieve the desired level of surface finish.

The high cost of computer-aided technologies has been reported as a limitation in patient-specific medical applications (Sanghera, Naique, Papaharilaou, & Amis, 2001). However, given the process and prosthesis improvements and the related indirect cost savings associated with patients undergoing shorter consultations, the computer-aided method has clear advantages. The equipment also has application in many other patient-specific devise design and surgical scenarios (Eggbeer et al., 2005, 2006; Evans et al., 2004; Sykes et al., 2004; Verdonck et al., 2003).

The application of computer-aided technologies may also be used to quantify the target prosthesis weight, by measuring the remaining breast volume and translating this to the prosthesis form. Body posture is known to be compromised post-mastectomy (Rostkowska, Bak, & Samborski, 2006) and providing a well-matched prosthesis may help to reduce injuries and the need for physiotherapy.

5.17.4   Conclusions

The technique presented provided a prosthesis that accurately fitted the defect chest wall of the post-mastectomy patient providing a comfortable fit and improved retention over an 'off the shelf' alternative. By designing the prosthesis digitally and using the mirrored contour of the breast within the brassier a better symmetry was achieved for the wearer.

Manufacturing a custom prosthetic appliance is more cost and labour intensive than an off the shelf solution. This technique has illustrated how computer-aided methods are able to offer a cost-effective alternative to the traditional labour intensive techniques by reducing the length of patient consultation, reducing the number of patient visits, reducing the quantity of materials used and providing a more flexible and repeatable method of working.

References

Available online from Health Solutions Wales. http://www.wales.nhs.uk/sites3/docmetadata. cfm?orgId=527&id=105899.

Available online from NHS Hospital Episode Statistics. http://www.hesonline.nhs.uk/Ease/ servlet/ContentServer?siteID=1937&categoryID=210.

Available online from UK National Statistics. http://www.statistics.gov.uk/downloads/ theme_health/MB1-37/MB1_37_2006.pdf.

Eggbeer, D., Bibb, R., & Evans, P. L. (2006). Towards identifying specification requirements for digital bone anchored prosthesis design incorporating substructure fabrication: a pilot study. *International Journal of Prosthodontics*, *19*(3), 258–263.

Eggbeer, D., Bibb, R., & Williams, R. (2005). The computer aided design and rapid prototyping of removable partial denture frameworks. *Proceedings of the Institution of Mechanical Engineers Part H*, *219*(3), 195–202.

Evans, P. L., Eggbeer, D., & Bibb, R. (2004). Orbital prosthesis wax pattern production using computer aided design and rapid prototyping techniques. *The Journal of Maxillofacial Prosthetics & Technology*, *7*, 11–15.

Gallagher, P., Buckmaster, A., O'Carroll, S., Kiernan, G., & Geraghty, J. (November, 2009). Experiences in the provision, fitting and supply of external breast prostheses: findings from a national survey. *European Journal of Cancer Care*, *18*(6), 556–568.

Glaus, S. W., & Carlson, G. W. (2009). Long-term role of external breast prostheses after total mastectomy. *The Breast Journal*, *15*(4), 385–393.

Kovacs, L., Eder, M., Hollweckb, R., Zimmermanna, A., Settlesc, M., Schneiderd, A., et al. (2007). Comparison between breast volume measurement using 3D surface imaging and classical techniques. *Breast*, *16*(2), 137–145.

Livingston, P. M., White, V. M., Roberts, S. B., Pritchard, E., Hayman, J., Gibbs, A., et al. (2005). Women's satisfaction with their breast prosthesis: what determines a quality prosthesis? *Evaluation Review*, *29*(1), 65–83.

Rostkowska, E., Bak, M., & Samborski, W. (2006). Body posture in women after mastectomy and its changes as a result of rehabilitation. *Advances in Medical Sciences*, *51*, 287–297.

Rozen, W. M., Rajkomer, A. K. S., Anavekar, N. S., & Ashton, M. W. (2009). Post-mastectomy breast reconstruction: a history in evolution. *Clinical Breast Cancer*, *9*(3), 145–154.

Sanghera, B., Naique, S., Papaharilaou, Y., & Amis, A. (2001). Preliminary study of rapid prototype medical models. *Rapid Prototyping Journal*, *7*(5), 275–284.

Sykes, L. M., Parrott, A. M., Owen, P., & Snaddon, D. R. (2004). Applications of rapid prototyping technology in maxillofacial prosthetics. *International Journal of Prosthodontics*, *17*(4), 454–459.

Thomas, K. F. (1994). *Prosthetic rehabilitation*. London: Quintessence Publishing Co. Ltd.

Verdonck, H. W. D., Poukens, J., Overveld, H. V., & Riediger, D. (2003). Computer-assisted maxillofacial prosthodontics: a new treatment Protocol. *International Journal of prosthodontics*, *16*(3), 326–328.

Weinberg, S. M., Naidoo, S., Govier, D. P., Martin, R. A., Kane, A. A., & Marazita, M. L. (2006). Anthropometric precision and accuracy of digital three-dimensional photogrammetry: comparing the genex and 3dMD imaging systems with one another and with direct anthropometry. *The Journal of Craniofacial Surgery*, *17*(3), 477–483.

Orthotic rehabilitation applications

5.18 Orthotic rehabilitation applications case study 1: a review of existing anatomical data capture methods to support the mass customisation of wrist splints

Acknowledgements

The work described in this case study was first report in the reference below and is reproduced here in part or in full with the permission of both CRDM and Taylor and Francis Publishers.

Paterson, A. M. J., Bibb, R. J., & Campbell, R. I. (2010). A review of existing anatomical data capture methods to support the mass customisation of wrist splints. *Virtual and Physical Prototyping*, 5(4), 201–207. http://dx.doi.org/10.1080/17452759.2010.528183

Paterson, A. M. J., Bibb, R. J., & Campbell, R. I. (2010). A review of existing anatomical data capture methods to support the mass customisation of wrist splints. In D. Jacobson, C. E. Bocking, Rennie (Eds.), *11th National Conference on Rapid Design*, Prototyping & Manufacture, AEW, CRDM, Ltd., High Wycombe, pp. 97–108, ISBN 978-0-9566643-0-3.

5.18.1 Introduction

Mass customisation (MC) is an increasingly common and popular manufacturing approach in many disciplines, particularly when combined with additive manufacture (AM). The MC–AM relationship is expanding and progressing, from the consumer market to the medical industry for development of surgical planning, teaching, preparation, aids and assistive technologies (Webb, 2000). MC is intended to collect the needs and requirements of the end user to make life easier and more comfortable through customised fit, function, performance and aesthetics. The European initiative project CUSTOM-FIT illustrates the need for customisation, with particular interest in assistive technologies for disabled users (Gerrits, Jones, & Valero, 2004). Therefore, anatomical data capture is essential for the manufacture of customised assistive devices.

According to E. Donnison (personal communication, 4 February 2010), the prescription of wrist splints inevitably requires a certain amount of customisation to suit the patient's individual fit and requirements, regardless of whether the splints are fully custom–fit or prefabricated off-the-shelf splints. This is particularly relevant for patients with degenerative diseases such as rheumatoid arthritis, because patients often need splints that will be comfortable, robust, and long-lasting. A wide range of literature suggests the suitability of anatomical imaging equipment such as Magnetic Resonance Imaging (MRI) and Computed Tomography (CT) for the design of prosthetics and orthotics (Bibb, 2006; Bibb & Winder, 2010; Minns, Bibb, Banks, & Sutton, 2003), but little research has been focused on the data acquisition of hand and wrist geometry for the mass customisation of wrist splints.

A US patent application by Fried (2007) stated ownership of a process for creating wrist splints, from data acquisition of anatomical features of wrists and hands to splint fabrication using AM. However, data acquisition of the wrists and hands is not a straightforward process. To date, there is no standardised method for collecting topographical skin surface data for the wrists and hands, and numerous problems emerge with various data collection methods. Therefore, this chapter evaluates the strengths and weaknesses of four different data acquisition methods: CT, MRI, three-dimensional (3D) laser scanning, and anthropometrics. The most suitable method will be identified to support the digitisation process of customised wrist splint design and manufacture for small-scale research. This report is not a comprehensive study of all data acquisition methods, nor does it imply that the chosen method is suitable for any or all clinical applications, but it reviews the most common data acquisition types discussed in other case studies within the field of medicine to support future doctorate research.

5.18.2   Data acquisition methods

5.18.2.1   Noncontact data acquisition

Computed tomography

Computed tomography is widely used within medicine, typically for diagnosis and surgical planning. It has the ability to generate both 3D and 4D images, along with quality volume-rendered imaging. The patient is placed between an X-ray tube and a detector array; X-rays pass through the patient and the level of attenuation is detected by the detector array and then logged by a computer (Goodenough, 2000). Measurements are taken from various angles to produce a series of axial slices and are illustrated in greyscale to demonstrate different densities within the body, ranging from black to represent air to white to represent the densest bone (Bibb, 2006). There are two main types of CT: sequential and spiral (Goodenough, 2000). Spiral CT can capture data in real time to produce high-quality 4D imagery and can be used for functional analysis procedures such as CT coronary angiography and diagnosis of joint instability (Feyter et al., 2007; Tay et al., 2007). This is particularly beneficial for patients who are unable to maintain a still position (Bibb, 2006; Feyter et al., 2007). Other advantages of CT are high-image resolution between soft tissue, bone, and air, and the ability to improve contrast and decrease structure noise. Users can also focus on a specified field of view to produce more accurate scans (Bibb, 2006; Goodenough, 2000). For these reasons, CT within anaplastology for the design of prosthetics is now standard practice and is often linked with AM for prosthetic fabrication (Bibb, 2006; Eggbeer, 2008).

However, there are several drawbacks to CT imaging. Radiation is the biggest concern; CT imaging was dismissed by Eggbeer (2008) (Chua, Chou, Lin, Lee, & Saw, 2000) for prosthetic design owing to radiation exposure. Radiation exposure is directly proportional to the duration of scanning, so higher image resolution and larger area coverage will require longer scanning times, which exposes patients to greater radiation dosage (Bibb & Winder, 2010; Tay et al., 2007). Bibb and Winder (2010) stated that radiologists should rationalise the acceptable level of accuracy and resolution from CT imaging so scanning times may be balanced with radiation exposure. Also, resolution between different soft tissues is often poor because collected images are divided by pixel shade; two different densities that share a pixel create an intermediate density known as the partial pixel effect, which can create a blurred boundary (Bibb, 2006). Minns et al. (2003) identified issues in lack of definition when combining CT imagery with AM; CT slice distances can range from 0.5 mm upward, whereas certain AM equipment is capable of creating 0.15-mm-thick layers (Bibb, 2006). Therefore, intermediate sections between CT data layers are required. Also, a large field of view can demonstrate poor resolution when scanning small, intricate detailing (Bibb, Eggbeer, & Evans, 2010). However, Tay et al. (2007) justified the use of CT for wrist function analysis because they claimed that the wrist is not sensitive to radiation.

Magnetic resonance imaging

Approximately 500 MRI machines are in use within the United Kingdom, and over 20,000 worldwide (Institute of Physics, 2008). Magnetic resonance imaging uses three different magnetic fields: static magnetic, switched gradient and pulsed radio-frequency. The combination of these fields causes hydrogen atoms within the body to align, so the equipment can differentiate between reactions and physical attributes (Bibb & Winder, 2010; Institute of Physics, 2008; Pickens, 2000). Magnetic resonance image slices can be used solely for diagnoses, or combined through suitable medical software into a 3D virtual form for design and fabrication of surgical guides and prosthetics (Bibb, 2006; Fitzpatrick et al., 1998). The captured data illustrate different densities; areas with a high concentration of hydrogen atoms, such as soft tissue, are represented in light greyscales, whereas areas with low hydrogen atom concentration appear darker (Bibb, 2006). Therefore, MRI is excellent at differentiating between different soft tissues and air in close proximity, and can be used for capturing skin surfaces (Bibb, 2006).

Various strengths are associated with MRI, such as high resolution of soft tissue and bone morphology, and unlike CT, patients are not exposed to radiation (NHS, 2009). However, there are several disadvantages. Over one million examinations are performed using MRI equipment in the United Kingdom every year. Therefore, the equipment is in high demand among a wide variety of medical specialities, and waiting times can be lengthy (NHS, 2009). This is particularly relevant when considering patients who require more urgent access to MRI equipment. The mechanical design of the equipment can also deter patients who experience claustrophobia, and the noise created may be unnerving (Bibb, 2006; Laurence, 2000; Wiklund & Wilcox, 2005). MRI scanning can also be time-consuming and movement can cause distortion and shadowing. Various studies have investigated the administration of general anaesthesia and patient sedation during MRI screening to gather undisturbed data imagery, but these methods may come with concerns, particularly for children and individuals with compromised health conditions (Laurence, 2000; Lawson, 2000; Low, O'Driscoll, MacEneaney, & O'Mahony, 2008). Patients with pacemakers, cardioverter-defibrillators and metal fabricated implants also face difficulties, because exposure to strong magnetic fields could be detrimental to patient health, although Roguin (2009) discussed these concerns and rationalised MRI usage after careful clinical evaluation.

To date, no literature has been found with regard to collecting skin surface topography of the wrist and hand using MRI. This may be because volume-rendered imaging is rarely performed on MRI scanning owing to the time involved in scanning and processing data, and most diagnoses can be performed with single accurate slice images.

Optical-based systems

Optical-based systems are becoming more prominent across a range of fields in medicine. Close-range triangulation laser scanning, for example, involves a laser source and one or more sensors (Figure 5.131). A laser is emitted from the source and pointed towards the artefact; the reflected light is then captured by the sensor(s). Because the angle of the source relative to the sensors is known, the distance of the reflected source can be calculated using triangulation to form a 3D virtual representation of the artefact.

Structured light systems also have a light-emitting source, but in this case patterns such as moiré are emitted across a wider area; these patterns alternate in shape, size, and position over a set period of time. The reflected data captured

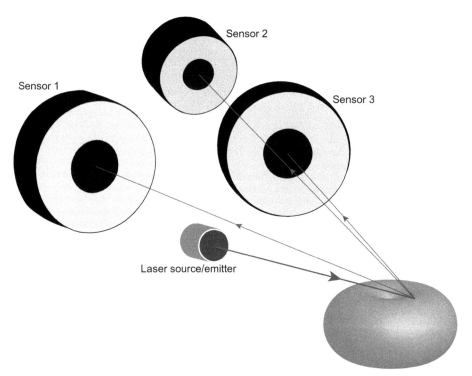

Figure 5.131 Three-dimensional (3D) laser scanning principles.

of the emitted patterns are compared in complementing software to create a 3D representation. Photogrammetry tends to refer to a single-mounted structured light system and will capture data only in its field of vision. Stereophotogrammetry, however, involves the use of multiple light-emitting sources and sensors strategically placed in different positions relative to one another ideally to surround the artefact to be scanned. Each unit captures a different angle of the object. These scans can then be combined together in supporting software, and depending on the positions of the light source and sensor, can give a complete 360° 3D representation of the artefact.

Uptake of the technology has been limited in the past owing to the cost of the equipment, particularly high-resolution stereophotogrammetry equipment. However, recent developments in technology show promise for opening up a new market for low-cost optical-based systems for the future. The Xbox Kinect, for example, can be used to analyse biomechanics (Bonnechère et al., 2014) but it can also be used to gather topographic data as a 3D scanner, and it has been used to compare breast reconstruction data (Henseler, Kuznetsova, Vogt, & Rosenhahn, 2014).

In terms of applications, Chua et al. (2000) chose the 3D laser scanning method over other conventional methods for prosthesis modelling, such as plaster of paris impressions, MRI and CT. Laser scanning has also been used to assess anatomical changes in facial morphology over time to demonstrate its suitability (Kau, Zhurov, Bibb, Hunter, & Richmond, 2005).

A significant benefit of optical-based data capture systems is that only the patient's skin surface is captured. This substantially reduces data file sizes compared with MRI and CT imagery, which also capture imaging of internal anatomy (Chua et al., 2000). Processing time is also significantly reduced; point cloud data can be converted to a polygonal mesh within 3D Computer-Aided Design (CAD)-based software. Another benefit is the speed of data collection. Kau et al. (2005) reported facial scanning times of 2.5 s and, depending on the optical equipment used, accuracy as high as 0.05 mm between data points (Bibb et al., 2000; Chua et al., 2000). The financial benefits of optical methods are also an incentive compared with CT and MRI because they offer affordable hardware and software and minimal training requirements, which correspond to ease of use, availability, accessibility, efficiency, validity and repeatability (Kau et al., 2005).

However, a significant problem with any optical-based system is the inability to capture wanted internal structures and intricate surfaces owing to line-of-sight limitations (Bibb et al., 2000; Surendran, Xu, Stead, & Silyn-Roberts, 2009). This is particularly relevant to the application of prosthetics and orthotics. Certain topographic sections of the human anatomy have intricate creases and folds, particularly between fingers and the thenar webbing when the hand is in a neutral position; scanning these areas can result in void data and unwanted point convergence, as demonstrated in Figure 5.132 (Bibb et al., 2010; Chua et al., 2000; Li, Chang, Dempsey, & Cai, 2008). Various sources suggest using reverse engineering software capable of post-processing to produce a watertight model by repairing and re-sculpturing void data (Bibb et al., 2000; Chua et al., 2000; Surendran et al., 2009). However, this process can be time-consuming and may not be a true representation of the scanned object once corrected (Bibb et al., 2000). Bibb et al. (2000) suggested another approach to capture

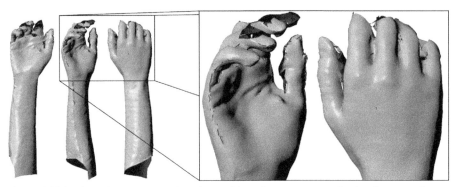

Figure 5.132 Direct 3D laser scanning: data voids and convergence of points between fingers.

shadowed data by collecting and combining several overlapping scan data files to produce a complete model. However, data point density would increase and alignment of point clouds may be difficult (Bibb et al., 2000; Surendran et al., 2009). Time and expertise in reverse engineering software are also significant factors for combining scans effectively; software such as Geomagic Studio (Geomagic Solutions, 430 Davis Drive, Suite 300, Morrisville, NJ) is available in which the process is simplified for the user through more complex software algorithms, but at a cost. Another limitation to 3D laser scanning is inaccurate data acquisition owing to involuntary movement, causing noise and distortion (Bibb, 2006); this concern is particularly relevant to this investigation. Two studies suggested using custom-made position jigs to prevent movement and achieve better scanning results, with limited success (Bibb et al., 2000; Li et al., 2008).

In contrast, Direct Dimensions (Direct Dimensions I, 2010) identified an accurate technique of capturing anatomical attributes to support the fabrication of a lifelike prosthetic hand using AM. The process involved scanning a plaster cast model of a patient's hand in a neutral position (Direct Dimensions I, 2010). Two laser scanners were mounted to a Faro Arm and a motorised precision coordinate measuring machine. The setup was capable of capturing intricate detail such as pores and creases (Direct Dimensions I, 2010). The case study demonstrates that scanning can be effective when scanning inanimate objects, because concerns over involuntary movement are diminished. This method was supported by Tikuisis, Meunier, and Jubenville (2001), who reported highly accurate resolution of scanned hand data. Based on the conclusion of Surendran et al. (2009), plaster of paris would be an effective material for scanning because of its matte appearance, even colour tone, and opacity, and its ability to capture fine detail, but casting can be messy, time-consuming, and uncomfortable for patients.

5.18.2.2   Contact data acquisition

Anthropometrics

A report by Greiner (1990) has been stated as a reliable resource for accurate anthropometric measurement techniques for hands (Greiner, 1990; Wilcox, 2005). Calipers and goniometers are often used within orthopaedics for measurement and assessment of joint range of movement (Gajdosik & Bohannon, 1987). Williams (Williams, 2007) designed a technique for creating customised gloves by collecting key anthropometric measurements using calipers and goniometers. The values were entered into a parameter table, and the parametric 3D CAD hand model was adapted to meet these dimensions. The parametric model was converted into a suitable format for AM. Li, Chang, Dempsey, Ouyang, and Duan (2008) investigated the concept of extracting anthropometric data from scanned resin hand casts and stated that this would be a successful approach in future anthropometrics.

However, collecting enough anthropometric data to construct a wrist splint would be costly in time and labour, and would potentially demonstrate high error if measurements were entered incorrectly into a parametric model within a 3D CAD program. Particular difficulties would come from measuring prominent bony landmarks and contours within the wrist, such as the radial styloid process, ulnar styloid process, and pisiform (Li, Chang, Dempsey, & Cai, 2008; Srinivas Reddy & Compson, 2005). Because the hand and wrist have many degrees of freedom, any movement within the joints could affect the measurements, which in turn could invalidate other dependent measurements.

5.18.3   Conclusion and future work

This review showed that there is no standardised method to capture the entire surrounding skin surface topography of the wrist and hand to support customised splint design and manufacture. Table 5.15 lists ideal characteristics for data acquisition methods targeted at splint design.

A common problem among all data acquisition methods is voluntary and involuntary movement, which may be particularly problematic for patients who have medical conditions such as Parkinson disease. Therefore, methods of immobilisation through jigging or casting may be the most effective method to support data collection. However, given the range of movement capable from the joints within the wrists and hands, jigging may not be the most appropriate method, because the jig itself would need to cover joints and may interfere with data collection, particularly with optical-based methods such as 3D laser scanning. Jigging would be most appropriate for capturing anatomical contouring of the wrist and forearm, but if the patient requires additional support around the digits, i.e. to support fingers from ulnar drift, a polygonal mesh between the digits may be necessary to create this additional support; this may be difficult and time-consuming to capture.

Findings conclude that in terms of accuracy, resolution, patient safety, cost, speed, and efficiency, optical-based acquisition methods appear to be the most suitable to meet all needs.

In the context of gathering test data for continued research and development of a digitised approach to splinting, one may combine optical methods with plaster casting to reduce collection of ambiguous data while providing a repeatable and reliable data source for validation purposes. From an orthopaedic aspect (E. Donnison, personal communication, 4 February 2010), hands should be cast in the neutral or intrinsic plus position if possible, or a rested comfortable position if the splint

Table 5.15 Comparative advantages and disadvantages of data acquisition methods

Data capture method / Ideal characteristics	CT/MRI	Anthropometry	Direct 3D laser scanning	Indirect 3D laser scanning (support jig)	Indirect 3D laser scanning (plaster cast)
No trained specialists required	✗	✓	✓	✓	✓
Affordable for low budget projects	✗	✓	✓	✓	✓
Quick, easy access	✗	✓	✓	✓	✓
Low/no waiting times	✗	✓	✓	✓	✓
Minimum preparation time	✓	✓	✓	✗	✓
Quick data collection	MRI = ✗ CT = ✓	✗	✓	✓	✓
Low/no discomfort for patients	✓	✓	✓	✓	✓
No health/safety risks to patients	✗	✓	✓	✓	✓
Low risk of data capture error	✓	✗	✗	Dependent on suitability of jig	✓
Suitable accuracy	✓	✓	✓	✓	✓
Suitable resolution	✓	✓	✓	✓	✓
Data capture unaffected by movement	MRI = ✗ CT = ✓	✗	✗	Dependent on suitability of jig	✗
No line of sight limitations	✓	✓	✗	✗	✗
Little/no concern over positioning	✓	✗	✗	✗	✓
Reliable repeatability	✓	✗	✗	Dependent on suitability of jig	✓
Direct export into 3D CAD software	✓	✗	✓	✓	✓

is for a patient with a degenerative disease such as rheumatoid arthritis. It is an accepted fact that collecting scanned data between fingers is difficult, particularly when the hand is in a neutral position. Therefore, plaster casts could be placed and scanned in several different orientations if using fixed position optical-based scanners, until acceptable data are collected between the digits. Handheld scanners will also allow free movement around the plaster cast to capture greater detail of intricate surfaces from different angles. This will form the focus for future research, because a cast can provide and maintain a constant reference position suitable for repeatability testing.

Also, by using the technique suggested by Li, Chang, Dempsey, Ouyang et al. (2008), anthropometric measurements could also be taken from a plaster cast if required, without further consultation with the patient. Although the casting process can be time-consuming and messy, physical and digital versions of the patient's hand could be kept on record, which may be useful for orthopaedics when assessing disease progression. The process may be similar to the case study by Kau et al. (2005) with regard to changes of facial structure captured using laser scanning. How accurate the scan data would need to be to fabricate a comfortable form around the patient's wrist and hand has yet to be established. However, it is assumed that point density should vary throughout the model, with a denser point count around complex geometry and a sparse point count to represent simple geometry. Reverse engineering software capable of point cloud manipulation could be used to tailor scan data to a suitable level of accuracy to reduce file sizes and processing time. It may also be unnecessary to capture anatomical data between the fingers, but consultation with different disciplines (e.g. occupational therapy, physiotherapy) may demonstrate different requirements for each individual patient depending on the severity of the condition and individual needs with regard to levels of wrist and hand immobilisation.

Acknowledgements

Special thanks to Ella Donnison and Lucia Ramsey for their help, advice, and contribution.

References

Bibb, R. (2006). *Medical modelling: the application of advanced design and development techniques in medicine*. Cambridge: Woodhead Publishing, ISBN-13: 978–1845691387.

Bibb, R., Eggbeer, D., & Evans, P. (2010). Rapid prototyping technologies in soft tissue facial prosthetics: current state of the art. *Rapid Prototyping Journal, 16*(2), 130–137. http://dx.doi.org/10.1108/13552541011025852.

Bibb, R., Freeman, P., Brown, R., Sugar, A., Evans, P., & Bocca, A. (2000). An investigation of three-dimensional scanning of human body surfaces and its use in the design and manufacture of prostheses. *Proceedings of the Institution of Mechanical Engineers Part H, 214*(6), 589–594. http://dx.doi.org/10.1243/0954411001535615.

Bibb, R., & Winder, J. (2010). A review of the issues surrounding three-dimensional computed tomography for medical modelling using rapid prototyping techniques. *Radiography, 16*(1), 78–83. http://dx.doi.org/10.1016/j.radi.2009.10.005.

Bonnechère, B., Jansen, B., Salvia, P., Bouzahouene, H., Omelina, L., Moiseev, F., et al. (2014). Validity and reliability of the Kinect within functional assessment activities: comparison with standard stereophotogrammetry. *Gait Posture*, *39*(1), 593–598. http://dx.doi.org/10.1016/j.gaitpost.2013.09.018.

Chua, C. K., Chou, S. M., Lin, S. C., Lee, S. T., & Saw, C. A. (2000). Facial prosthetic model fabrication using rapid prototyping tools. *Integrated Manufacturing Systems*, *11*(1), 42–53. http://dx.doi.org/10.1108/09576060010303668.

Direct Dimensions I. (2010). *Case study: using 3D imaging to create high-res prosthetic hand*. Available at http://directdimensions.blogspot.com/2010/01/case-study-using-3d-imaging-to-create.html. Accessed 22.03.10.

Eggbeer, D. (2008). *The computer aided design and fabrication of facial prostheses (Ph.D. thesis)*. University of Wales.

Feyter, P. J. D., Meijboom, W. B., Weustink, A., Van Mieghem, C., Mollet, N. R. A., Vourvouri, E., et al. (2007). Spiral multislice computed tomography coronary angiography: a current status report. *Clinical Cardiology*, *30*(9), 437–442. http://dx.doi.org/10.1002/clc.16.

Fitzpatrick, J. M., Hill, D. L., Shyr, Y., West, J., Studholme, C., & Maurer, C. R. J. (1998). Visual assessment of the accuracy of retrospective registration of MR and CT images of the brain. *IEEE Transactions on Medical Imaging*, *17*(4), 571–585. http://dx.doi.org/10.1109/42.730402.

Fried, S. (2007). *Splint and or method of making same. US patent application 20070016323 A1*.

Gajdosik, R. L., & Bohannon, R. W. (1987). Clinical measurement of range of motion: review of goniometry emphasizing reliability and validity. *Physical Therapy*, *67*(12), 1867–1872.

Gerrits, A., Jonesr, C. L., & Valero, R. (2004). Custom-fit: a knowledge-based manufacturing system enabling the creation of custom-fit products to improve the quality of life. *Rapid Product Development Conference*, October 12–13, 2004. Portugal.

Goodenough, D. J. (2000). Tomographic imaging. In J. Beutel, H. L. Kundel, & R. L. Van Metter (Eds.), *Handbook of medical imaging, volume 1. Physics and psychophysics* (pp. 511–554). Bellingham, Washington: SPIE - The International Society for Optical Engineering. ISBN-13: 978-0819477729.

Greiner, T. (1990). *Hand anthropometry of U.S. military personnel*. Washington, DC: United States Department of Defense.

Henseler, H., Kuznetsova, A., Vogt, P., & Rosenhahn, B. (2014). Validation of the Kinect device as a new portable imaging system for three-dimensional breast assessment. *Journal of Plastic, Reconstructive & Aesthetic Surgery*, *67*(4), 483–488. http://dx.doi.org/10.1016/j.bjps.2013.12.025.

Institute of Physics. (2008). *MRI and the physical agents (EMF) directive*. London: Institute of Physics.

Kau, C. H., Zhurov, A., Bibb, R., Hunter, L., & Richmond, S. (2005). The investigation of the changing facial appearance of identical twins employing a three-dimensional laser imaging system. *Orthodontics & Craniofacial Research*, *8*(2), 85–90. http://dx.doi.org/10.1111/j.1601-6343.2005.00320.x.

Laurence, A. S. (2000). Sedation, safety and MRI. *British Journal of Radiology*, *73*(870), 575–577. http://dx.doi.org/10.1259/bjr.73.870.10911777.

Lawson, G. R. (2000). Sedation of children for magnetic resonance imaging. *Archives of Disease in Childhood*, *82*(2), 150–153. http://dx.doi.org/10.1136/adc.82.2.150.

Li, Z., Chang, C., Dempsey, P. G., & Cai, X. (2008). Refraction effect analysis of using a hand-held laser scanner with glass support for 3D anthropometric measurement of the hand: a theoretical study. *Measurement*, *41*(8), 842–850. http://dx.doi.org/10.1016/j.measurement.2008.01.007.

Li, Z., Chang, C. C., Dempsey, P. G., Ouyang, L., & Duan, J. (2008). Validation of a three-dimensional hand scanning and dimension extraction method with dimension data. *Ergonomics*, *51*(11), 1672–1692. http://dx.doi.org/10.1080/00140130802287280.

Low, E., O'Driscoll, M., MacEneaney, P., & O'Mahony, O. (2008). Sedation with oral chloral hydrate in children undergoing MRI scanning. *Irish Medical Journal*, *101*(3), 80–82.

Minns, R. J., Bibb, R., Banks, R., & Sutton, R. A. (2003). The use of a reconstructed three-dimensional solid model from CT to aid the surgical management of a total knee arthroplasty: a case study. *Medical Engineering & Physics*, *25*(6), 523–526. http://dx.doi.org/10.1016/S1350-4533(03)00050-X.

NHS. (2009). *MRI-scan: advantages and disadvantages*. Available at http://www.nhs.uk/Conditions/MRI-scan/Pages/Advantages.aspx. Accessed 02.02.10.

Pickens, D. (2000). Magnetic resonance imaging. In J. Beutel, H. L. Kundel, & R. L. Van Metter (Eds.), *Handbook of medical imaging, volume 1. Physics and Psychophysics* (pp. 373–461). Bellingham, Washington: The International Society for Optical Engineering. ISBN-13: 978-0819436214.

Roguin, A. (2009). Magnetic resonance imaging in patients with implantable cardioverter-defibrillators and pacemakers. *Journal of the American College of Cardiology*, *54*(6), 556–557. http://dx.doi.org/10.1016/j.jacc.2009.04.047.

Srinivas Reddy, R., & Compson, J. (2005). (i) Examination of the wrist—surface anatomy of the carpal bones. *Current Orthopaedics*, *19*(3), 171–179. http://dx.doi.org/10.1016/j.cuor.2005.02.008.

Surendran, N. K., Xu, X. W., Stead, O., & Silyn-Roberts, H. (2009). Contemporary technologies for 3D digitization of Maori and Pacific Island artifacts. *International Journal of Imaging Systems and Technology*, *19*(3), 244–259. http://dx.doi.org/10.1002/ima.20202.

Tay, S. C., Primak, A. N., Fletcher, J. G., Schmidt, B., Amrami, K. K., Berger, R. A., et al. (2007). Four-dimensional computed tomographic imaging in the wrist: proof of feasibility in a cadaveric model. *Skeletal Radiology*, *36*(12), 1163–1169. http://dx.doi.org/10.1007/s00256-007-0374-7.

Tikuisis, P., Meunier, P., & Jubenville, C. (2001). Human body surface area: measurement and prediction using three dimensional body scans. *European Journal of Applied Physiology*, *85*(3), 264–271. http://dx.doi.org/10.1007/s004210100484.

Webb, P. A. (2000). A review of rapid prototyping (RP) techniques in the medical and biomedical sector. *Journal of Medical Engineering & Technology*, *24*(4), 149–153. http://dx.doi.org/10.1080/03091900050163427.

Wiklund, M. E., & Wilcox, S. B. (2005). Human factors roundtable. In *Designing usability into medical products* (pp. 31–54). Boca Raton, FL: CRC Press. ISBN-13: 978-0849328435.

Wilcox, S. B. (2005). Finding and using data regarding the shape and size of the user's body. In M. E. Wiklund, & S. B. Wilcox (Eds.), *Designing usability into medical products* (pp. 77–83). Boca Raton, FL: CRC Press. ISBN-13: 978-0849328435.

Williams, G. L. (2007). *Improving fit through the integration of anthropometric data into a computer aided design and manufacture based design process* (Ph.D. thesis). Loughborough University.

5.19   Orthotic rehabilitation applications case study 2: comparison of additive manufacturing systems for the design and fabrication of customised wrist splints

Acknowledgements

The work described in this case study was first reported in the reference below and is reproduced here in part or in full with the permission of the Rapid Prototyping Journal.

Paterson, A. M., Bibb, R. J., Campbell, R. I., & Bingham, G. A. (2014). Comparison of additive manufacturing systems for the design and fabrication of customised wrist splints. *Rapid Prototyping Journal*, Vol.21, Issue 3, in press.

5.19.1   Introduction

Wrist splints provide multifaceted treatment outcomes to patients, including pain relief through immobilisation of affected joints (Callinan & Mathiowetz, 1996; Jacobs, 2003). Wrist immobilisation splints, for example, are designed to immobilise the wrist while allowing mobility of all digits to promote endurance for everyday tasks (Pagnotta, Korner-Bitensky, Mazer, Baron, & Wood-Dauphinee, 2005).

There are two main categories of splints: prefabricated off-the-shelf and custom-made. Prefabricated splints can be bought from a variety of stores such as pharmacies, but may also be prescribed by splinting practitioners, such as occupational therapists or physiotherapists. Prefabricated splints may come in a range of sizes (e.g. small, medium, and large), which assumes a one-size-fits-all strategy and are not necessarily tailored to suit an individual unless adjusted by the user or a splinting practitioner. Alternatively, custom-made splints are produced and distributed exclusively by splinting practitioners to suit each individual patient's lifestyle, as well as anatomical demands relative to their condition. Custom-made splints offer superior fit and comfort and in many circumstances can be less bulky than off-the-shelf items. They can also be made to accommodate extremes of size and deformity that are not always possible with off-the-shelf items, which inevitably have limits on their adjustability. Custom-made splints maintain their shape at all times whereas off-the-shelf items need to be adjusted each time they are put on and it is not always possible to replicate the adjustment precisely on each occasion. Consequently, off-the-shelf splints cannot accommodate every patient and there will always be a need for custom-made splints.

This chapter focuses on the creation of custom-made splints, because its end use is considered synonymous with fundamental benefits of AM in terms of mass customisation with regard to anatomical fit, function, and appearance. Although splints can be designed for many purposes, this study focuses on wrist immobilisation splints intended to alleviate the symptoms of Rheumatoid Arthritis as (RA) an exemplar application. An image of an immobilisation splint can be found in Figure 5.133. If prescribed by a splinting practitioner, patients are typically provided with wear

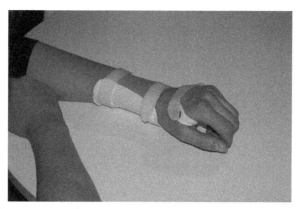

Figure 5.133 Custom-made wrist immobilisation splint.

instructions, including how to don and doff the splint and the expected wear schedule. Each wear schedule is tailored to suit each individual patient, their lifestyle, and the suggested optimal treatment approach to their condition, determined by the therapist (Lohman, Poole, & Sullivan, 2001).

A process model for designing and fabricating a custom-made wrist immobilisation splint can be found in Figure 5.134, deduced from several sources (Austin, 2003a; Jacobs & Austin, 2003; Lohman, 2001). In summary, custom-made wrist immobilisation splints are typically handmade; they are formed from sheets of thermoplastic that are cut, heated, moulded to the patient, adjusted, and then finished with fasteners (such as Velcro) to ensure a secure fit to the patient. Fundamentally, the splinting process is inherently a combination of designing and making in a single process. As a result, limitations of materials and fabrication processes impede the design for fit and function. Consequently, several factors affect patient compliance, such as discomfort and poor aesthetics, often resulting in a reduced willingness to wear splints to match the prescribed wear regime. The aesthetics of splints can have implications for the suggested duration and location of wear (Veehof, Taal, Willems, & van de Laar, 2008), because splints typically look clinical and unattractive despite the best efforts of clinicians to finish splints to a high standard and to suit patient preference. For example, patients are encouraged to choose different Velcro colours in a bid to improve compliance (Austin, 2003b). However, choice is limited to the material stock available to the clinic and the associated properties of the thermoplastic.

Because the manufacturing process is entirely manual and skills-dependent, the splint may also be poorly fitted, resulting in shear stress, directional misalignment, and pressure over bony prominences, which in turn can induce pressure sores (Coppard, 2001). Furthermore, the presence of a thermoplastic splint with uniform thickness and limited perforation can induce excessive perspiration, which can collect within the porous elements of padding (if present). In turn, this harbours bacteria, resulting in an odorous, unhygienic, yet often compulsory form of treatment for patients (Coppard & Lynn, 2001). Furthermore, splints are difficult to keep clean, particularly if a padded lining is present.

In response to these issues, the opportunity for using AM for upper extremity splinting was considered a viable option for future splint fabrication. Campbell (2006) stated that AM can account for functional, environmental, ergonomic, aesthetic,

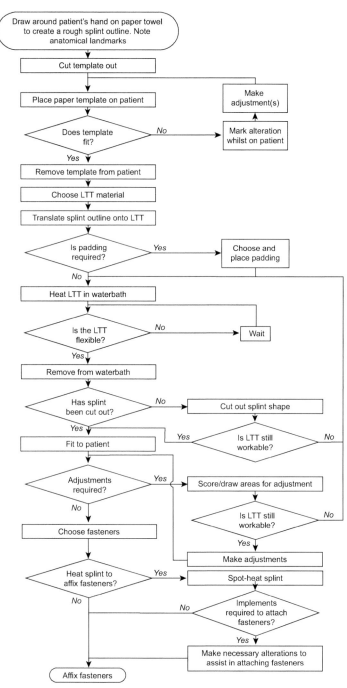

Figure 5.134 Traditional splint fabrication process (Paterson, Bibb, & Campbell, 2012). Image courtesy of SFF Symposium.

emotional, and user-fit requirements, and as such is a proven viable method for the design and fabrication of customised body-fitting items. The scope for AM applications continues to widen in a broad range of disciplines; use in medicine and dentistry, for example, continues to be the world's third largest serving industry (15.1%) in the AM sector for the past 11 years (Wohlers, 2012). Additive manufacturing has been explored in a range of exoskeletal assistive devices ranging from clubfoot treatment methods (Cook, Gervasi, Rizza, Kamara, & Xue-Cheng, 2010; Gervasi, Cook, Rizza, Kamara, & Liu, 2009; Smith, 2011a) to spinal braces (Summit & Trauner, 2010a) and ankle–foot orthoses (Faustini, Neptune, Crawford, & Stanhope, 2008; Gibson, Woodburn, Porter, & Telfer, 2014; Mavroidis et al., 2011; Pallariet al., 2010); the latter is the focus of a Framework 7 European project called A-Footprint. Most of these works focused on the use of laser sintering (LS), the benefits of which include relatively low part cost compared with other AM processes and the ability to retrieve unsintered powder for future use. Furthermore, the fact that unsintered powder subsequently behaves as support for down-facing structures is also a significant advantage, because this enables the creation of complex geometries without incurring significant cleanup time and subsequent costs associated with manual post-processing, while reducing the build time. Additive Manufactured Textiles (AMT) described by Bingham et al. (2007) demonstrate benefits of AM in terms of part consolidation and assembly builds while enabling functional articulating wearable structures. Various AMT linkage designs and arrays have been explored for stab-resistant body armour, which demonstrates its capabilities in generating functional constructs to enable movement of the intended wearer (Johnson, Bingham, & Wimpenny, 2013). Furthermore, AMTs look visually appealing, and although their integration into artefacts to date remains a niche topic, the scope for integrated aesthetic yet discrete functional AM textiles in custom-fitting wearable devices is entirely feasible.

Extending the use of AM in context of assistive devices, Gibson et al. (2014) investigated the suitability of a low cost extrusion-type 3D printer; their justification was a low-cost, in-clinic approach to fabrication through the use of a RapMan 3D print extrusion system (3D Systems, Rock Hill, SC, USA). Furthermore, Mavroidis et al. (2011) used Stereolithography (SLA) (3D Systems, USA) for ankle foot orthoses, because a range of materials could be used to offer different properties; for example, Arptech's DSM Somos 9120 Epoxy (Arptech Pty Ltd, Victoria, Australia) offered a flexible solution while being biocompatible. Stereolithography also demonstrates a high level of surface quality and resolution compared with original STL data. In the context of using material jetting systems for fabrication of assistive devices, Smith (2011b) described the developments of the Miraclefeet organisation and North Design Labs with Objet Connex systems into custom-made paediatric clubfoot orthoses. The most significant benefit of the approach was through exploiting the multimaterial build capabilities on offer, enabling heterogeneous builds including a range of Shore hardness alongside more rigid materials to incorporate functional parts, such as flexible hinges and soft edge features for improved functionality and comfort.

Despite the previous studies into AM for lower limb prosthetics and orthotics, the suitability of AM technologies in the context of upper extremity (i.e. hand, wrist and forearm) splinting has yet to be compared and evaluated, with suitable courses of action established for future development. To date, there has only been limited research and development into AM upper extremity splinting. The application has been implied by several additional sources (Fried, 2007; Fried, Michas, & Howard, 2005; Summit & Trauner, 2010b), and although no commercial approach has been proposed yet, a number of research institutes and individuals have explored the feasibility of AM splinting. For example, Fraunhofer IPA used an EOS P100 LS system with PA2200 powder (EOS GmbH Electro Optical Systems, Krailling, Germany) to offer a single example of an attractive solution to traditional splinting (Grzesiak, 2010). The benefit of LS in this instance was the ability to introduce intricate locking mechanisms that would have required less post-processing to remove unsintered powder acting as support structures. Fraunhofer IPA explored the integration of Voronoi patterns to their prototype (Breuninger, 2010), and Evill (Cortex, 2013) proposed the use of LS for splinting wrist fractures, incorporating a honeycomb structure with interlocking fasteners. Palousek, Rosicky, Koutny, Stoklasek, and Navrat (2013) also explored AM for splinting, but the intent was limited to reverse engineering a single existing splint that had been designed for traditional manufacturing techniques, and this attempt failed to recognise or explore the fundamental opportunities of design for AM principles such as improved aesthetics and or lightweight lattice-type structures.

In terms of material jetting technologies, Carpal Skin by Oxman (2010) explored the multiple material build capabilities available with Objet Connex systems (Stratasys, Eden Prairie, MN, USA). A range of Shore hardness was incorporated alongside stiffer materials in one build, resulting in a heterogeneous splint. The term 'synthetic anisotropy' was established to communicate the effect of directional influencers in the form of a specified pattern distribution to allow or restrict movement, the dispersion of which was dictated by an algorithm sourced from a pain map defined by an individual (Oxman, 2011; PopTech, 2009). This approach explores the opportunities of design for AM, but no clinical validity has been published to date. Taking an alternative stance to performance, functionality, and design strategies of integrated features, Paterson et al. (2012) explored the potential integration of multiple materials into wrist splints under the direction of qualified and experienced clinicians who specialise in the design and fabrication of custom-fitted wrist splints within the National Health Service (NHS) in the United Kingdom. The research focused on developing a specialised workflow designed for splinting practitioners to allow them to design splints in a virtual environment to support AM; the report focused specifically on the intent for placing multiple materials to behave as hinges or cushioned features as opposed to traditional fabrication processes, in which a similar approach would be impossible to replicate.

Having identified the relative strengths of four different AM systems along with existing work into other custom-fit devices, the authors chose to explore opportunities in upper extremity splinting for improved fit, functionality, and aesthetics. Features that were considered potentially beneficial included a best-fit first time approach to provide a customised fit for a patient relative to the anatomical and rehabilitation needs, and AMT elements for hinges using LS for easier donning and doffing. Furthermore, the authors chose to explore the opportunities for integrating varying levels of Shore hardness into wrist splints using the Objet Connex system, to include functional features such as integrated elastomer hinges or cushioned features over bony prominences. These new opportunities were impossible to deliver in traditional splinting, but are now entirely feasible as a result of AM. By exploring new, novel integrated applications for upper extremity splinting, future avenues for product development could be pursued. It was anticipated that with these potential benefits, patient compliance would be improved.

5.19.2   Aim and objectives

The research reported in this chapter represents one stage of a long-term research project exploring the whole process of digital splint design and manufacture, from data acquisition to data manipulation in 3D CAD to support AM. This chapter describes the exploration of AM prototypes through comparison and subsequent suitability of different AM processes based on the digital design workflow developed and described by Paterson (2013). By investigating previous research activity and potential AM process benefits, several design characteristics were planned for upper extremity splint integration for this investigation. Particular focus was placed on improving aesthetics, fit, and function of splints, and on exploring existing design features in the context of upper extremity splinting.

The aim of this chapter was to demonstrate and evaluate the suitability of a range of AM processes to deliver custom-fit wrist splints, using the following objectives:

Objective 1. To evaluate the design and fabrication of homogeneous AM splints.
Objective 2. To evaluate the design and fabrication of heterogeneous (multiple material) AM splints with functional features, e.g. an elastomer hinge and soft edges.
Objective 3. To demonstrate varied part consolidation, resulting from AM processes through integrated fastener features.
Objective 4. To evaluate the integration of a textile hinge to demonstrate assembly build capabilities.
Objective 5. To establish relative strengths and limitations of each AM process relative to recognised best practice in splint design.
Objective 6. To establish areas for future research and development related to the design of wrist splints for AM.

5.19.3   Method

One fundamental requirement of the AM splinting approach was to deliver a customised fit to the patient, because this is a standard requirement in traditional custom-made splinting. To deliver this, the patient's skin surface topography would be required to capture the patient's anatomical data. The data could then be used to extract and subsequently generate a 3D virtual form of a splint to match the topography.

Before AM splints could be fabricated for evaluation, 3D CAD splint models had to be generated. The organic topography of a forearm and hand was required to generate an accurate profile for CAD manipulation. Scan data were acquired from a healthy volunteer by 3D laser scanning a plaster cast, as described by Paterson, Bibb, and Campbell (2010). A plaster cast was used to eradicate concerns with noise that would otherwise be collected through involuntary movement and tremor during scanning. The plaster cast provided a repeatable, static data source in case a repeat scan was required later in the research process (Paterson, 2013). This form of data acquisition was to enable this research only, and is not suggested for future clinical practice. The plaster cast was scanned with a Z Corp ZScanner 800 handheld 3D laser scanner (3D Systems, USA). The same scan data were used for all subsequent splint designs, and therefore each demonstrated almost identical anatomical topographies (excluding integrated features such as lattices and fasteners) that would therefore fit the participant used to generate the plaster cast. Using these scan data, six different splints were designed in a range of 3D CAD software packages and plug-ins:

- Autodesk® Maya® 2011 (Autodesk, San Rafael, CA, USA)
- Geomagic Studio 2012 (3D Systems, Rock Hill, SC, USA)
- McNeel Rhinoceros® Version 4.0 (Robert McNeel and Associates, Seattle, WA, USA)
- Grasshopper plug-in (Robert McNeel and Associates, Seattle, WA, USA)
- PTC Pro/Engineer Wildfire 5.0 (PTC, Needham, MA, USA)
- FreeForm Modelling Plus (Geomagic, 3D Systems, USA)

The justification for using a wide range of CAD programs was related to the underlying research; a separate and previous stage in the overall research project was to develop a specialised CAD workflow that effectively replicated the splint design process used by splinting practitioners/clinicians. Therefore, there was a need to establish suitable tools that could ideally replicate and/or improve on traditional fabrication methods and techniques in a virtual environment within 3D CAD programs. Several different CAD packages were used during this phase of the research because each program had different tools, offering different strengths and limitations that were considered appropriate for splinting applications. Tools and CAD strategies were developed in that phase of the research to create a customised software workflow, and these methods were taken forward to the design and fabrication of the customised AM splints described in this chapter. The design workflow and exploration of CAD approaches are described fully by Paterson (2013) and are the subject of pending publications. In short, this workflow translates into a 3D CAD environment with the same design intent as the traditional design approach. Therefore, the design rules for the design of splints are adhered to in the digital workflow; this has been evaluated and approved by a number of qualified splinting professionals. The design steps include, for example, defining the boundary of the splint following anatomical landmarks, defining the material thickness, rounding edges, and alleviating pressure on sensitive areas. Over and above the traditional approach, the choice of lattice pattern is introduced to reduce weight, improve ventilation, and enhance aesthetics. This workflow has been developed to facilitate recognised clinical practice within a digital environment and has been critically evaluated by experienced and qualified clinical professionals.

Four different 3D CAD models were created in total, designed for homogeneous AM fabrication using:

- LS: EOS P100 Formiga, made with EOS PA2200 50:50 powder (EOS GmbH Electro Optical Systems, Krailling, Germany). This process was used to explore opportunities to integrate textile elements into a splint for added functionality.

- Fused Deposition Modelling (FDM): Stratasys Dimension SST1200es, made with ABS (Stratasys, Eden Prairie, MN, USA).
- SLA: 3D Systems 250, made with Accura® Xtreme resin (3D Systems, USA)
- PolyJet matrix material jetting: Objet Connex 500, made with FullCure® 515 and FullCure® 535 to generate RGD5160-DM (Stratasys, Eden Prairie, MN, USA), displaying ABS-like properties.

Finally, a heterogeneous splint was designed to exploit the multiple material capabilities of Objet Connex technologies, using TangoBlackPlus and VeroWhitePlus to generate DM9840FLX and DM9850FLX material ranges (Stratasys, Eden Prairie, MN, USA). Although outputs were similar to Oxman (2012), the combination with integrated aesthetic lattice structures was also targeted as an output in this context. With the exception of the Objet Connex heterogeneous build, material choice was not considered important within this investigation, because the research focused on highlighting potential differences in AM systems and their specific capabilities. However, the authors acknowledge that material choice is crucial in many aspects in assessing part quality and delivering for the intended need. However, to the extent that it was able to vary properties in the heterogeneous build, material choice was important for the multi-material splint because this was considered the most important characteristic to display in this context.

In the next section, each of these designs will be described in more detail, corresponding to the AM process considered most appropriate for the design. Different designs were fabricated on different AM systems to demonstrate the versatility of AM systems relative to their particular strengths. Furthermore, LS was only used for the fabrication of a textile splint, because previous studies described in Section 5.19.1 already demonstrated homogeneous splint prototypes using LS; the development of a textile element in this case was therefore considered a potential novel contribution to knowledge.

5.19.3.1 Homogeneous AM textile splint

The AMT splint featured an AMT element along the ulnar aspect. This was incorporated to consolidate splint parts into one assembly. The AMT element was designed to behave as a hinge to enable the user to open the splint for easier donning and doffing. Furthermore, the AMT element was designed to follow contours of the upper extremity geometry, demonstrating the drapability and free movement described by Bingham et al. (2007). The repeating units in the AMT element were formed to follow the topography of the scan data and were aligned using a custom mesh array algorithm devised by Bingham and Hague (2013) in MATLAB® (MathWorks, Natick, MA, USA). The textile element was then incorporated into the remaining splint geometry using uniform spaced links generated in Grasshopper; a generative modelling plug-in for Rhinoceros to enable quick adjustments within set parameters. The automated linkages were united with the main structures of the splint using a Boolean Union function.

The AMT element was specifically designed to exploit the freedom of form available through LS (Figure 5.135), because support structures commonly required

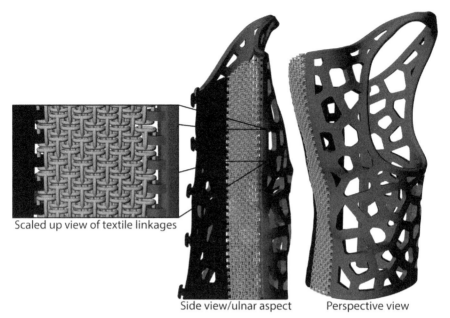

Scaled up view of textile linkages

Side view/ulnar aspect   Perspective view

Figure 5.135 Splint prototype with textile hinge, modelled using Matlab, Rhino, and Grasshopper for linkage integration.

by other AM systems would not be required in this instance. Because un-sintered powder provides support for overhangs, LS was considered the most effective approach, as overhangs in AMT linkages were abundant. Cleanup time was also considered, because the un-sintered powder would only require removal via high-pressure air jets and vibration to remove excess powder between linkages. Other processes could be used, such as 3D jetting/printing techniques. However, depending on the exact process and material combination used, support material removal could result in additional time and in some cases could damage the AMT linkages. SLA would also prove ineffective for this approach because supports would be required for links within the textile element, which would prove difficult and time-consuming to remove.

In addition, mushroom-like fasteners were integrated into the design to demonstrate part consolidation of fasteners within the build, as opposed to detached fasteners used on traditional splints such as Velcro.

5.19.3.2   Homogeneous circumferential build designs

Two circumferential splint designs were modelled in 3D CAD (Figure 5.136). The splints were designed in two corresponding parts to enable donning and doffing. The design was intended to behave as a pinch-splint, in which the user

Figure 5.136 Circumferential, homogeneous two-part splints for SLA (Voronoi) and Objet (Swirl) builds, respectively.

could pinch the palmar region laterally to separate the two halves when taking the splint off. This pinch design demonstrated the ability to integrate subtle, discrete fasteners into the splint while still being functional, which subsequently highlights part consolidation compared with traditional fastening methods (e.g. Velcro, D-rings).

5.19.3.3   *Heterogeneous splint using Objet Connex technologies*

A splint was designed for heterogeneous AM system fabrication, as described by Paterson et al. (2012). The underlying intent of this approach was to enable the practitioner to specify and localise areas where softer materials might benefit the patient to relieve pressure (e.g. elastomer elements over bony prominences or areas prone to pressure sores). In traditional splinting, practitioners may have to create cavities over bony prominences or integrate separate gel discs (Coppard & Lynn, 2001), but this approach affects the topography and subsequent aesthetics of the splint. However, the use of the Objet Connex would enable subtle integration of elastomer features that would not drastically affect the topography of the splint, therefore creating a less cumbersome appearance that may be more conducive to complying with wearing regimes for the patient.

Figure 5.137 shows the 3D CAD model developed specifically for the Objet Connex 500 system, to exploit its multimaterial capabilities. Various closed shells were required within the 3D CAD model to allocate different digital materials before fabrication. The shells were created by trimming the initial scan mesh (generated from cloud data) in Geomagic Studio before manipulating and thickening the geometry in other CAD software such as McNeel Rhinoceros. Elements labelled 1 were soft elastomer edges to provide a comfortable interface between the skin and the rigid splint structures (labelled 2). A soft elastomer cushion inspired by Pagnotta et al. (2005) was located over a bony prominence (pisiform) and a flexible hinge (element 4) was integrated along the ulnar aspect of the splint to enable donning and doffing.

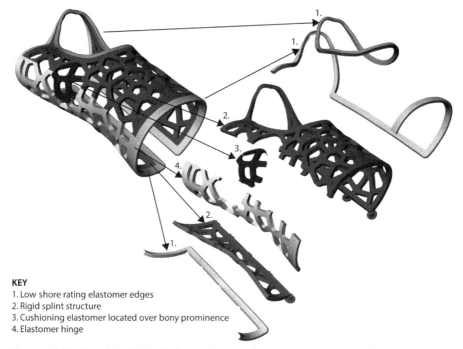

KEY
1. Low shore rating elastomer edges
2. Rigid splint structure
3. Cushioning elastomer located over bony prominence
4. Elastomer hinge

Figure 5.137 Closed shell distribution and intent of heterogeneous wrist splint.

5.19.4   Results

All of the splint prototypes were built according to the suppliers' recommended parameters using commercially available materials and machines. Each AM process and subsequent build is reviewed in the following sections.

5.19.4.1   Homogeneous AM textile splint

The LS splint shown in Figure 5.138 proved successful in capturing its intended outcome of an integrated AMT hinge to enable easier donning and doffing; the links offered sufficient freedom to enable this. The links proved strong enough during a preliminary wearer trial to maintain their structure without failing when the splint was worn. Furthermore, the AMT element added a unique aesthetic quality to the splint. The union of AM textiles and upper extremity splinting with the aim as a medical intervention was considered a world-first.

However, a small number of links remained fused together as a result of residual un-sintered powder trapped between linkages. It is anticipated that the porous nature of the surfaces would also inherently affect the hygiene of the splint by absorbing dirt, sweat, sebum, dead skin cells, etc., as described by Bibb (2006). Furthermore,

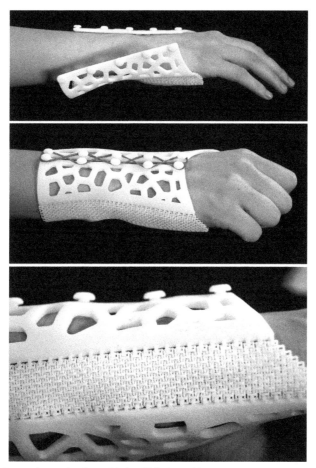

Figure 5.138 Laser sintered splint with AMT linkage hinge.
Image courtesy of Dr. C. Majewski, University of Sheffield, UK.

the AMT element exacerbates these concerns. The small links and tight textile design used in this example could also potentially catch on vellus and/or terminal arm hair, causing discomfort if extracted from hair follicles (i.e. trapping and pulling out body hairs). However, larger links could reduce this risk. Cleaning such a splint would also prove problematic unless immersed in a detergent or washed with an automated process/ system such as a dishwasher, as proposed by Fried (Fried, 2007). Laser sintering parts can withstand dishwashing; this has been discussed with reputable LS suppliers, although to date this has not been rigorously tested or reported. As a preliminary test, the researchers placed the splint in a dishwasher at various temperatures (45 °C, 50 °C and 65 °C) along with branded dishwasher detergent, with no visible after-effects.

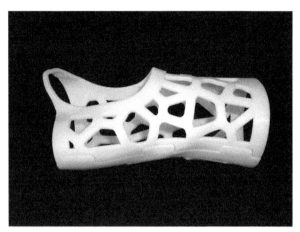

Figure 5.139 Fused deposition modelling splint.

Figure 5.140 Accura Xtreme® splint, built on a 3D Systems 250 (Paterson et al., 2012). *Prototype courtesy of Dr. D. Eggbeer, Product Design Development Research (PDR), Cardiff.* Image courtesy of SFF symposium.

5.19.4.2 *Homogeneous circumferential splints*

The FDM splint shown in Figure 5.139 demonstrated comparatively poor surface quality, with obvious layering and stepping; these factors affected the aesthetics, which also could affect the comfort of the splint at the edges. Pitted areas between layers and tracks could collect waste products as described earlier, and therefore could be an unhygienic solution for a splinting application. However, the ABS material is relatively robust and a widely used material in domestic and wearable products such as frames of eyeglasses; subsequently, it is reasonable to anticipate that the ABS splint can withstand mechanical cleaning with mild detergents.

The SLA splint shown in Figure 5.140 was oriented to reduce the requirements for supports, as can be seen in Figure 5.141. Similar to the homogeneous ABS-like splint, the design of the splint was effective in allowing donning and doffing by pinching the palmar element. Overall, the surface quality of the SLA splint was considered the highest of all the AM processes used in this investigation. The smooth surfaces facilitate cleaning and minimises hygiene risks. However, although side and up-facing surfaces are smooth, down-facing surfaces demonstrated abrasive

Figure 5.141 Stereolithography build, showing support lattice structures.
Image courtesy of S. Peel, PDR.

imperfections where supports had been removed (Figure 5.141). Such imperfections could cause discomfort for a patient if left untreated. Manual post-processing with abrasives would be required, as described by Bibb, Eggbeer, Evans, Bocca, and Sugar (2009), adding cost in labour and resources if the approaches were implemented for clinical application. However, this could be minimised by using patterns that formed self-supporting structures that would eliminate most of the supports and the issues encountered in their removal. A simple example is shown in Figure 5.142. The disadvantage of this approach would be a reduction in aesthetic possibilities and limited patient choice of pattern.

Finally, a fourth homogeneous splint built on the Objet Connex is shown in Figure 5.143. An initial visual and tactile interpretation of surface quality was considered acceptable compared with other AM processes described in this chapter. In addition, the fastener design and overall splint structure were able to perform and withstand its intended function in flexing to allow the user to don and doff by pinching the palmar aspect of the splint, and subsequently demonstrated part consolidation in the context of AM splint fabrication.

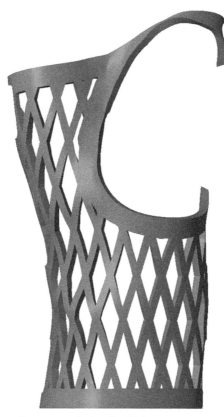

Figure 5.142 Self-supporting pattern design for SLA.

However, a significant amount of support material (FullCure 705) was required, but was removed easily and relatively quickly with a high-pressure water jet.

5.19.4.3   Heterogeneous splint using Objet Connex technologies

Before building this example, material choices were specified in Objet Studio™ (Stratasys, Eden Prairie, MN, USA). The software interface was used to define variables for Objet Connex build systems. The splint made with the DM98 material range provided a stronger colour contrast, as shown in Figure 5.144.

The 9850 Shore 50 elastomer placed over the bony prominence (pisiform) expanded when pressure was applied within the splint, which demonstrated that a small amount of expansion could accommodate swelling if required. The hinge was also functional, although several failures occurred over 12 months (Figure 5.145). This result was repeated from flexion when opening and closing the splint. The splints were used as proof-of-concept prototypes, and therefore have been handled by a large number of people to demonstrate the intent of the research. This particular splint prototype was handled by a large number of individuals in presentations and demonstrations and consequently underwent more than 50 openings and closings over a 12-month period. Many people

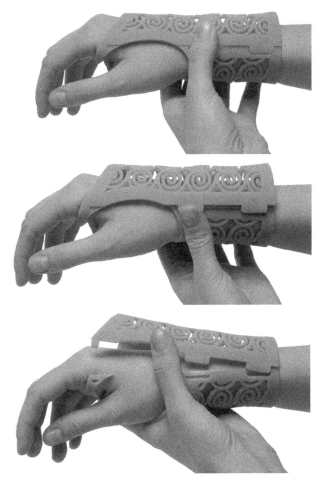

Figure 5.143 FullCure® 515/FullCure® 535 with ABS-like properties, built on the Objet Connex 500 model.

handled the prototype with little knowledge of its physical limitations and it is possible that its physical limits were exceeded by careless handling. In addition, the prototype was displayed in exhibitions and was subject to extended periods of exposure to strong light sources, which may have affected its physical properties over time. It was also noted that the position and shape of the flexible hinge were not ideal. The position was along the ulnar aspect of the forearm, and the intended wearer had to use significant force while adjusting the posture of the hand and wrist to don the splint correctly. Simultaneously, a significant amount of force was applied to the hinge while in an open position in putting it on; therefore, the elastomer elements were more susceptible to compression and tension forces, resulting in split lines. An additional hinge could have been placed along the radial aspect so both sides of the splint could have been opened to help donning. Similarly, the shape of the splint hinge was inappropriate because it was formed in 3D CAD by two parallel planes intersecting the splint geometry. If one considers a living

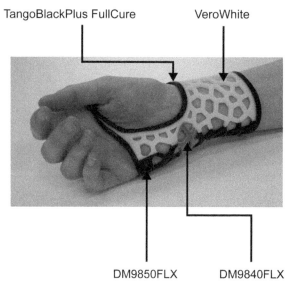

Figure 5.144 Heterogeneous splint; assorted materials within one AM build, using the Objet Connex.

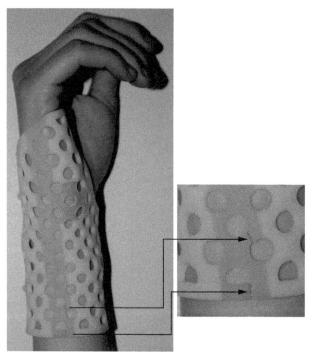

Figure 5.145 Failures in the multi-material elastomer hinge (Paterson et al., 2012). Image adapted courtesy of SFF Symposium.

hinge in a polypropylene DVD case, for example, rotation around a single axis results in a uniform distribution of compression and tension exerted throughout the length of the hinge structure. However, because of the organic topography of the splint, compressive and tensile stresses varied throughout the structure, with subsequent higher tensile concentration towards the borders of the splint, whereas a higher compression concentration was demonstrated on the inner region of the hinge. Therefore, the shape of the hinge element may have benefitted from being a varied shape to suit the topography. Although not formally documented, a slight level of creep was also observed over time, resulting in a slight twist in the splint. This is most likely because the splint had been stood upright on display for extended periods and it would be reasonable to assume that were the splint worn for extended periods it would be more likely to retain the intended shape. Therefore, future research would be required to assess the extent of creep, as well as to establish design rules to reduce local strain and therefore avoid failures at the hinge. However, surface quality and resolution were considered adequate compared with FDM and LS.

Unfortunately, the multimaterial build required a large volume of support material (FullCure 705), as shown in Figure 5.146. Much like the ABS-like splint, the support material was removed with a high-pressure water jet. This increased the cost in terms of material consumption, but also those related to labour time. Prototype properties and costs are summarised in Table 5.16.

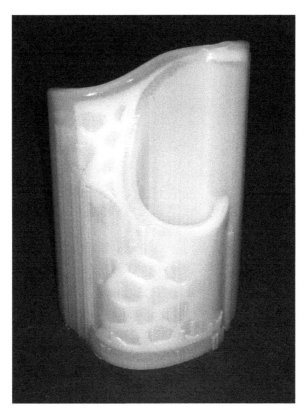

Figure 5.146 Support material required for the Objet multi-material build in upright position.

Table 5.16 Summary of prototype splints

AM process					
Evaluation criteria	SLA	LS (textile splint)	FDM	PolyJet (homogeneous)	PolyJet (heterogeneous)
System	3D systems 250	EOS P100 Formiga	Stratasys Dimension SST1200es	Object Connex 500	Object Connex 500
Material(s)	Accura® Xtreme	PA2200	ABS	FullCure® 515/ FullCure® 535 (ABS-like)	TangoBlackPlus VeroWhitePlus DM9850FLX DM9840FLX
Unique benefits		Textile element			Heterogeneous
Build speed	Poor (draining and post-curing required)	Poor (heat-up and cool-down periods required)	Poor	Adequate	Adequate
Surface quality	Good	Poor	Poor	Adequate	Adequate
Accuracy and resolution	Good	Good	Poor	Good	Good
Z plane build weaknesses?	Mild	Mild	Severe	Mild	Mild
Rigid build supports required?	Yes	No	Yes	Yes	Yes
Post-curing required?	Yes	No	No	No	No
Rigid support removal required?	Yes: removal by hand, followed by surface sanding	No, although removal of powder between linkages is required	Yes, although Water-Works supports are soluble	Yes, using high-pressure water jet	Yes, using high-pressure water jet
Comparative material cost (£/1 kg) (excluding VAT and shipping)	High £179.00	Lowest £55	High £290	Highest £246 and £267, respectively	High TangoBlackPlus = £245 VeroWhite = £200

5.19.5   Conclusions and future work

Each process displayed benefits and limitations in the context of upper extremity splinting. When compared, the most inappropriate AM process was considered to be FDM. Despite the advantages of robust materials, improvements in surface quality would be needed. Stereolithography proved to have good surface quality and reasonably robust materials, the effects of cleaning notwithstanding. Laser Sintering (LS) and PolyJet material jetting each displayed unique advantageous characteristics made feasible by the AM process. Previous studies and prototypes that used LS such as Fraunhofer IPA (Grzesiak, 2010) and Evill (Cortex, 2013) demonstrated the ability to integrate aesthetically pleasing structures, but incorporating a textile element has not previously been reported in the context of splinting. This chapter further describes additional features that could be beneficial in the future. In addition, the multi-material splints demonstrate similarities to Oxman (2010). However, the underlying ethos of placing different materials was a different approach, allowing clinicians to specify materials with varying Shore hardness that they would consider clinically appropriate, whereas Oxman (2010) developed an automated approach to integrate materials to direct or restrict the patient's movement. Both strategies demonstrate strengths, and ideally both would be made available features in specialised 3D CAD for clinicians to explore in the future.

In terms of integrated functionality, the heterogeneous splint was the most versatile and could be exploited in a range of situations, as highlighted by Paterson et al. (2012). If the digitised splinting approach were to be introduced as a realistic option for clinics, the choice of AM process would depend on the needs of patients, as prescribed by their therapists, but a thorough understanding of different AM processes and relevant AM materials would be needed by the therapists.

Despite the aesthetic and functional advantages displayed in the results, several developments would be required before such processes would be feasible for adoption in clinical situations. First, the development of suitable materials would need to be explored further, taking into consideration long-term exposure to the skin. Although a number of AM materials such as Objet's MED610™ transparent, rigid material and Stratasys's ABS-M30i claim ISO 10993 (biological evaluation of medical devices) and/or USP 23 Class VI approval in a range of conditions such as irritation and hypersensitivity, regulations involve standardised tests that may not necessarily take into account specific design and manufacturing processes. This certification demonstrates that the materials are inherently low in toxicity, but the European Medical Device Directive and various international equivalents require clinical trials that prove the safety of the entire design, materials and manufacturing process.

Cost analysis must also be performed to determine which process, if any, can be cost-effective in terms of clinical demands. Although current practice involves the use of cheap materials, the labour costs could be cut dramatically. This is especially the case when fabricating duplicate splints, for example. In current practice, the entire crafting process must be repeated for each and every splint – there are no economies of scale. When using AM, repeat splints would incur materials costs only. It is also important to consider the hidden costs of time and travel involved in clinic appointments for the patient as well as the clinician. The AM process can eliminate repeat prescription time; if patients require a replacement or duplicate splint as a result of previous failure or the desire for a different aesthetic design, they would not need a repeat clinic session with the practitioner to fabricate a new splint. Instead, a request could be logged and a duplicate splint ordered instantaneously. Such an order could be added to a queue for manufacture and then dispatched when the build is complete, a similar approach to the latter stages of the proposed automated process by Fried (Fried, 2007). This reduction in clinic time could reduce demands on the clinic, potentially reducing waiting times and patient waiting lists. In turn, this could improve patient satisfaction in the health care system. The economic advantages of an AM approach are predicated on replacing a high-labour cost manual crafting process with a much more efficient design then manufacture process. Consequently, the economic factors are context dependent. It can be envisioned that in regions of high labour cost, the AM approach may greatly reduce labour costs to the extent that the higher material costs of AM are more than compensated for. This work was conducted within a UK National Health Service context but future work on costing will enable a more direct calculation of economic benefit in different contexts around the world. Similar arguments for the potential benefits of a digital design and manufacture process have been explored in maxillofacial prosthetics by Eggbeer, Bibb, Evans, and Ji (2012). Another implication in cost-effectiveness is improved compliance, leading to improved patient outcomes in the longer term. The impact of patient involvement (i.e. choosing patterns and colours) and enhanced aesthetics on compliance is the subject of current research by the authors.

An aesthetic consideration available in clinics that could not be demonstrated fully was the customisation of colour choice. Colour ranges are currently limited in AM processes; the 3D Systems ZPrinter 450 (3D Systems, Rock Hill, SC, USA) can provide multicolour builds, whereas a range of FDM machines can offer various single colour builds. However, within the scope of this investigation, colour choice amongst LS, SLA, FDM and PolyJet material jetting is limited. The Objet Connex gave the widest variety, allowing for an integrated range from black to white. Other colours are achievable with the Objet Connex, such as VeroBlue (RGD840), green (ABS-like RGD810), and transparent (VeroClear RGD810) (Stratasys, Eden Prairie, MN, USA), provide multicolour builds with vibrant colours, as demonstrated by Oxman (2012). However, at the time of publication, these facilities were not commercially available for exploration. Some service providers are able to supply colour-dyed LS parts, but typically colour choice is limited and colour fastness is an issue.

It was hypothesised that the mechanical failures of the multi-material splints were the result of a number of factors, including:

- Limited tensile strength, elongation, and tear resistance of flexible Objet materials, e.g. 0.5–1.5 MPa, 150–170%, and 4–6 kg/cm, respectively for FLX9840-DM material (figures according to (Stratasys, 2013)). Similar issues have been highlighted elsewhere (Eggbeer et al., 2012; Moore & Williams, 2012) (Figure 5.147).
- Degradation of material over a period of time (including creep).
- Suboptimal location/shape of the flexible hinge relative to the organic geometry of the splint, and the force exerted when opening and closing the splint when donning and doffing.

However, further research is required to explore these areas further, with opportunities to establish interventions to overcome these limitations. Because the underlying research focused on 3D CAD processes to design splints for AM, performing mechanical tests on the prototypes was outside the focus of this investigation. The

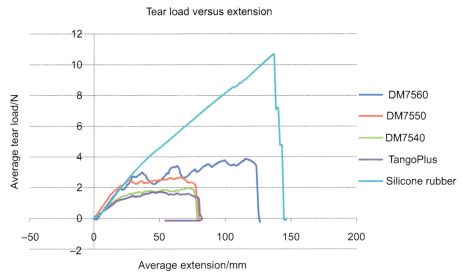

Figure 5.147 Tear load versus extension of Objet Connex materials and silicone rubber for maxillofacial prosthetics.
Adapted from Eggbeer et al. (2012).

authors have performed research using finite element analysis on a number of the proposed designs to enable comparison of the mechanical properties with traditional practice; this work will be published in due course. Structural analysis of homogeneous and heterogeneous splint builds will be performed by the authors to compare results with splints made with traditional fabrication methods and techniques. Concerns relating to ultraviolet (UV) exposure of photopolymer resin processes such as material jetting and SLA builds also need further exploration in an attempt to resolve breakdown of elastomer elements, as suggested by Eggbeer et al. (2012). Similar efforts should be performed on prevention of creep and discolouration of such materials and the AM processes used, particularly for PolyJet builds. Work is also being performed on the effects of material exposure to mechanical cleaning and everyday household chemicals, such as washing powders, liquids and other detergents, because this could potentially lead to suggestions regarding cleaning techniques for splints. As described previously, the authors have begun preliminary testing of cleaning splints in a dishwasher, and as such will form further research to determine suitable variables for extending the life of a splint.

In Section 5.19.1 (Introduction), the authors showed that previous researchers had limitations relating to clinical validity for the use of upper extremity splinting. Similarly, the authors also recognise and acknowledge that significant future research is required with respect to their own documented work before conducting clinical trials to consider the efficacy of the approach for end use applications. This includes exploring analytical studies relating to all design work to consider material properties and structural integrity. The authors also acknowledge that further work must be explored regarding suitable data capture methods to support the digitised splinting approach for suggested improvements to fit and function. In addition, comments relating to surface finish and surface roughness were not quantified but preliminary in this case. Therefore, further research is required to quantify the surface roughness of AM splints, and the potential effect on the skin if worn by a patient, such as abrasion. Finally, the current research is limited in terms of the usability of splints, because developments to date have not allowed for clinical trials. Further work will be required to assess the usability of different splint designs relative to a range of manual tasks (e.g. driving), against which patients can compare and contrast their previously prescribed splints made with traditional methods.

Acknowledgements

Many thanks to Dr. Candice Majewski of the University of Sheffield and Dr. Dominic Eggbeer and Sean Peel of the National Centre for Product Design and Development Research, Cardiff, for building the SLS® and SL prototype splints, respectively. Thanks to Nigel Bunt and Sarah Drage of HK Rapid Prototyping Ltd. for building the black and white Objet Connex splint, to Mark Tyrtania at LaserLines for the FDM splint, and to Phil Dixon, Loughborough University, for assistance with the building of numerous Objet Connex prototypes. The research was carried out as part of a PhD research project at Loughborough University.

References

Austin, N. M. (2003a). Process of splinting. In M. Jacobs, & N. M. Austin (Eds.), *Splinting the hand and upper extremity: principles and process* (pp. 88–99). Baltimore: Lippincott Williams & Wilkins. ISBN-13: 978-0683306309.

Austin, N. M. (2003b). Equipment and materials. In M. Jacobs, & N. M. Austin (Eds.), *Splinting the hand and upper extremity: principles and process* (pp. 73–87). Baltimore: Lippincott Williams & Wilkins. ISBN-13: 978-0683306309.

Bibb, R. (2006). *Medical modelling: the application of advanced design and development techniques in medicine.* Cambridge: Woodhead Publishing, ISBN-13: 978–1845691387.

Bibb, R., Eggbeer, D., Evans, P., Bocca, A., & Sugar, A. (2009). Rapid manufacture of custom-fitting surgical guides. *Rapid Prototyping Journal, 15*(5), 346–354. http://dx.doi.org/10.1108/13552540910993879.

Bingham, G. A., & Hague, R. J. M. (2013). Efficient three dimensional modelling of additive manufactured textiles. *Rapid Prototyping Journal, 19*(4), 269–281. http://dx.doi.org/10.1108/13552541311323272.

Bingham, G. A., Hague, R. J. M., Tuck, C. J., Long, A. C., Crookston, J. J., & Sherburn, M. N. (2007). Rapid manufactured textiles. *International Journal of Computer Integrated Manufacturing, 20*(1), 96–105. http://dx.doi.org/10.1080/09511920600690434.

Breuninger, J. (2010). *Voronoi wrist splint [conversation]* (personal communication, 26 November 2010).

Callinan, N. J., & Mathiowetz, V. (1996). Soft versus hard resting hand splints in rheumatoid arthritis: pain relief, preference and compliance. *American Journal of Occupational Therapy, 50*(5), 347–353. http://dx.doi.org/10.5014/ajot.50.5.347.

Campbell, R. I. (2006). Customer input and satisfaction. In N. Hopkinson, J. M. Hague, & P. M. Dickens (Eds.), *Rapid manufacturing: an industrial revolution for the digital age* (pp. 19–38). Chichester: John Wiley & Sons, Ltd. ISBN-13: 978-0470016138.

Cook, D., Gervasi, V., Rizza, R., Kamara, S., & Xue-Cheng, L. (2010). Additive fabrication of custom pedorthoses for clubfoot correction. *Rapid Prototyping Journal*, *16*(3), 189–193. http://dx.doi.org/10.1108/13552541011034852.

Coppard, B. (2001). Anatomical and biomechanical principles of splinting. In B. M. Coppard, & H. Lohman (Eds.), *Introduction to splinting: a clinical-reasoning & problem-solving approach* (2nd ed.) (pp. 34–72). St Louis, MO: Mosby, Inc. ISBN-13: 978-0323009348.

Coppard, B. M., & Lynn, P. (2001). Introduction to splinting. In B. M. Coppard, & H. Lohman (Eds.), *Introduction to splinting: a clinical reasoning & problem-solving approach* (2nd ed.) (pp. 1–33). St Louis, MO: Mosby, Inc. ISBN-13: 978-0323009348.

Cortex, E. J. (2013). Available at http://jakevilldesign.dunked.com/cortex. Accessed 01.08.13.

Eggbeer, D., Bibb, R., Evans, P., & Ji, L. (2012). Evaluation of direct and indirect additive manufacture of maxillofacial prostheses. *Proceedings of the Institution of Mechanical Engineers, Part H: Journal of Engineering in Medicine*, *226*(9), 718–728. http://dx.doi.org/10.1177/0954411912451826.

Faustini, M. C., Neptune, R. R., Crawford, R. H., & Stanhope, S. J. (2008). Manufacture of passive dynamic ankle-foot orthoses using selective laser-sintering. *IEEE Transactions of Biomedical Engineering*, *55*(2), 784–790.

Fried, S. (2007). *Splint and or method of making same. US patent application 20070016323 A1.*

Fried, S., Michas, L., & Howard, J. (2005). *Method of providing centralized splint production. US patent application US20050015172 A1.*

Gervasi, V., Cook, D., Rizza, R., Kamara, S., & Liu, X. (2009). Fabrication of custom dynamic pedorthoses for clubfoot correction via additive-based technologies. In Bourell D., (Ed.), *Proceeding of the 20th annual international solid freeform fabrication symposium*, Austin, TX, pp. 652–661.

Gibson, K. S., Woodburn, J., Porter, D., Telfer, S. (2014). Functionally optimised orthoses for early rheumatoid arthritis foot disease: a study of mechanisms and patient experience. *Arthritis Care & Research*, *66*(10), 1456–1464. http://dx.doi.org/10.1002/acr.22060.

Grzesiak, A. (July 08, 2010). Fraunhofer additive manufacturing alliance–highlights, current RTD activities and strategic topics. In Fraunhofer (Ed.), *Proceedings of the additive manufacturing international conference* (p. 14). Loughborough, UK.

Jacobs, M. (2003). Splint classification. In M. Jacobs, & N. Austin (Eds.), *Splinting the hand and upper extremity: principles and process* (pp. 2–18). Baltimore: Lippincott Williams & Wilkins. ISBN-13: 978-0683306309.

Jacobs, M., & Austin, N. (2003). Splint fabrication. In M. Jacobs, & N. Austin (Eds.), *Splinting the hand and upper Extremity: principles and process* (pp. 98–157). Baltimore: Lippincott Williams & Wilkins. ISBN-13: 978-0683306309.

Johnson, A., Bingham, G. A., & Wimpenny, D. I. (2013). Additive manufactured textiles for high-performance stab resistant applications. *Rapid Prototyping Journal*, *19*(3), 199–207. http://dx.doi.org/10.1108/13552541311312193.

Lohman, H. (2001). Wrist immobilisation splints. In B. M. Coppard, & H. Lohman (Eds.), *Introduction to splinting: a clinical reasoning & problem-solving approach* (2nd ed.) (pp. 139–184). St Louis, MO: Mosby, Inc. ISBN-13: 978-0323009348.

Lohman, H., Poole, S. E., & Sullivan, J. L. (2001). Clinical reasoning for splint fabrication. In B. M. Coppard, & H. Lohman (Eds.), *Introduction to splinting: a clinical & problem-solving approach* (2nd ed.) (pp. 103–138). St Louis, MO: Mosby, Inc. ISBN-13: 978-0323009348.

Mavroidis, C., Ranky, R., Sivak, M., Patritti, B., DiPisa, J., Caddle, A., et al. (2011). Patient specific ankle-foot orthoses using rapid prototyping. *Journal of Neuro Engineering and Rehabilitation*, *8*(1), 1. http://dx.doi.org/10.1186/1743-0003-8-1.

Moor, J. P. and Williams, C. B. (2012). Fatigue characterisation of 3D printed elastomer material. In D. Bourell, R. H. Crawford, C. C. Seepersad, J. J. Beaman, & H. Marcus (Eds.), *Proceedings of the twenty third annual international solid freeform fabrication symposium - an additive manufacturing conference, August 6–8, 2012* (pp. 641–655).

Oxman, N. (2010). *Material-based design computation. (Ph.D. thesis), Massachusetts Institute of Technology.*

Oxman, N. (2011). Variable property rapid prototyping. *Virtual and Physical Prototyping*, 6(1), 3–31. http://dx.doi.org/10.1080/17452759.2011.558588.

Oxman, N. (2012). *Imaginary beings @ Centre Pompidou.* Available at http://materialecology.blogspot.co.uk/2012/05/imaginary-beings-centre-pompidou.html. Accessed June 18, 2012.

Pagnotta, A., Korner-Bitensky, N., Mazer, B., Baron, M., & Wood-Dauphinee, S. (2005). Static wrist splint use in the performance of daily activities by individuals with rheumatoid arthritis. *Journal of Rheumatology*, 32(11), 2136–2143.

Pallari, J. H. P., Dalgarno, J., Munguia, J., Muraru, L., Peeraer, L., Telfer, S., et al. (2010). Design and additive fabrication of foot and ankle-foot orthoses. In D. Bourell (Eds.), *Proceedings of the twenty first annual international solid freeform fabrication proceedings - an additive manufacturing conference, August 9–11* (pp. 834–845).

Palousek, D., Rosicky, J., Koutny, D., Stoklasek, P., & Navrat, T. (2013). Pilot study of the wrist orthosis design process. *Rapid Prototyping Journal*, 20(1), 27–32. http://dx.doi.org/10.1108/RPJ-03-2012-0027.

Paterson, A. M. (2013). *Digitisation of the splinting process: exploration and evaluation of a computer aided design approach to support additive manufacture* (Ph.D. thesis). Loughborough University.

Paterson, A. M., Bibb, R. J., & Campbell, R. I. (2010). A review of existing anatomical data capture methods to support the mass customisation of wrist splints. *Virtual and Physical Prototyping Journal*, 5(4), 201–207. http://dx.doi.org/10.1080/17452759.2010.528183.

Paterson, A. M., Bibb, R. J. & Campbell, R. I. (2012). Evaluation of a digitised splinting approach with multiple-material functionality using additive manufacturing technologies. In D. Bourell, R. H. Crawford, C. C. Seepersad, J. J. Beaman, & H. Marcus (Eds.), *Proceedings of the 23rd annual international solid freeform fabrication symposium – an additive manufacturing conference, Austin, TX, August 6–8, 2012* (pp. 656–672).

PopTech. (2009). *Neri Oxman: on designing form.* Available at http://www.youtube.com/watch?v=txl4QR0GDnU. Accessed 17.04.10.

Smith, S. (2011a). *Miraclefeet and objet bring innovation to braces.* Available at http://www.deskeng.com/articles/aabamt.htm. Accessed 25.05.11.

Smith, S. (2011b). *Micraclefeet and objet bring innovation to braces.* Available at http://www.deskeng.com/articles/aabamt.htm. Accessed 07.08.12.

Stratasys. (2013). *Objet digital materialsTM data sheets [online PDF].* Available at http://www.stratasys.com/materials/polyjet/~/media/879CBF2F6582406C9C1B629F4F9E05D5.ashx. Accessed 04.04.13.

Summit, S., & Trauner, K. B. (2010a). *Custom braces, casts and devices having limited flexibility and methods for designing and fabricating. US patent application 2010/0268138A1.*

Summit, S., & Trauner, K. B. (2010b). *Modular custom braces, casts and devices and methods for designing and fabricating. US patent application 20100268135A1.*

Veehof, M. M., Taal, E., Willems, M. J., & van de Laar, M. A. F. J. (2008). Determinants of the use of wrist working splints in rheumatoid arthritis. *Arthritis Care & Research*, 59(4), 531–536. http://dx.doi.org/10.1002/art.23531.

Wohlers, T. T. (2012). *Wohlers report 2012: additive manufacturing and 3D printing state of the industry. Annual worldwide progress report.* Colorado: Wohlers Associates, Inc.

5.20 Orthotic rehabilitation applications case study 3: evaluation of a digitised splinting approach with multiple-material functionality using additive manufacturing technologies

Acknowledgements

The work described in this case study was first reported in the references below and is reproduced here in part or in full with the permission of the University of Texas at Austin.

Paterson, A. M., Bibb, R. J., & Campbell, R. I. (2012). Evaluation of a digitised splinting approach with multiple material functionality using additive manufacturing technologies. In D. Bourell, R. H. Crawford, C. C. Seepersad, J. J. Beaman & H. Marcus (Eds.), *Proceedings of the Twenty-Third Annual International Solid Freeform Fabrication Symposium – An Additive Manufacturing Conference. Austin, TX: University of Texas at Austin* (pp. 656–672).

5.20.1 Introduction

Rheumatoid arthritis (RA) is a chronic, systemic autoimmune disease that typically affects joints within the hands, wrists, ankles and feet (Biese, 2007; Melvin, 1982). Symptoms can include inflamed synovia and tendon sheaths and destruction of cartilage and bone, resulting in pain and discomfort (Melvin, 1982). Borenstein, Silver, and Jenkins (1993) stated that the approach to RA treatment is a multilayered pyramid consisting of 'education, physical and occupational therapy, rest and nonsteroidal anti-inflammatory drugs' (p. 545). Occupational therapy in particular addresses limitations that patients may encounter during everyday activities in an attempt to circumvent the limitations, and to improve well-being and quality of life. One method of intervention is splint prescription, the perceived benefits of which are multifaceted (Colditz, 1996; Jacobs, 2003; Melvin, 1982; Taylor, Hanna, & Belcher, 2003):

1. Relieves pain through immobilisation and protection of affected joints.
2. Protects painful contractures from impacts, scarring, and excessive movement.
3. Promotes movement of stiff joints through immobilisation of more mobile joints.
4. Encourages healing of fractures and contractures.
5. Prevents or corrects deformities and contractures.
6. Rests affected joints.
7. Provides support.

Practitioners such as occupational therapists and physiotherapists may prescribe either custom-made or off-the-shelf prefabricated splints. However, this chapter describes and discusses the design and fabrication methods of custom-made Static Wrist Immobilisation (SWI) splints in particular (shown in Figure 5.148), because they are one of the most commonly prescribed splints among a range of conditions (Stern, 1991).

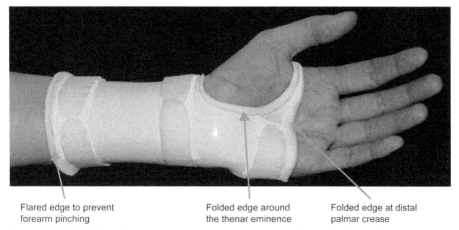

Flared edge to prevent   Folded edge around   Folded edge at distal
forearm pinching   the thenar eminence   palmar crease

Figure 5.148 Traditionally manufactured static wrist immobilisation splint and common characteristics.

Therapists must take into account the likelihood of their patients adhering to their splint-wearing regimens. Unfortunately, Sandford, Barlow, and Lewis (2008) found that two-thirds of patients reported non adherence, and advised therapists to be aware of and acknowledge low adherence levels. There are several reasons for poor patient adherence in terms of wear duration and frequency (Callinan & Mathiowetz, 1996; Melvin, 1982; Sandford et al., 2008; Spoorenberg, Boers, & Linden, 1994; Taylor et al., 2003):

- The splint does not address the patient's condition.
- The patient has received insufficient information about the condition.
- The patient has not had sufficient information justifying the need for the splint.
- The splint is unattractive.
- The splint is difficult to don or doff.
- The splint is uncomfortable to wear.
- The splint is impractical in certain environments or for certain tasks.
- The patient may not be interested or informed about the potential beneficial outcome of wearing the splint.

Hygiene issues may also contribute to low adherence; open-cell padding within splints can absorb moisture, such as perspiration, which can lead to odours and collection of bacteria (Coppard & Lynn, 2001). More generally and in terms of assistive devices, Louise-Bender Pape, Kim, and Weiner (2002) reviewed literature linking perceived social stigma to the association of assistive devices, which can contribute to nonuse of such devices. In addition, 78% of 27 participants reported immobilisation splints as unwieldy (Spoorenberg et al., 1994).

5.20.1.1   The splinting process

Please refer to Figure 5.134 in the previous case study, which depicts the splint design and fabrication workflow, deduced from the literature (Jacobs & Austin, 2003; Kiel, 1983; Lohman, 2001). The therapist must choose the low-temperature thermoplastic

(LTT) based on a variety of properties before splint forming, including contour conformability, thickness and colour. Colour selection by the patient may be encouraged in a bid to improve adherence (Coppard & Lynn, 2001). However, choice is often dictated by stock available in clinics or properties that are deemed more important by the therapist, such as thickness. In addition, therapists are advised to consider and apply padding before forming the LTT, to avoid inducing pressure to prone areas (Cooper, 2007; Coppard & Lynn, 2001; Taylor et al., 2003). Padding types can vary from sheets and stockinettes to silicone gel discs or pads, which can be placed over bony prominences for cushioning (Coppard & Lynn, 2001). Other common SWI splint characteristics are proximal edge flaring to avoid forearm pinching, and folding of both the distal and thenar edges to provide a more comfortable edge against the skin while adding rigidity to the splint (Figure 5.148).

Although many of the characteristics described are integrated to address functional needs, results may appear unwieldy, voluminous, and unsightly. For example, if cavities are integrated over bony prominences to relieve pressure, the appearance of the altered topology can be compromised. In addition, folded or rolled edges add volume to the palmar region, potentially affecting palmar grasp capacity. In terms of the fabrication process, there is little room for error; the further the therapist progresses down the fabrication workflow, the more challenging adjustments may be to make. If additional or replacement splints are required, the whole process must be repeated.

5.20.1.2 Additive manufacturing for upper extremity splint fabrication

Several research studies have explored the use of AM for upper extremity splint fabrication. Fraunhofer IPA fabricated a prototype SWI splint using an EOS P100 LS system (EOS GmbH, Krailling, Germany) with polyamide (PA2200) powder (Breuninger, 2012); the benefits were improved aesthetics, bespoke fit, and integral fasteners to exploit the AM capabilities of part consolidation. A hand immobilisation (HI) splint prototype developed by Materialise (fabricated using an EOS P730 LS machine with PA2200 powder) gave a reduction in weight (Pallari, 2011). Oxman (2011) fabricated *Carpal Skin* using Objet Connex500™ technologies (Objet Geometries, Rehovot, Israel), by exploiting the system's capabilities to enable multiple-material builds (Objet Ltd, 2012a). The structures within the splint-like gloves were dictated by a predefined pain map to allow or restrict movement via a reaction-diffusion pattern (Oxman, 2011). Not only does the use of AM in the context of upper extremity splinting allow for exploitation of bespoke fit and function, both prototypes by Oxman (2011) and Fraunhofer IPA demonstrated improved aesthetics as a result of geometric complexity and freedom, synonymous characteristics that are only viable as a combination through AM.

Sketches shown in Figure 5.149 by Bibb (2009) show the intent of AM for splint fabrication with lattice feature integration. The lattice structures were intended to look aesthetically pleasing while promoting airflow to the skin, in a bid to reduce perspiration. An additional intent was to enable patients to personalise splints by choosing their own perforation patterns in an attempt to improve patient adherence.

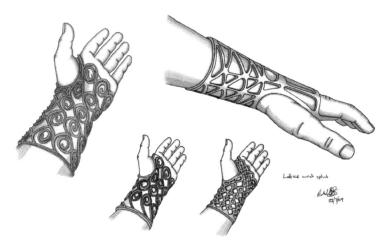

Figure 5.149 Additive manufacturing splint concepts.

The investigators were also interested in exploring additional and novel approaches for multiple-material integration into splints, particularly the use of elastomeric regions for a variety of new design features as a result of Objet Connex capabilities. For example, radial and/or ulnar-based elastomer hinges could be integrated into the splint to allow for easier donning and doffing. In addition, specific elastomer regions could be placed over bony prominences (similar to gel discs and pads in traditional splinting). Both concepts could be achieved without affecting the overall topology of the splint. It was anticipated that regional patches could protect and cushion bony prominences while providing dynamic pressure over areas prone to fluctuating oedema with an aim to prevent oedema pooling.

However, if such a design and manufacturing approach is to be realised for upper extremity splinting in a clinical domain, supporting software technologies need to be developed and tailored to suit the target user. Knoppers and Hague (2006); Pallari, Dalgarno, and Woodburn (2010); Rogers, Stephens, Gitter, Bosker, and Crawford (2000) supported the need for product-specific Computer-Aided Design (CAD) software for practitioners, to enable the intended users to adopt quickly and easily without dedicating time to learn. In addition, Smith (1991) justified the need for customised software, because it gives the intended users the tools to achieve what they want and need while removing unnecessary functions.

5.20.2   Research aim and objectives

A number of weaknesses in traditional splinting have been identified, as well as the strengths and feasibility of AM in splint fabrication. The need for customised 3D CAD software to support the use of AM has also been noted. Therefore, the aim of this chapter was to evaluate the feasibility of new design features for upper extremity splinting only made viable through AM, as well as the digitisation of traditional splint characteristics. Objectives to achieve the aim are:

Objective 1. The replication of key splint features/characteristics in a virtual environment.

Objective 2. Exploration of new features only viable through AM, which could potentially address concerns regarding aesthetics, hygiene, comfort and form. Features include:

2.1. Aesthetically appealing and personalised lattice integration.
2.2. Multiple-material integration (using Objet Connex technologies) to
2.2.1. Imitate gel-discs/pads.
2.2.2. Potentially replace the need to create bulbous features in splints.
2.2.3. Provide dynamic pressure over areas susceptible to fluctuating oedema.
2.2.4. Integration of flexible hinges for easier donning/doffing.

Objective 3. Evaluation of the feasibility of the digitised splinting approach via a specialised CAD software prototype:

3.1. Refine replicated characteristics into a virtual workflow.
3.2. Devise a representation of the digitised approach via a software prototype.
3.3. Evaluate the digitised approach, specifically new features as a result of AM.

The scope of this investigation was purely related to the data manipulation of acquired patient scan data within 3D CAD to support AM splint fabrication. The investigation does not address data acquisition methods suitable for clinical environments, although Paterson, Bibb, and Campbell (2010) reviewed data acquisition methods to generate test data in support of this investigation. Similarly, the process does not investigate the use of Finite Element Analysis (FEA) for assessing the structural integrity of splints, although it can be assumed that this would be a crucial feature if the digitised approach were to be realised. Also, the digitised splinting approach is not intended to make the splinting profession obsolete, nor does it imply the redundancy of splinting practitioners. The intended proposition is to provide therapists with an additional toolset to capture the design intent quickly and easily, while addressing the needs and concerns of patients through improved aesthetics, fit, and functionality.

A systematic approach was adopted to develop a sequence of functions to meet the objectives. Each objective will be addressed individually.

5.20.3 Methods

5.20.3 1 Objective 1: replication of key splint features/ characteristics in a virtual environment

Critical splint characteristics required for the specialised splinting workflow were tested iteratively with a range of 3D CAD software packages, including Geomagic Studio (Geomagic, 2010) and McNeel Rhinoceros (Robert McNeel & Associates, 2011), using an action research strategy. The approach was intended to test and refine different CAD tools and CAD strategies to capture and replicate the design intent of typical splint features, such as pressure relief cavities. An example of the iterative testing is described by Paterson et al. (Paterson, Bibb, & Campbell, 2010).

5.20.3.2 Objective 2: explore new features only viable through AM

Objective 2.1: aesthetically appealing and personalised lattice integration

Similar to addressing Objective 1, an action research strategy was adopted to explore lattice applications to a previously defined splint surface. After iterative testing, it became apparent that an additional feature within the workflow was required to form a splint border, which would encase the lattice structure. This would address inherent issues of sharp protrusions, which could otherwise cause lacerations if borders were not implemented into the proposed splints.

Objective 2.2: multiple-material integration/heterogeneous structures

As mentioned previously, the purpose of using Objet Connex technologies was to explore the viability of new features in the context of splinting, making the transition from homogeneous to heterogeneous splints. Objet Connex technologies can deposit materials in predefined combinations, which dictate regional material proportions to form so-called Digital Materials™ (DM). For example, to create variations in shore hardness among the Objet flexible (FLX) FLX97-DM range, the primary material (TangoPlus) and secondary material (VeroWhitePlus) are interspersed by varying dual-jet distribution using Objet's PolyJet Matrix Technology. The material depositions are then cured using UV light (Objet, 2012; Objet Ltd, 2012b).

There were two main approaches to consider for multiple-material integration: continuous functional grading (CFG) versus stepped functional grading (SFG) (Figure 5.150). Continuous functional grading proposes a gradual transition of one or more properties across a defined volume, and has been of significant interest across a range of disciplines. However, the implementation of such a feature in terms of CAD modelling strategies, file export, and AM systems is still under development (Erasenthiran & Beal, 2006). Knoppers and Hague (2006) and Siu and Tan (2002) highlighted the difficulties of representing functionally graded geometries in boundary representation (B-rep) modelling, because the geometries were defined by a series of surfaces to determine the topology of a closed volume. If functional grading were to be a viable feature

	Continuous functional grading (CFG)	Stepped functional grading (SFG)
TangoPlus VeroWhite DM9740		
TangoPlus VeroWhite DM9750		
TangoPlus VeroWhite DM9760	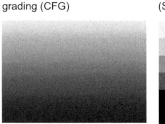	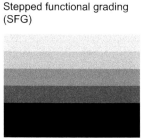
TangoPlus VeroWhite DM9770		
TangoPlus VeroWhite DM9785		
TangoPlus VeroWhite DM9795		

Figure 5.150 Continuous functional grading (CFG) versus specific boundary representation (SBR) for SFG.

in splints, an alternative or additional modelling strategy within the CAD methodology (e.g. voxel-based modelling) would need to be adopted. A consequential limitation, however, is the increased memory power, processing consumption and modelling complexity during the design phase (Erasenthiran & Beal, 2006; Oxman, 2011). This would add unnecessary complexity to the geometry and CAD strategies. In addition, current Objet Connex technologies only offer a finite number of predefined deposition variations, which limits the CFG approach. As a result, CFG was considered excessively complex and unnecessary for this particular application, and it was decided that SFG would be sufficient to deliver the finite variation that Objet Connex technologies offer.

The disadvantage of SFG is that it creates boundary lines between adjacent materials properties, which, depending on the differences in the adjacent properties, could lead to stress concentrations at the boundary or abrupt changes in flexural modulus that could cause unwanted folds or creases at the boundary. The worst-case scenario can be imagined to be a choice of only two materials: for example, very rigid and very flexible. In that circumstance, the abrupt change at the boundary is likely to present problems. This would be mitigated by the ability to specify a greater number of regions and a sufficiently high number of materials property choices. It can be assumed that for a given application, a sufficiently high number of materials choices could satisfactorily approximate continuous variation.

Objective 2.2.1: imitate gel-discs/pads

This objective was to explore the possibility of using multi-material capability to reproduce the function of gel pads. In existing practice, the rigid shell of the splint is raised in local areas to provide space to insert soft pads. These are required to reduce pressure on bony prominences and improve comfort. The ability to define material properties within a localised region enables the body of the splint to be made soft in the required area. This eliminates the pads and reduces bulk of the splint.

Objective 2.2.2: potentially replace the need to create bulbous features in splint

This is essentially the same function as 2.2.1 except that in existing practice some areas of the splint may be raised to provide space for sensitive areas or bony prominences but without placing a gel disc/pad into the space.

Objective 2.2.3: provide dynamic pressure over areas susceptible to fluctuating oedema

This function is currently difficult to accommodate with a single material rigid splint. The ability to alter the flexibility of the splint, possibly in specific regions, would enable a conformal fit while allowing some degree of expansion that would accommodate temporary or fluctuating swelling. This would be a unique advantage of the AM approach over existing practice.

Objective 2.2.4: integration of flexible hinges for easier donning/doffing

With existing practice, the rigid splint has to incorporate large openings to allow donning and doffing. This leaves large regions unsupported. In addition, straps and fasteners have to be added into or onto the splint, which adds construction time, increases bulk, and adds features that will be more susceptible to damage and wear. The ability to build in flexible joints or hinges enables a splint design with greater coverage and therefore increased support. The ability to build in hinges and fasteners reduces the bulk and the time required to add those features after forming the splint. This approach has many other potential advantages in eliminating purchasing and stocking fasteners, straps, etc. In addition, frequently straps and fasteners such as Velcro wear, fail, or become unacceptably soiled in use earlier than the splint itself.

In terms of 3D CAD strategy testing to support the digitised splinting approach, the refined strategy for enabling multiple-material builds was to create a curve on the splint surface to act as a trim boundary before pattern application. After applying the pattern to the surface, the splint surface could then be trimmed (Figure 5.151(a)). Separate patches could then be thickened (Figure 5.151(b) and (c)), and exported as separate shells in a single STL file. Each shell can subsequently be assigned a digital material in the Objet preparation software. For example, in Figure 5.151(c) there are three shells. The predominant grey shell could be a rigid material, the green shell could be an intermediate flexible material, and the red shell could be soft.

5.20.3.3 Objective 3: evaluation of approach via a specialised splinting CAD software prototype

The requirements established after CAD testing were refined into a specialised CAD software workflow (Objective 3.1) (Figure 5.152). The order was determined by considering the traditional splinting workflow depicted in Figure 5.134 (see Case study 2 in Section 5.19) as well as constraining/best practice CAD approaches (e.g. detailed application to surfacing before thickening). An important feature to note in

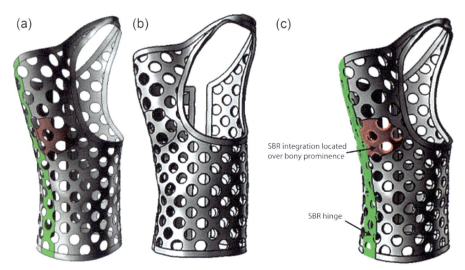

Figure 5.151 Example of elastomer integration over a bony prominence and medial/ulnar hinge, (a) three separate surfaces present, (b) thickened splint and (c) regions of splints, depicting different materials.

the digitised approach was the intent for traversal operation, the ability to move back and forth between different features independent of one another. Fundamentally, this was anticipated to be an important step for splint fabrication compared with traditional splint fabrication, in which the further one progresses through design and fabrication, the more difficult and time-consuming it may be to make adjustments.

Having established a refined workflow, a high-fidelity concept software prototype was developed within Microsoft Access 2010 (Microsoft Corporation, 2010a) using

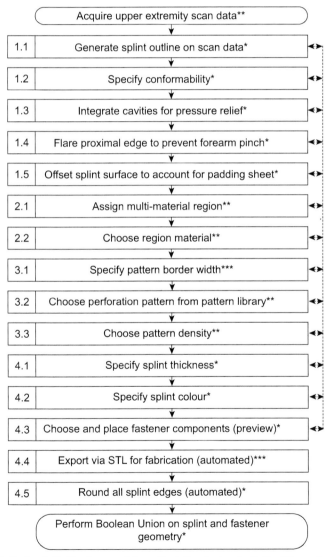

Figure 5.152 Refined digitisation workflow: *existing feature/activity, **new feature/activity, ***necessary feature as a result of a new feature/activity.

Visual Basic for Applications 2010 (Microsoft Corporation, 2010b), which could be used to depict the workflow for final evaluation (Objective 3.2). The prototype featured a viewport and a series of slider controls and drop-down boxes to control features such as pattern density (Figure 5.153). When controls were adjusted, the image in the viewport would change to suit the recent adaptation to provide direct visual feedback to the user. The prototype also featured a pattern library and a separate fastener library, which allowed users to browse different perforation patterns and fasteners. The user

Figure 5.153 Prototype interface example, and resulting changes from altering pattern perforation density.

would then be able to select the desired options from the libraries for automatic place-ment to the splint geometry in the viewport.

Ten splinting practitioners within the United Kingdom were invited to participate in evaluation sessions (physiotherapist: $n = 2$; occupational therapist: $n = 8$). Eight participants had one-to-one sessions, whereas two joined in one session for their con-venience. Each evaluation session was composed of four activities: a briefing into the intent of AM for splint fabrication, a demonstration of the prototype, user trials of the software prototype, and a semi-structured interview. Qualitative data were captured regarding the digitised approach, together with specialist feedback regarding multi-ple-material integration into splints. Participants completed a demographic question-naire to gather data including the number of years of splinting experience. In addition, participants signed informed consent forms agreeing to the capture of audio and com-puter screen-capture recordings while testing the prototype. The recordings were used to identify trends in opinions and were transcribed into NVivo (NVivo, 2011), for cod-ing and trend identification. In addition, proof-of-concept splints were manufactured using various AM processes to give physical, tactile representations of the intended output (Figure 5.154). The purpose was to show participants what could be achieved using AM, to ensure that the intent was conveyed effectively to participants in the context of upper extremity splinting.

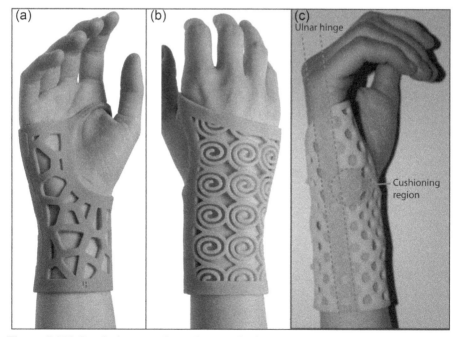

Figure 5.154 Proof-of-concept AM splints, (a) fabricated using 3D Systems SLA 250 (3D Systems Xtreme material), (b) fabricated using Objet Connex500 (RGD5160-DM), (c) fabricated using Objet Connex500, with multi-material regions (ulnar hinge: FLX9760-DM; cushioning region: FLX9740-DM; remainder of splint: VeroWhitePlus FullCure 835).

5.20.4 Results and discussion

A wide range of positive comments and suggestions were gathered during the evaluation sessions. The approach not only highlighted many exciting new avenues for exploration within splint design and fabrication, it expanded the window of opportunity for Objet Connex technologies. Participants mentioned that the integration of multiple materials within a single splint was a completely new and exciting toolbox for therapists that could not be achieved using existing fabrication methods. Participants were interested in new applications as a result of multiple-material capabilities, such as protection and cushioning of bony prominences without compromising the splint topography. Of the three participants who were asked, all participants favoured mono-splints with elastomer hinges (as shown in Figure 5.154(c)), to make donning and doffing easier for patients with restricted dexterity; two participants even suggested having two hinges placed along the radial and ulnar aspects of the wrists, to assist further. Three participants were interested in using elastomer materials to apply dynamic pressure over areas prone to fluctuating oedema. In extension to these applications, one participant suggested the use of elastomer materials in splints for treatment or management of burns and scars, and protection of post-surgical metalwork (i.e. alignment screws and brackets). Another participant suggested integration of elastomer borders on splint edges, which could aid in pressure distribution and provide a softer interface between the more rigid structure of the splint and the skin. Therapists also expressed interest in allowing partial dynamic features to allow or restrict movement, similar to *Carpal Skin* by Oxman (2011).

Many participants suggested additional creative features to be integrated within the digitised workflow, such as multicolour ability; one participant asked whether photographs could be applied to splints. Multicolour and multi-material single builds are possible with the newest release of the Objet Connex3 (Stratasys, 2014), but the material properties require further development before they may be suitable for this particular application. Therapists were also keen on part consolidation with regard to fastener integration. Participants felt that the workflow depicted in the software prototype demonstrated that used in traditional fabrication, with the exception of new features, and placement of colour and thickness controls. In response to traversal movement within the workflow, eight participants thought this was a useful feature, particularly because variables could be altered independently from one another; a desirable novelty compared with traditional splinting.

The intent of edge filleting in the CAD workflow (Figure 5.152, flowchart item 4.5) was to relieve the need for rolling of splint edges, consequently reducing cumbersome structures presented in traditional splinting. Therapists were surprised at the strength and rigidity of the prototype splints without having folded or rolled edges within the palm. One participant was interested to know whether this would improve palmar grasp capacity, because material volume was reduced. However, some participants thought they would still want the opportunity to produce the effect of rolled and folded edges within the splinting software.

Despite the abundance of positive responses, participants also raised concerns regarding data acquisition methods, structural integrity, fabrication time, and cost of the approach. Two participants were concerned that oedematous, inflamed regions might protrude through the lattice structure, potentially causing discomfort for the patient. Another participant thought that not only would patient adherence be improved as a result of improved aesthetics, but the level of improved wear duration could potentially be detrimental; they were concerned that patients would be wearing splints for longer durations than is necessary or recommended, potentially resulting in splint dependency. Cooper stated that static splints should not be worn more than necessary because they contribute to stiffness, atrophy, and joint disuse (Cooper, 2007). Therefore, a more stringent wearing regimen might be required to combat this concern if AM splints were to be eventually prescribed.

5.20.5   Future work

Although the investigation highlighted many new positive and exciting applications within AM splint fabrication, there is much to be done before the approach can be used as a clinically viable treatment method. To date, the software prototype described in this chapter is purely a visual representation for evaluative purposes. Future development is required to create fully functional specialised splinting CAD software. New features must also be investigated for feasibility into workflow integration, such as edge folding. The ability to reuse previously defined splint files to make replacement splints must also be an integral feature in the CAD approach, because this would allow therapists to fabricate duplicate splints without incurring additional design time. The opportunity to adjust previously defined splint files should also be explored, so therapists may alter one or many design variables where required. Alternative file format exports are also required to support depiction of complex organic geometry, while allowing for multiple-material and mono/multicolour capabilities; the development of STL 2.0 or AM file format is promising, particularly for the composition of complex geometries and multiple-material builds in support of the Objet Connex capabilities (Hiller & Lipson, 2009). In terms of CAD process efficiency, Objet have developed CADMatrix™, a plug-in for material assignment within selected CAD applications before automated STL creation (Objet Ltd, 2009). If the proposed software prototype were developed, this approach could be applied within the custom software.

Cost was a concern expressed by participants that demands further work. It is too early in the development of the CAD approach to predict costs realistically. However, the advantages of the CAD/AM approach for future cost reduction is based on two premises. First, there is little scope for cost reduction in existing practice. The materials costs incurred in current practice are minimal and by far the greater proportion of cost is attributed to time and salary costs for the professionals involved. There is little scope for cost reduction of the existing materials and significant cost reduction would only be possible by speeding up work or reducing salaries, neither of which is likely to be acceptable, and any savings would be small and incremental. A shift to a CAD/AM approach would achieve cost savings by eliminating physical work (especially for remakes or replacements) and enabling a much faster design stage to be completed by professionals. Separation of the manufacture from the design enables designs to be done at any time, potentially without the patient being physically present, whereas manufacture does not incur salary costs. Although AM machine and materials costs are high, it is reasonable to assume that these costs will reduce significantly over time, as has already been demonstrated by other technologies. The potential advantages of a CAD/AM approach in the provision of custom-fitting medical devices have been recognised in other research, such as the provision of maxillofacial prosthetics (Bibb, Eggbeer, & Evans, 2010).

Further development of medical-grade materials conforming to ISO 10993 category standards will also be necessary for this application, in preparation for clinical trials. In response to burns and scar treatment, further research into suitable materials would be required. The multiple-material splint shown in Figure 5.154(c) also had a significant amount of warping and deformation at room temperature; these issues will also need to be addressed.

Although not within the focus of this investigation, FEA should also be investigated to address concerns regarding structural integrity, both the way in which it may be applied to the splint geometry in 3D CAD and the way in which the user would interact with the feature. The approach would need to be easy to use with the potential for automated alterations through analysis and regeneration of problem areas. Therapists expressed the interest into multiple-material splints to either permit or restrict movement, termed Synthetic Anisotropy by Oxman (2011). Therefore, research needs to be performed into how to achieve such results in terms of 3D CAD modelling and specification of Objet Connex materials, but also into the most efficient way to integrate such features within the specialised splinting CAD software.

In response to therapists' concerns into oedematous regions protruding through lattice structures, displacement mapping could be used on the outer splint surface to overcome such problems. This could be used instead of or in addition to material-to-perforation ratio alterations. This would resolve issues related to oedema and would also be a step forward in resolving concerns regarding structural integrity.

Acknowledgements

Our sincerest thanks to Lucia Ramsey at the University of Ulster, Belfast, Ella Donnison at Patterson Medical Ltd., Sutton-in-Ashfield, Jason Watson at Queens Medical Centre, Nottingham, and Dr. Dominic Eggbeer at The National Centre for Product Design and Development Research, Cardiff for their help, support and advice throughout the project. Extended thanks to all participants involved in the study, for their time, expertise, and valued feedback. This research was funded by Loughborough University.

References

Bibb, R. J. (2009). *Lattice within splints. Drawing. July 22.*
Bibb, R., Eggbeer, D., & Evans, P. (2010). Rapid Prototyping technologies in soft tissue facial prosthetics: current state of the art. *Rapid Prototyping Journal, 16*(2), 130–137. http://doi.org/10.1108/13552541011025852.
Biese, J. (2007). Arthritis. In C. Cooper (Ed.), *Fundamentals of hand therapy: clinical reasoning and treatment guidelines for common diagnoses of the upper extremity* (pp. 349–375). St Louis, MO: Mosby Inc., Elsevier Inc.

Borenstein, D. G., Silver, G., & Jenkins, E. (1993). Approach to initial medical treatment of rheumatoid arthritis. *Archives of Family Medicine*, 2(5), 545–551. http://doi.org/10.1001/archfami.1993.01850060113018.

Breuninger, J. (2012). *Wrist splint [email]* Message to A. M. Paterson (a.m.paterson@lboro.ac.uk). Sent 11 July 2012, 14:31(Accessed 11 July 14.53).

Callinan, N. J., & Mathiowetz, V. (1996). Soft versus hard resting hand splints in rheumatoid arthritis: pain relief, preference and compliance. *The American Journal of Occupational Therapy*, 50(5), 347–353. http://doi.org/10.5014/ajot.50.5.347.

Colditz, J. C. (1996). Principles of splinting and splint prescription. In C. A. Peimer (Ed.), *Surgery of the hand and upper extremity* (pp. 2389–2409). New York: McGraw-Hill.

Cooper, C. (2007). Fundamentals of clinical reasoning: hand therapy concepts and treatment techniques. In C. Cooper (Ed.), *Fundamentals of hand therapy: clinical reasoning and treatment guidelines for common diagnoses of the upper extremity* (pp. 3–21). St Louis, MO: Mosby Inc., Elsevier Inc.

Coppard, B. M., & Lynn, P. (2001). Introduction to splinting. In B. M. Coppard, & H. Lohman (Eds.), *Introduction to splinting: a clinical reasoning & problem-solving approach* (2nd ed.) (pp. 1–33). St Louis, MO: Mosby, Inc. ISBN-13: 978-0323009348.

Erasenthiran, P., & Beal, V. (2006). Functionally graded materials. In N. Hopkinson, R. J. M. Hague, & P. M. Dickens (Eds.), *Rapid manufacturing: an industrial revolution for the digital age* (pp. 103–124). Chichester: John Wiley & Sons, Ltd.

Geomagic. (2010). *Geomagic studio version 12 [software]*. Durham, NC: Geomagic.

Hiller, J.D., & Lipson, H. (2009). STL 2.0: a proposal for a universal multi-material additive manufacturing file format. In D. Bourell (Ed.), *Proceedings of the twentieth annual international solid freeform fabrication symposium* (pp. 266–278). University of Texas, Austin, TX, August 3–5.

Jacobs, M. (2003). Splint classification. In M. Jacobs, & N. Austin (Eds.), *Splinting the hand and upper extremity: principles and process* (pp. 2–18). Baltimore: Lippincott Williams & Wilkins. ISBN-13: 978-0683306309.

Jacobs, M., & Austin, N. (2003). Splint fabrication. In M. Jacobs, & N. Austin (Eds.), *Splinting the hand and upper extremity: principles and process* (pp. 98–157). Baltimore: Lippincott Williams & Wilkins. ISBN-13: 978-0683306309.

Kiel, J. H. (1983). *Basic hand splinting techniques: a pattern-designing approach* (1st ed.). Lippincott Williams & Wilkins. ISBN-13: 9780316491778.

Knoppers, R., & Hague, R. J. M. (2006). CAD for rapid manufacturing. In N. Hopkinson, R. J. M. Hague, & P. M. Dickens (Eds.), *Rapid manufacturing: an industrial revolution for the digital age* (pp. 39–54). West Sussex: John Wiley & Sons, Ltd.

Lohman, H. (2001). Wrist immobilisation splints. In B. M. Coppard, & H. Lohman (Eds.), *Introduction to splinting: a clinical reasoning & problem-solving approach* (2nd ed.) (pp. 139–184). St Louis, MO: Mosby, Inc. ISBN-13: 978-0323009348.

Louise-Bender Pape, T., Kim, J., & Weiner, B. (2002). The shaping of individual meanings assigned to assistive technology: a review of personal factors. *Disability & Rehabilitation*, 24(1–3), 5–20.

Melvin, J. L. (1982). *Rheumatic disease: occupational therapy and rehabilitation* (2nd ed.). Philadelphia: F. A. Davis Company.

Microsoft Corporation. (2010a). *Microsoft® Access® 2010 [software]* (Version 14.0.6112.5000 (64bit) ed.). Redmond, WA: Microsoft Corporation.

Microsoft Corporation. (2010b). *Microsoft® visual Basic® for applications 2010 [software]* Version 7.0 1625.

NVivo. (2011). *Software* (Version 9 ed.). Doncaster, VIC: QSR International Pty Ltd.

Objet L. (2012). *New objet materials: the power behind your 3D printer. [Datasheet PDF].* Available at http://www.objet.com/Portals/0/docs2/New%20materials%20data%20sheets_low%20res.pdf. Accessed 12.07.12.

Objet Ltd. (2009). *CADMatrixTM - Objet add-in for CAD solidworks®.* Available at http://objet.com/sites/default/files/CADMatrix_solidworks_A4_IL_low.pdf. Accessed 02.12.12.

Objet Ltd. (2012a). *Objet connex family.* Available at http://www.objet.com/3D-Printer/Objet_Connex_Family/. Accessed 02.06.12.

Objet Ltd. (2012b). *PolyJet MatrixTM 3D printing technology.* Available at http://www.objet.com/products/polyjet_matrix_technology/. Accessed 22.06.12.

Oxman, N. (2011). Variable property rapid prototyping. *Virtual and Physical Prototyping, 6*(1), 3–31. http://doi.org/10.1080/17452759.2011.558588.

Pallari, J. (2011). *A-footprint project [Email].* Sent to A. M. Paterson (a.m.paterson@lboro.ac.uk). Sent 9 August 2011, 16:13. Former Technical Product Manager at Materialise, Leuven, Belgium. All correspondence now to J. Pallari, Peacocks Medical Group, Newcastle upon Tyne, UK, 2011.

Pallari, J. H. P., Dalgarno, K. W., & Woodburn, J. (2010). Mass customisation of foot orthoses for rheumatoid arthritis using selective laser sintering. *IEEE Transactions on Biomedical Engineering, 57*(7), 1750–1756.

Paterson A. M., Bibb, R. J., & Campbell, R. I. (2012). Digitisation of the splinting process: development of a CAD strategy to support splint design and fabrication. In A. E. W. Rennie, C. E. Bocking (Eds.), *12th conference on rapid design, prototyping and manufacturing, June 17, 2011, Lancaster University* (pp. 97–104).

Paterson, A. M., Bibb, R. J., & Campbell, R. I. (2010). A review of existing anatomical data capture methods to support the mass customisation of wrist splints. *Virtual and Physical Prototyping Journal, 5*(4), 201–207. http://doi.org/10.1080/17452759.2010.528183.

Robert McNeel & Associates. (2011). *McNeel Rhinoceros® [software].* Version 4.0 Service Release 9. Seattle, Washington: Robert McNeel & Associates.

Rogers, B., Stephens, S., Gitter, A., Bosker, G., & Crawford, R. (2000). Double-wall, trans-tibial prosthetic socket fabricated using selective laser sintering: a case study. *Journal of Prosthetics and Orthotics, 12*(3), 97–100.

Sandford, F., Barlow, N., & Lewis, J. (2008). A study to examine patient adherence to wearing 24-hour forearm thermoplastic splints after tendon repairs. *Journal of Hand Therapy, 21*(1), 44–52. http://doi.org/10.1197/j.jht.2007.07.004.

Siu, Y. K., & Tan, S. T. (2002). Representation and CAD modeling of heterogeneous objects. *Rapid Prototyping Journal, 8*(2), 70–75.

Smith, M. F. (1991). *Software prototyping: adoption, practice and management.* Maidenhead: McGraw-Hill Book Company (UK) Ltd.

Spoorenberg, A., Boers, M., & Linden, S. (1994). Wrist splints in rheumatoid arthritis: a question of belief? *Clinical Rheumatology, 13*(4), 559–563.

Stern, E. B. (1991). Wrist extensor orthoses: dexterity and grip strength across four styles. *The American Journal of Occupational Therapy, 45*(1), 42–49. http://doi.org/10.5014/ajot.45.1.42.

Stratasys. (2014). *Objet500 Connex3-vivid color and multi-material 3D printing.* Available at http://www.stratasys.com/3d-printers/design-series/precision/objet500-connex3. Accessed 05.07.14.

Taylor, E., Hanna, J., & Belcher, H. J. C. R. (2003). Splinting of the hand and wrist. *Current Orthopaedics, 17*(6), 465–474. http://doi.org/10.1016/j.cuor.2003.09.001.

5.21 Orthotic rehabilitation applications case study 4: digitisation of the splinting process – development of a CAD strategy for splint design and fabrication

Acknowledgements

The work described in this case study was first reported in the references below and is reproduced here in part or in full with the permission of CRDM, Ltd.

Paterson, A. M., Bibb, R. J., & Campbell, R. I. (2011). Digitisation of the splinting process: development of a CAD strategy to support splint design and fabrication. In C. Bocking, & A. E. W. Rennie (Eds.), *Twelfth conference on Rapid Design, Prototyping and Manufacturing*, *CRDM Ltd.*: High Wycombe, pp. 97–104.

5.21.1 Introduction

Designing and fabricating custom-made splints is a highly skilled, creative process. Orthotists must understand anatomical and biomechanical principles of the upper extremities, the best strategy for addressing different ailments, and the biomechanical capabilities of each patient (Coppard, 2001). Splint designs and materials can vary depending on a patient's condition(s), daily lifestyle and activity levels (Lohman, Poole, & Sullivan, 2001). However, there are direct and indirect factors resulting in poor patient compliance in terms of wear duration and frequency.

The splinting process can be time-consuming and awkward for both the patient and the orthotist, particularly because the complexity of the splint increases; mistakes or weakened areas of splints often require additional support material to maintain the desired structural integrity, but inevitably reduce the aesthetic appeal of the splint (Coppard & Lynn, 2001). Splints should be low-profile, light, and finished professionally, because poor-looking splints will deter patients from wearing them, particularly in social environments, owing to the desire to fit in (Doman, Rowe, Tipping, Turner, & White, 2005; Louise-Bender Pape, Kim, & Weiner, 2002).

Poorly fitted splints also result in poor compliance, because they can cause discomfort and pain during everyday activities (Lohman et al., 2001). The splinting design and fabrication process can also be painful for some patients, because the orthotist may have to apply force onto thermoplastic templates while draping it onto the patient's extremities to fit it correctly. The design and fabrication of other custom-fitting assistive devices such as hearing aids have been digitised using 3D patient scan data acquisition methods and AM. However, the success of these advances is partially the result of the development of suitable 3D CAD software strategies specifically developed for specialised applications. To date, no CAD strategy for splint design has been published. It is anticipated that CAD use would reduce splinting time and provide a more comfortable clinic experience. An appropriate CAD strategy must be able to cater to the creative skills and clinical decision

Figure 5.155 Proposed digitised splinting process. This chapter addresses the sub-stages assign a splint outline and make alterations if necessary.

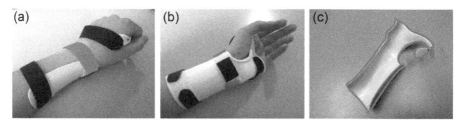

Figure 5.156 A custom-made static wrist immobilisation splint in use (a and b) and the outlined border of a static wrist immobilisation splint (c).

making that orthotists currently possess. However, it should be simple and intuitive and not require additional training to be used effectively (Pallari, Dalgarno, & Woodburn, 2010). Therefore, this chapter outlines part of a CAD strategy (Figure 5.155) that will directly manipulate 3D patient scan data to design and fabricate a more aesthetically pleasing, comfortable, and bespoke wrist splint using AM. In particular, the chapter focuses on the most suitable CAD tools to assign a splint shape for a static wrist immobilisation splint (Figure 5.156). The most suitable CAD methods had to replicate or potentially improve conventional splinting methods that are currently used to form a splint outline, as highlighted in Figure 5.156(c). Various CAD methods have been analysed, with strengths and weaknesses identified.

5.21.2   Current splinting techniques

To understand the needs of orthotists, a thorough understanding of the splinting processes was required. Conventional splint fabrication involves a trial-and-error approach involving various steps depending on the splint required for the condition. A static wrist immobilisation splint can be fabricated in 19 steps (Lohman, 2001). In summary, the patient's hand is placed palm-down on a piece of A4 paper and the orthotist draws around the patient's hand and forearm. The orthotist marks the anatomical landmarks on the paper (Figure 5.157), such as the radial styloid (marked A), ulnar styloid (B), second metacarpophalangeal (MCP) joint head (C), and the fifth MCP joint head (D). The remaining points in Figure 5.157 are offset from other anatomical landmarks. By intersecting the points, a rough splint template is drawn. The template is cut out and drawn around on a sheet of thermoplastic. Next, the orthotist places the thermoplastic material in a heated water bath until it reaches its Glass Transition Temperature (GTT). At this point, the thermoplastic is removed, cut to shape, and then draped on the patient's extremity.

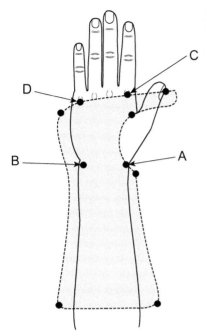

Figure 5.157 Volar-based static wrist immobilisation splint pattern.
Adapted from Lohman (2001).

Depending on the thermoplastic characteristics, the orthotist may have to apply pressure to form it correctly. While in its flexible state, excess material is removed from the thermoplastic using scissors or other suitable implements. Adjustments must be made until the orthotist is satisfied with the fit, which may require repeated heating in the water bath. The orthotist can then add fastenings such as Velcro and/or pop rivets.

As described, assigning a splint outline requires structural consideration, but the main pitfall of conventional splinting is that the orthotist must approximate the overall form of the splint using only a few guidelines on a 2D plane, not within 3D space. As a result, the orthotist must make many adjustments during the fitting process. This is often costly in terms of time, labour, materials, and funds available. A common problem when designing static wrist immobilisation splints on a 2D plane is that the drawing of the forearm trough can often be sketched too narrow (Lohman, 2001). Consequently, when the 2D thermoplastic sheet is cut out and fitted, the trough will be too narrow for the patient and will not provide adequate support. If this is the case, it must be discarded and started again. However, by digitising the splinting process, the limitations of conventional splinting may be reduced. Orthotists can manipulate the scan data and view alterations directly to produce a suitable form in fewer stages, without approximation or material waste. The splinting experience may also be significantly less painful for patients, because the orthotist will not need to apply pressure to the thermoplastic during fitting, nor will the patient be expected to place the hand palm-down on paper (depending on the scan data acquisition method), because this is a painful experience if severe deformities are present. In fact, at this design stage, the patient need not be present at all.

5.21.3   Experimental procedures

Various modelling strategies were explored, including mesh modelling, surface modelling, solid modelling, and haptic modelling. The following software packages were used during testing:

- Geomagic Studio®, Version 12 (Geomagic, 3200 East Hwy 54, Cape Fear Building, Suite 300, Research Triangle Park, NC, 27,709, USA)
- Pro Engineer® Wildfire 4.0 (PTC Corporate Headquarters, 140 Kendrick Street, Needham, MA 02,494, USA)
- Rhinoceros® 4.0 (McNeel North America, 3670 Woodland Park Ave N, Seattle, WA 98,103, USA)
- Maya® 2011 (Autodesk, Inc., 111 McInnis Parkway, San Rafael, CA 94,903, USA)
- FreeForm Modelling Plus®, Version 10 (Geomagic, 3D Systems, USA)

Different tools within the software packages were trialed and compared for suitability and similarity to conventional splinting techniques. The methods offered potentially desirable splint fabrication characteristics, such as easy assignment of a splint outline, while offering additional benefits such as being able to adjust the outline position and size interactively.

The trial CAD methods were divided into three categories:

1. *Curve on Surface Category*
   Method 1. Sketch on Polygon Mesh and Trim tools (Figure 5.158(a)). This method was performed in Rhinoceros. The user draws freehand sketches on the scan mesh surface by using the left mouse button and dragging the mouse across the mesh. This creates a non uniform rational B-spline (NURBS) curve. Initially the curve automatically attaches to the mesh. However, the position of the curve is view orientation–dependent, so the user may have to make more than one curve when the view orientation is changed to optimise curve position. The user must then manually close any open loops by using the Match command to attach anchor points or nearby NURBS curves. Alterations can be made by moving the control points/knots, but the curve detaches from the surface as a result; if this is the case, the user must re-attach the edited curve back onto the scan mesh. Once a closed loop is created, the user can trim the scan mesh with the curve using the Trim tool.

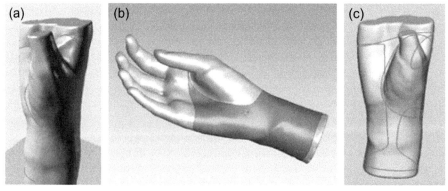

Figure 5.158 Rhinoceros Sketch on Polygon Mesh (a), Geomagic Studio Trim with Curve (b) and FreeForm Modelling Plus Trim Mesh with Curve (c).

Method 2. Trim with Curve tool (Figure 5.158(b)). This is a tool within Geomagic Studio. The user creates an NURBS curve on the scan mesh by placing numerous control points on the mesh. The control points are constrained to the surface, so the user can alter the points easily if required. Once the user is satisfied with the shape and position of the NURBS curve, the NURBS curve is used to trim the mesh (providing the curve is a closed loop).

Method 3. Trim Mesh with Curve (Figure 5.158(c)). This method combines two tools within FreeForm Modelling Plus; the user must create a closed loop on the scan mesh, and then use a mesh trim tool. The method is a similar approach to Trim with Curve within Geomagic Studio, except the user must use a PHANTOM® Haptic stylus to place the points on the scan data.

2. *Import category*

Method 1. Boolean intersection (Figure 5.159(a) and (b)). This method was tested using the Pro Engineer Interactive Surface Design Extension (ISDX). Four imported surfaces

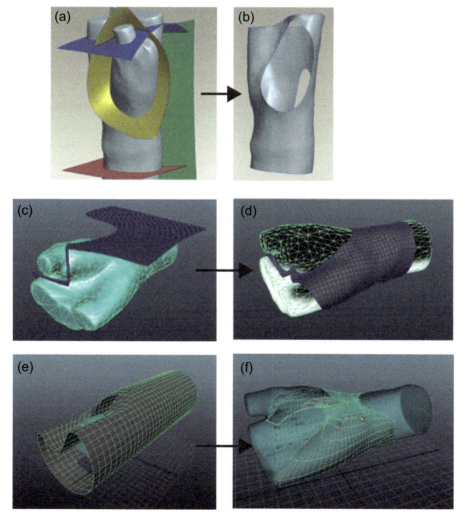

Figure 5.159 Pro Engineer ISDX Boolean Intersection (a and b), Maya nCloth Drape (c and d) and Maya Shrinkwrap plug-in (e and f).

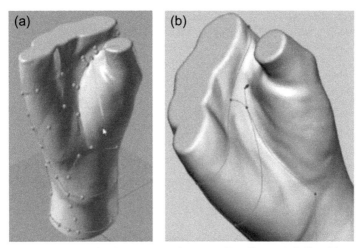

Figure 5.160 Rhinoceros RhinoReverse plug-in (a) and FreeForm Modelling Plus curve tool for surface boundary creation (b).

are placed accordingly to intersect the scan data. The user must then use the Intersect tool to create an intersecting curve for each intersecting region. Each curve is then used to trim the scan data, and the unwanted surface(s) can be deleted.

Method 2. Drape (Figure 5.159(c) and (d)). This method uses nCloth nDynamics in Maya. A template cloth is imported and draped under gravitational and magnetic influence onto the scan data, to replicate the thermoplastic draping techniques in current splinting methods. The virtual cloth is then developed and the scan data are redundant.

Method 3. Shrinkwrap (Figure 5.159(e) and (f)). This method uses the MultiTool 1.0.1 Maya plug-in (M. Richter, http://www.creativecrash.com/maya/downloads/scripts-plugins/utility-external/export/c/multitool). A surrounding mesh (i.e. cylinder) is imported, re-orientated in a suitable position around the scan mesh, and then shrinkwrapped onto the scan data. Once complete, the scan data are redundant and the shrinkwrapped mesh is developed further.

3. *Surface patching category*

Method 1. RhinoReverse (Figure 5.160(a)): The user places numerous curve control points on the scan mesh to construct a series of four-sided NURBS curve boundaries that are then surfaced automatically. The curves are constrained to the scan mesh, and once they are surfaced with NURBS patches the scan mesh is redundant and the NURBS surfaces are developed instead.

Method 2. Patching (Figure 5.160(b)). This method is similar to RhinoReverse patching except the method uses a PHANTOM® haptic stylus and FreeForm Modelling Plus.

5.21.4   Results

Each category was assessed and compared; strengths and weaknesses were identified and evaluated.

The Curve on Surface category displayed many strengths; the user can locate anatomical landmarks that are used in current splinting methods and tailor the splint shape directly around them. The user also has freedom over the placement of control points for a more controlled fit. The resulting advantage of this is that the user can see the shape of the splint as it progresses and can adjust accordingly, and can see directly where the mesh will be trimmed before performing a Trim command. This means the user can capture the design intent of the splint outline easily and quickly. The scan data can also be used directly as the splint surface, so the contours of the patient's upper extremities can be complemented closely by the proposed splint. There are also no limits on the hand position from the scan data acquisition method, because the user can place points on any part of the scan mesh and from any angle (view orientation–dependent). As a result, the scan data do not need re-orienting with respect to a reference. No weaknesses were encountered compared with conventional splinting techniques.

The Import category strengths included a potential for automation; assuming a library of different splint templates is available, the user could simply browse through the library and then apply a suitable template to the scan mesh. The process would also save the specialist from redesigning the same splint from scratch.

However, Import weaknesses were significant. The methods result in a limited view of the scan data, restricting the ability to identify anatomical landmarks and the ideal splint placement. In addition, the shape of the splint outline is difficult to interpret until the import geometry has been used. Therefore, it is highly likely that the applied import feature will need adjusting to suit each individual. In some cases, the templates may not suit the needs of the patient and their condition, so the user would need to adapt an existing template. Also, to account for anthropometric variations of different patients, the import features may require rescaling using anthropometric measurements to account for extremes of size. Success of the import features also depends on the quality of the scan data in terms of digit position; the thumb orientation in particular can obstruct the correct operation of the import features, which can lead to overlapping/intersecting geometry. Re-orientation of either the scan data or the import features will be required to ensure the correct regions of the scan data are used and manipulated. This will ultimately result in another stage in the digitisation process.

Surface Patching category strengths are similar to those of the Curve on Surface category, in which curves are easy to apply and adjust on the scan mesh. However, the weaknesses are also significant. Each curve-on-surface must have the same number of control points, and each surface boundary must consist of four sides only. As a result, there is no reason obvious to novice users as to why each surface must have a four-sided boundary. The process can also be repetitive and time-consuming, particularly if the orthotist requires a high-quality replication of the scan mesh; a higher number of four-sided NURBS surfaces will offer higher accuracy but requires more time in curve construction. The user may also have difficulty constructing a splint with four-sided boundaries, either because they forget the fundamental limitations or are unable to link up boundaries with four sides to create a smooth splint outline. Another limitation is that the scan data are not used directly; more time is required to gain better scan accuracy, because the user must create more patches. However, the fundamental disadvantage of this category is that the methods are an unnecessary extension to the Curve on Surface category.

5.21.5   Conclusion

After identifying the strengths and weaknesses of each category, the Curve on Surface category would be the most suitable for orthotists to allow the level of creativity with which they are familiar because it has no weaknesses.

The main limitation of the Import category is that the organic geometry of every patient is different; with regard to precut splint patterns, 'generic patterns rarely fit persons correctly without adjustments' (Coppard & Lynn, 2001). The same principle applies in this digitised context; having generic patterns/templates with a one size fits all approach to manipulate the scan data would not necessarily be a faster alternative to the orthotist. Anthropometric measurements could be taken to adjust the template to a more suitable size, but this would add another stage before the splinting process, in taking the anthropometric measurements. The Import geometry may not satisfy the needs of the patient's condition, either. The traditional splinting fabrication method allows the orthotist to adapt splints to suit patients' individual needs and cater for more than one ailment in some cases. To account for this, a large database would be required with many different splint types with minor differences. In this case, the orthotist would have to search through many import options, and may end up selecting an unsuitable splint by mistake. Alternatively, the orthotist may have to adapt existing templates using more advanced CAD tools, which would add time and complexity to the process.

Of all three categories, Surface Patching would be the most inappropriate, because of the surfacing requirements. If the fundamentals of NURBS surfacing are met, i.e. four-sided surfaces from curves with equal control point count, the reasons for this are likely to be unclear to orthotists. In addition, the greatest weakness to the Patching category is that it is an extension to the Curve on Surface category. Not only does it take longer to construct the patch boundaries, but another stage is required to surface boundaries. Conformability of the topological detail using Patching is also significantly reduced.

Of the three methods within the Curve on Surface category, Sketch on Polygon Mesh would be unsuitable, because the curves must be drawn freehand. Trimming a freehand curve would create a rough splint edge and could be uncomfortable for the patient. Therefore, the Trim with Curve method in Geomagic Studio and Trim Mesh with Curve method in FreeForm Modelling Plus remain. As mentioned previously, the main difference between these methods is the haptic hardware required for FreeForm Modelling Plus, allowing the user to feel anatomical landmarks. However, the haptic hardware is costly and the tactile sensation of the PHANTOM® stylus may not make splint design easier; it can be assumed that many orthotists are familiar with how to use a computer mouse, but have had little to no interaction with haptic hardware. Therefore, if a haptic system were to be implemented, orthotists would have to dedicate time to learn how to use the haptic equipment effectively, which is an undesirable consequence. After consulting with a senior hand therapist,[1] the favoured method of assigning the splint outline is the Trim with Curve tool within Geomagic Studio. The method requires only two

---

[1] Ella Donnison, clinical specialist and senior hand therapist, Patterson Medical Ltd (Homecraft Roylan), Nottingham. Interview date 17 March 2011, 5:00 pm.

actions: place a curve on the surface using control points and trim the scan data using that curve, reducing the splinting process from 19 steps to two in its entirety.

Acknowledgements

Many thanks to Ella Donnison for time and contribution to the research project, and to Loughborough University for the funding and support provided.

References

Coppard, B. (2001). Anatomical and biomechanical principles of splinting. In B. M. Coppard, & H. Lohman (Eds.), *Introduction to splinting: a clinical-reasoning & problem-solving approach* (2nd ed.) (pp. 34–72). St Louis, MO: Mosby, Inc.

Coppard, B. M., & Lynn, P. (2001). Introduction to splinting. In B. M. Coppard, & H. Lohman (Eds.), *Introduction to splinting: a clinical reasoning & problem-solving approach* (2nd ed.) (pp. 1–33). St Louis, MO: Mosby, Inc.

Doman, C., Rowe, P., Tipping, L., Turner, A., & White, E. (2005). Tools for living. In E. Turner, M. Foster, & S. E. Johnson (Eds.), *Occupational therapy and physical dysfunction: principles, skills and practice* (5th ed.) (pp. 165–210–178). London, Churchill Livingstone, Elsevier.

Lohman, H. (2001). Wrist immobilisation splints. In B. M. Coppard, & H. Lohman (Eds.), *Introduction to splinting: a clinical reasoning & problem-solving approach* (2nd ed.) (pp. 139–184). St Louis, MO: Mosby, Inc.

Lohman, H., Poole, S. E., & Sullivan, J. L. (2001). Clinical reasoning for splint fabrication. In B. M. Coppard, & H. Lohman (Eds.), *Introduction to splinting: a clinical & problem-solving approach* (2nd ed.) (pp. 103–138). St Louis, MO: Mosby, Inc.

Louise-Bender Pape, T., Kim, J., & Weiner, B. (2002). The shaping of individual meanings assigned to assistive technology: a review of personal factors. *Disability & Rehabilitation, 24*(1–3), 5–20.

Pallari, J. H. P., Dalgarno, K. W., & Woodburn, J. (2010). Mass customisation of foot orthoses for rheumatoid arthritis using selective laser sintering. *IEEE Transactions on Biomedical Engineering, 57*(7), 1750–1756.

5.22   Orthotic rehabilitation applications case study 5: evaluation of a refined 3D CAD workflow for upper extremity splint design to support AM

Acknowledgements

The work described in this case study was first reported in the references below and is reproduced here in part or in full with the permission of CRDM, Ltd.

Paterson, A. M., Bibb, R. J. & Campbell, R. I. (2012). Evaluation of a refined three-dimensional Computer Aided Design workflow for upper extremity splint design to support Additive Manufacture. In C. Bocking, & A. E. W. Rennie (Eds.), *Thirteenth Conference on Rapid Design, Prototyping and Manufacturing, High Wycombe: CRDM Ltd.*, pp. 61–70.

5.22.1   Introduction

Patients with chronic conditions such as Rheumatoid Arthritis (RA) are often prescribed custom-made Static Wrist Immobilisation (SWI) splints (commonly known as cock-up splints) (refer to Figure 5.156 for an example). Intended outcomes of such splints include reducing pain and inflammation, preventing deformities and contractures, and promoting movement of stiffer joints along with lost motor function (Cooper, 2007; Jacobs & Austin, 2003; Lohman, 2001; Schultz-Johnson, 1996). SWI splints are designed to enable full dexterity of the digits to allow perseverance in various everyday tasks (Lohman, 2001). Various fundamental features are built into each splint, some of which are bespoke to the patient's needs. For example, cavities are often integrated over pressure-prone areas to avoid forming pressure sores (Coppard & Lynn, 2001). In addition, the proximal edges of splints are flared for a more comfortable edge against the skin (Jacobs & Austin, 2003).

However, evidence suggests that patient adherence is affected by:

- *Fit and Function*. If the splint is not fabricated correctly (such as insufficient attention given to pressure-prone areas), the splint can be uncomfortable to wear. Pressure can result in soft tissue damage such as pressure sores, ulceration, and persistent oedema (Coppard, 2001; Lohman et al., 2001). As a result, patients may discard the splints (Lohman et al., 2001). Similarly, if donning and doffing of the splint is difficult or painful for patients, they may be reluctant to wear them. In addition, dissatisfaction occurs because of the fasteners used and their tendencies to snag on clothing (Veehof, Taal, Willems, & van de Laar, 2008). Velcro is commonly used as a fastener for SWI splints. If the splint is worn for the specified amount of time (dictated by the therapist) or worn for rigorous, dirty activities, Velcro deteriorates in terms of its ability to latch correctly.
- *Aesthetics*. Despite the efforts of practitioners to fabricate splints to a high quality, the functional appearance of splints can affect patients' willingness to wear them in social situations, such as visiting friends and relatives (Veehof et al., 2008). Louise-Bender Pape, Kim, and Weiner (2002) describe patients' reluctance to use assistive devices within the

public domain, because they think assistive devices threaten their sense of fitting in. In addition to Velcro's tendency to snag fabrics, it attracts and collects other loose items such as hair and fabric fibres within the hooks, reducing the aesthetics further. Co-design of the splint is also limited between the patient and therapist owing to the materials available. Patients may have the option to pick the colour of Velcro straps and Low Temperature Thermoplastic (LTT) wherever possible in an attempt to encourage adherence (Austin, 2003a; Coppard & Lynn, 2001).

In addition, the fabrication process can be laborious and hands-on. There is heavy emphasis within splinting textbooks that prior planning is critical to avoid wasting time and resources; for example, therapists are encouraged to consider padding before forming the splint, because delaying padding integration can result in pressure points (Coppard & Lynn, 2001). This can lead to further discomfort if issued to the patient. Alternatively, the therapist may have to re-form the splint to account for padding, but this is costly in terms of labour and clinic time.

Austin (2003b) uses the acronym PROCESS to describe splint design as a systematic workflow:

1. **P**attern creation – up to and including evaluation of patients, their condition(s), and treatment choice. The splint pattern is created by drawing around a patient's forearm and hand on a piece of paper/towel and noting relevant landmarks (e.g. bony prominences).
2. **R**efine pattern – the fit of the paper pattern is tested on the patient and adjustments are made when necessary.
3. **O**ptions for materials – Low-temperature thermoplastic, padding solutions, and suitable fasteners are chosen. At this point, the paper/towel template is transferred onto the chosen LTT.
4. **C**ut and heat – The LTT is placed into a heated water bath until pliable. The LTT is then cut to shape using scissors. Padded features are also applied during this stage.
5. **M**oulding and **E**valuating fit – The LTT is fitted to the patient and adjustments are made.
6. **S**trapping and components – Fasteners are fixed to the formed splint.
7. **S**plint finishing touches – Austin (Austin, 2003b) claimed that edge-flaring/rolling should be within this phase. However, Lohman (Lohman, 2001) stated it should be before adding fasteners.

In a bid to resolve issues with adherence, an alternative fabrication approach has been investigated to exploit the geometric freedom offered through Additive Manufacture (AM). Concept sketches by Bibb (2009) (refer to Figure 5.149) suggest integration of aesthetically pleasing lattice structures that could also improve ventilation to reduce perspiration and its effects (such as maceration). In addition, the approach had scope to encourage co-design between the practitioner and the patient; the patient would be able to personalise the splint by specifying a bespoke pattern to promote user-centred design.

However, to exploit the geometric freedom of AM, geometric manipulation within 3D CAD software is necessary. Introducing mainstream CAD software to splinting practitioners would be an inappropriate solution because practitioners would need to dedicate a significant amount of time learning how to use the software effectively. In response to this issue, Rogers, Stephens, Gitter, Bosker, and Crawford (2000) and Pallari et al. (2010) suggested the need for specialised software for physicians, to allow them to capture the design intent quickly and with minimal training for prosthetic and orthotic fabrication using AM.

Therefore, the aim of this investigation was to devise a way to improve patient adherence by introducing AM for splint fabrication, but more important, by providing the CAD tools required to achieve this. The objectives required to meet the aim were to:

1. Complete a task analysis of conventional splinting to identify key heuristics displayed in traditional custom-made splint design and fabrication.
2. Capture the key fabrication characteristics in a virtual environment using suitable CAD packages and tools, by manipulating virtual upper extremity scan data.
3. Identify the most suitable CAD tools/strategies for new features (i.e. lattices).
4. Reform the strategies identified in 2. and 3. into a logical workflow for practitioners.
5. Translate the workflow devised in 4. into a specialised CAD software prototype.
6. Evaluate the software prototype generated in 5. through user trials and interviews.

The proposed approach is not intended to replace the expertise of splinting practitioners, nor is it intended to make the profession obsolete. Instead, the investigation aims to develop a new toolset to empower therapists in future splint design and fabrication. In addition, the scope of this chapter and associated PhD research is to address the development of a specialised splinting CAD workflow, with the intent to represent it as specialised 3D CAD software. The authors acknowledge that a suitable data acquisition method has yet to be identified to support the digitised splinting approach within a clinical environment. However, Paterson, Bibb, and Campbell (2010) explored a range of data acquisition methods to capture test data to support this investigation. Furthermore, this chapter focuses on a small yet vital element within a larger explorative study into the macro-digitisation of the splinting process, as described by Paterson, Bibb, and Campbell (2011).

5.22.2   Method

The objectives set out in the section were addressed in the following manner:

Objective 1. Initially, the needs and requirements of practitioners were identified through different media, attendance of a splinting class, by clinical observation, and through a literature survey. The intention was to identify key fabrication methods shared among therapists for the design and fabrication of SWI splints. In addition, the decision-making process of clinicians was vital to ensure the proposed workflow corresponded with their expectations in the design process and to make the transition from conventional to digital design and fabrication easier. As a result, the devised task analysis workflow of traditional splinting is shown in Figure 5.134 (please refer to Case study 2 in Section 5.19) as a baseline for future development. It became apparent that the main disadvantage to the workflow was the need to repetitively heat the LTT to make adjustments, and that making alterations in the latter stages of the workflow was more costly in time and resources.

Objective 2. Having identified the key features and behaviours of SWI splint fabrication, a variety of CAD tools and approaches were tested to capture the design intent of required features in a virtual environment. Features such as creating the splint outline (Paterson et al., 2011) flaring edges and adding cavities over bony prominences were investigated. An action research strategy was adopted to plan, test, evaluate, and refine each approach. Testing was performed on STL scan data acquired through indirect 3D laser scanning of an upper extremity plaster cast (Paterson et al., 2010), using a Z Corp ZScanner® 800 3D laser Scanner (3D Systems, USA) (3D Systems, 2012). A range of 3D CAD software packages were used during testing, as described by Paterson et al. (2011).

Objective 3. To facilitate the integration of lattice structures in the workflow, CAD tools and approaches were investigated for lattice pattern repetition. A similar action research strategy to Objective 2 was adopted for exploring lattice integration.

Objective 4. Where applicable, different CAD tools and approaches devised in Objectives 2 and 3 were compared and evaluated against set criteria: namely, being able to capture the initial design intent. Successful CAD strategies were taken forward for workflow refinement. The workflow is shown in Figure 5.152 (please refer to Case study 3 in Section 5.20).

Three key points about the refined workflow were:

1. Integration of lattice feature capabilities.
2. Discretionary workflow operation. Traditional splinting involves several mandatory stages to make both major and minor adjustments. As a solution, the workflow gives users the ability to alternate to and from different stages, so they are not restricted if they feel the need to make adjustments.
3. Change in colour and thickness option location. Choosing the colour and thickness of LTT in traditional splinting is often one of the first decisions to be made (as shown in Figure 5.134, 'Choose LTT material'). However, there is no opportunity to change the LTT colour or thickness further down the process chain. In addition, the colour of the LTT was viewed as a minor concern compared with key features such as flaring edges. Therefore, it was placed at the end of the refined workflow as a final adjustment.

Objective 5. Once the workflow had been refined, qualitative feedback was required from experienced splinting practitioners. The most appropriate data collection method was to introduce therapists to the workflow via an interactive software prototype (Figure 5.161). The prototype was a robust method to introduce splinting practitioners to features typically found in CAD software, such as a viewport. By making the prototype interactive, direct feedback could be obtained on the effect of participants' decision making by viewing the changes in the viewport. The prototype also replicated the likely interface features if the workflow were to be developed into fully functional software. A high-fidelity throwaway prototype was developed within Microsoft Access, with customised features created via Visual Basic for Applications. Most parameters were adjustable using slider controls. In addition, the pattern integration stage featured a library of different lattices; the concept intended to show that both the therapist and patient could browse for a pattern that would suit the patient's preference.

Objective 6. Ten participants were involved in the evaluation phase. All participants were certified splinting practitioners within the United Kingdom, within the field of either occupational therapy ($n = 8$) or physiotherapy ($n = 2$), with splinting experience ranging from 1 year to over 10 years. None of the participants had prior CAD experience. Participants were given a briefing on the aims and objectives of the investigation, supporting information about the benefits of AM, and the need for specialised CAD software to create AM splints. To explain the capabilities of AM, proof-of-concept splint prototypes were created (Figure 5.162). Participants

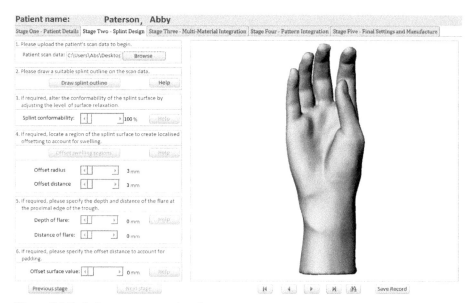

Figure 5.161 Software prototype interface.

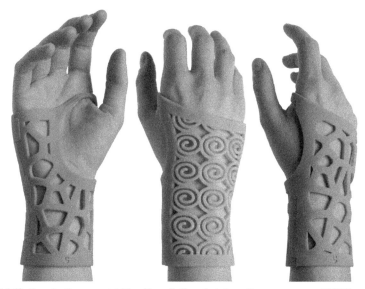

Figure 5.162 Proof-of-concept AM splints (left and right splints courtesy of PDR).

were then invited to test the software prototype, followed by a semi-structured interview. Probing questions were asked concerning the workflow through direct comparison with traditional splinting methods, as well as the extent of capturing design intent. Voice recordings were used to capture verbal feedback, which were then used to identify trends in opinions. Participants were asked to sign an informed consent form before the evaluation session.

5.22.3   Results and discussion

Although none of the participants had previous experience with using 3D CAD software, participants commented on how easy the prototype was to use. All participants were fully engaged with the software prototype; many suggested additional creative tools to be integrated, such as multicolour options and lattice perforation motifs in the shape of branded logos. Therapists were in favour of the digitised approach and their ability to alternate to and from different stages in the prototype to make adjustments. This compares favourably with conventional splint fabrication, in which the further down the process chain one goes, the more complex and potentially more expensive and time-consuming it is to make adjustments. Most participants were not concerned by the overall workflow order demonstrated in the prototype; when discussing the change in location of the colour and thickness options, eight participants were not concerned with the change. They thought that colour was better suited towards the end of the workflow because it was seen as a final touch and less important than key features such as edge flaring or cavity integration, which appeared earlier in the workflow. In response to perceived benefits offered through AM, all participants were excited by the prospect of integrating lattice patterns. One participant thought that cartoon character perforations, for example, would be ideal for paediatrics. However, one participant was concerned that issues with adherence might be reversed, to the point where patients would be wearing the splints too much. This could lead to muscle atrophy, so a more stringent wearing regimen would be required to avoid this issue. Two participants had concerns regarding window oedema as a result of the lattice structures, with a chance that oedema could protrude through perforations. On a related note, all participants had concerns about the structural integrity of the splints resulting from material removal. Most participants expressed concerns regarding initial data acquisition of the upper extremity and the way in which they could capture desired positions without therapist intervention. AM fabrication time and cost of the approach were also concerns.

5.22.4   Conclusions and further work

The introduction of a new splinting toolset was well received by all participants, providing a method to promote co-design between the patient and therapist. In addition, the prospect of decision-making during splint fabrication made the digitised approach a welcoming prospect to participants, with many suggestions for future feature integration.

In response to participants' concerns regarding window oedema and compromised structural integrity, further research is required into material-to-perforation ratios and the potential for surface displacement mapping instead. Finite Element Analysis (FEA) should be investigated for integration within the workflow, either to highlight problem areas or to automatically adjust the geometry to make splints structurally sound. This could be achieved through varying thicknesses, similar to work demonstrated by Rogers et al. (2007), or alterations in pattern material-to-perforation ratio. However, it is likely to be a challenging development to map because of the effects of different pathologies on extremity mobility, particularly on an individual basis with unique deformities and range of movement. In addition, a major concern among most participants was the ability to capture the desired posture of the patient's upper extremity, because the most appropriate data collection method in this context is ill-defined for clinical applications. Therefore, future research of data acquisition in a clinical context is vital to support the use of AM for upper extremity splint fabrication.

Acknowledgements

Many thanks to Dr Dominic Eggbeer, Sean Peel and staff at The National Centre for Product Design and Development Research (PDR), Cardiff, for manufacturing proof-of-concept stereolithography splints. In addition, our sincerest thanks to Lucia Ramsey (University of Ulster), Ella Donnison (Pulvertaft Hand Clinic, Derby) and all participants for their contribution throughout the investigation.

References

3D Systems. (2012). *ZScanner® 800*. Available at http://www.zcorp.com/documents/182_ZScanner800-tearsheet-v05wb.pdf. Accessed 12.06.12.

Austin, N. M. (2003a). Equipment and materials. In M. Jacobs, & N. M. Austin (Eds.), *Splinting the hand and upper extremity: principles and process* (pp. 73–87). Baltimore: Lippincott Williams & Wilkins.

Austin, N. M. (2003b). Process of splinting. In M. Jacobs, & N. M. Austin (Eds.), *Splinting the hand and upper extremity: principles and process* (pp. 88–99). Baltimore: Lippincott Williams & Wilkins.

Bibb, R. J. (July 22, 2009). *Lattice within splints. Drawing*.

Cooper, C. (2007). Fundamentals. In C. Cooper (Ed.), *Fundamentals of hand therapy: clinical reasoning and treatment guidelines for common diagnoses of the upper extremity* (pp. 3–21). St. Louis, MO: Mosby Inc., Elsevier Inc.

Coppard, B. (2001). Anatomical and biomechanical principles of splinting. In B. M. Coppard, & H. Lohman (Eds.), *Introduction to splinting: a clinical-reasoning & problem-solving approach* (2nd ed.) (pp. 34–72). St Louis, MO: Mosby, Inc.

Coppard, B. M., & Lynn, P. (2001). Introduction to splinting. In B. M. Coppard, & H. Lohman (Eds.), *Introduction to splinting: a clinical reasoning & problem-solving approach* (2nd ed.) (pp. 1–33). St Louis, MO: Mosby, Inc.

Jacobs, M., & Austin, N. (2003). Splint fabrication. In M. Jacobs, & N. Austin (Eds.), *Splinting the hand and upper extremity: principles and process* (pp. 98–157). Baltimore: Lippincott Williams & Wilkins.

Lohman, H. (2001). Wrist immobilisation splints. In B. M. Coppard, & H. Lohman (Eds.), *Introduction to splinting: a clinical reasoning & problem-solving approach* (2nd ed.) (pp. 139–184). St Louis, MO: Mosby, Inc.

Lohman, H., Poole, S. E., & Sullivan, J. L. (2001). Clinical Reasoning for splint fabrication. In B. M. Coppard, & H. Lohman (Eds.), *Introduction to splinting: a clinical & problem-solving approach* (2nd ed.) (pp. 103–138). St Louis, MO: Mosby, Inc.

Louise-Bender Pape, T., Kim, J., & Weiner, B. (2002). The shaping of individual meanings assigned to assistive technology: a review of personal factors. *Disability & Rehabilitation*, *24*(1–3), 5–20.

Pallari, J. H. P., Dalgarno, J., Munguia, J., Muraru, L., Peeraer, L., Telfer, S., & Woodburn, J. (2010). Design and additive fabrication of foot and ankle-foot orthoses. In *Proceedings of the twenty first annual international solid FreeForm fabrication proceedings - an additive manufacturing conference, August 9–11* (pp. 834–845).

Paterson, A. M., Bibb, R. J., & Campbell, R. I. (2010). A review of existing anatomical data capture methods to support the mass customisation of wrist splints. *Virtual and Physical Prototyping Journal*, *5*(4), 201–207.

Paterson, A. M., Bibb, R. J., & Campbell, R. I. (2011). Digitisation of the splinting process: development of a CAD strategy to support splint design and fabrication. In *12th conference on rapid design, prototyping and manufacturing, June 17, 2011*. CRDM Ltd.

Rogers, B., Bosker, G. W., Crawford, R. H., Faustini, M. C., Neptune, R. R., Walden, G., et al. (2007). Advanced trans-tibial socket fabrication using selective laser sintering. *Prosthetics and Orthotics International*, *31*(1), 88–100.

Rogers, B., Stephens, S., Gitter, A., Bosker, G., & Crawford, R. (2000). Double-Wall, transtibial prosthetic socket fabricated using selective laser sintering: a case study. *Journal of Prosthetics and Orthotics*, *12*(3), 97–100.

Schultz-Johnson, K. (1996). Splinting the wrist: mobilization and protection. *Journal of Hand Therapy*, *9*(2), 165–177.

Veehof, M. M., Taal, E., Willems, M. J., & van de Laar, M. A. F. J. (2008). Determinants of the use of wrist working splints in rheumatoid arthritis. *Arthritis Care & Research*, *59*(4), 531–536.

Dental applications

5.23 Dental applications case study 1: the computer-aided design and rapid prototyping fabrication of removable partial denture frameworks

Acknowledgments

The work described in this case study was first reported in the references below and is reproduced here in part or in full with the permission of the Council of the Institute of Mechanical Engineers and Quintessence Publishing Ltd.

Eggbeer D, Bibb R, Williams R, "The Computer Aided Design and Rapid prototyping of Removable Partial Denture Frameworks", Proceedings of the Institute of Mechanical Engineers Part H: Journal of Engineering in Medicine, 2005, Volume 219, Issue Number H3, pages 195–202. http://dx.doi.org/10.1243/095441105X9372

Eggbeer D, Williams RJ, Bibb R, "A Digital Method of Design and Manufacture of Sacrificial Patterns for Removable Partial Denture Metal Frameworks", Quintessence Journal of Dental Technology, 2004, Volume 2, Issue Number 6, pages 490–9.

The authors would like thank Frank Cooper at the Jewellery Industry Innovation Centre in Birmingham, UK, who kindly supplied the Perfactory® and Solidscape® rapid prototyping (RP) patterns and Kevin Liles at 3D Systems, Inc., who supplied the Amethyst® RP pattern.

5.23.1 Introduction

Computer-aided design (CAD), computer-aided manufacture (CAM), and rapid prototyping (RP) techniques have been extensively employed in the product development sector for many years and extensively used in maxillofacial technology and surgery (Bibb & Brown, 2000; Hughes, Page, Bibb, Taylor, & Revington, 2003). In addition, CAD/CAM technologies have been introduced into dentistry, particularly for the manufacture of crowns and bridges (Van der Zel, Vlaar, de Ruiter, & Davidson, 2001), but there has been little research into the use of such methods in the field of removable partial denture (RPD) framework fabrication. This may be partly attributed to the lack of suitable, dedicated software. Recent pilot studies showed that CAD/RP methods of designing and producing a sacrificial pattern for the production of metal components of RPD metal frameworks could have promising applications (Williams, Bibb, & Rafik, 2004; Williams, Eggbeer, & Bibb, 2004).

These studies explored the application of computer-aided technologies to the surveying of digital casts and pattern design and the subsequent production of sacrificial patterns using RP technologies.

Potential advantages offered by the introduction of advanced CAD/CAM/RP in the field of RPD framework fabrication include automatic determination of a suggested path of insertion, almost instant elimination of unwanted undercuts (re-entry points), and the equally rapid identification of useful undercuts. At another stage, components of a removable partial denture could be stored in a library and dragged and dropped in place on a scanned and digitally surveyed cast from icons appearing on screen, allowing virtual pattern making to be carried out in a much faster time than is achieved by current techniques. The quality assurance of component design can also be built into the software. Because RP machines build the object directly, scaling factors may also be precisely imposed to compensate for shrinkage in casting. In addition to the potential time savings, the CAD/RP process delivers inherent repeatability, which may help eliminate operator variation and improve quality control in the dental laboratory.

The current chapter reports on the application of CAD/RP methods to achieve the stages of surveying and design using a software package that provides a virtual sculpting environment, Geomagic FreeForm® Plus (Geomagic Solutions, 430 Davis Drive, Suite 300, Morrisville, NC 27560, USA). It also discusses the application of RP technologies to produce sacrificial patterns for casting the definitive chromium-cobalt framework component. The advantages, limitations, and future possibilities of these techniques are considered.

5.23.2   Materials and methods

5.23.2.1   Three-dimensional scanning

A three-dimensional (3D) scan of a partially dentate patient's dental cast was obtained using a structured white light digitizer (Comet 250; Steinbichler Optotechnik GmbH, AM Bauhof 4, D-83115 Neubeuern, Germany). This particular type of scanner is used in high-precision engineering applications and in maxillofacial technology (Bibb et al., 2000). Multiple overlapping scans were used to collect point cloud data that were aligned using Polyworks software (InnovMetric Software, Inc., 2014 Jean-Talon Blvd. North, Suite 310, Sainte-Foy, Quebec, Canada G1N 4N6). Spider software (Alias-Wavefront, Inc., 210 King Street East, Toronto, Ontario, Canada M5A 1J7) was used to produce a polygon surface, a stereolithography (STL) (Manners, C. R., 1993, 'STL File Format' available on request from 3D Systems, Inc., 333 Three D Systems Circle Rock Hill, SC 29730, USA) model file that could be imported into FreeForm®.

5.23.2.2   FreeForm® modelling

FreeForm® is a CAD package with tools analogous to those used in physical sculpting. A haptic interface (Phantom® Desktop haptic interface) incorporates positioning in 3D space and allows rotation and translation in all axes, while translating hand movements into the virtual environment (Figure 5.163). It also allows the operator to feel

Figure 5.163 The Phantom® stylus.

the object being worked on in the software. The combination of tools and force feedback sensations mimic working on a physical object and allows shapes to be designed and modified in an arbitrary manor.

Objects being worked on are referred to as virtual clay, which can be rotated and viewed from any angle on the screen. A buck setting prevents a model being unintentionally modified but allows clay to be added or copied. The STL cast was imported into FreeForm® as a buck model.

5.23.2.3 Surveying

Surveying is undertaken in dental technology laboratories to identify useful dental features for the RPD design to be retained in the oral cavity effectively. The parting line (also known as split line) function within FreeForm® was used to delineate up- and down-facing surfaces, thus identifying areas of undercut in a colour different from the buck model. The effect is identical to the physical technique of using dental survey lines to identify and mark the most bulbous areas of teeth with a pencil line (highlighted in Figure 5.164(a)). The undercuts were assessed to establish the best path of insertion and possible points for active clasp termination and the model was rotated accordingly.

A visual comparison (Figure 5.164(a) and (b)) was made between the physically surveyed cast and the same model cast surveyed using the software. Once a suitable angle was chosen, the model was re-exported as an STL file.

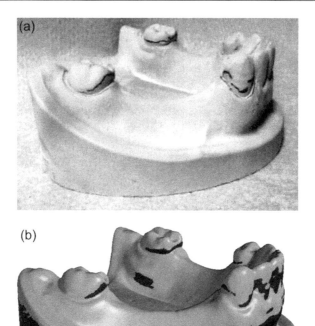

Figure 5.164 (a) The physically surveyed cast. (b) The digitally surveyed, buck cast. Undercuts are shown as dark areas.

5.23.2.4 Removing unwanted undercuts

When creating an RPD, most undercuts are removed so that the resulting framework can be inserted and removed in a comfortable manner. The STL file of the rotated cast was imported into FreeForm®, but this time using the extrude to plane option. When the cast was viewed from above, this option took the maximum extents of the profile and extruded them down by a user-defined distance. This effectively removed undercuts and replaced them with vertical surfaces (Figure 5.165(a) and (b)).

5.23.2.5 Identifying useful undercuts

FreeForm's ruler tool was used to measure the distance between the original cast model and the version with undercuts removed. The useful undercuts were marked with a line for use in the design stages. Removable partial dentures provide firm location on the existing dentition by using flexible clasps. The clasp components of the

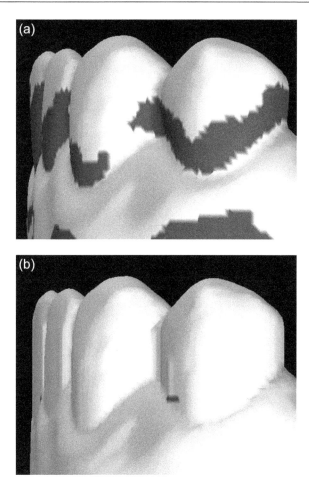

Figure 5.165 (a) Undercuts are shown as dark areas. (b) Undercuts have been removed and replaced by vertical surfaces.

RPD open on initial contact during insertion and removal, and return to their original position within the undercut on final seating, thus providing secure retention.

5.23.2.6 Creation of relief

Areas without teeth require a spacer known as relief, to prevent the framework from resting on the surface of the gums. Relief was created by selecting and copying an area from the cast with undercuts removed and then pasting this as a new piece of clay. This was then offset to the outside by 1 mm. The results of this process are highlighted in Figure 5.166.

The entire modified model was saved as an STL file and then re-imported using the buck setting to avoid unintentional modification during the next stages of RPD design.

Figure 5.166 Relieved edentulous areas are shown in the lighter colour on the dark cast.

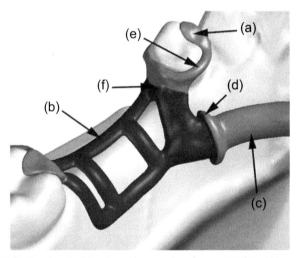

Figure 5.167 (a) Occlusal rest, (b) polymeric retention frame, (c) lingual bar, (d) acrylic line (e) nonactive clasp and (f) guide plate.

5.23.2.7 *Framework design*

The RPD design employed in this study was based on recognized dental technology methods emphasizing simplicity, aesthetics, and patient comfort (Budtz-Jorgensen & Bocet, 1998). Some of the key design features outlined in the design stages are labelled in Figure 5.167.

The entire framework was designed on the relieved buck cast with undercuts removed, with the exception of the clasp components. The clasps use the undercuts to function and were therefore designed on the original buck cast. The following techniques were used in the framework design.

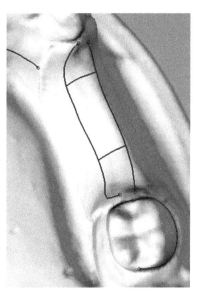

Figure 5.168 Construction curves.

Occlusal rests (Label (a) in Figure 5.167): A combination of 2D drawing, 3D creation, and manipulation tools was used to create pieces of clay that were copied and located where required on the teeth.

Polymeric retention framework (Label (b) in Figure 5.167), lingual bar (Label (c) in Figure 5.167), acrylic line (Label (d) in Figure 5.167), and nonactive clasps (Label (e) in Figure 5.167): The drawing tool was used to locate curves directly onto the cast surface. These formed the centre of the frameworks profile (Figure 5.168). The groove tool was used to define and create the exact oval and square sectional dimensions as clay.

Guide plates: (Label (f) in Figure 5.167) Guide plates were created using the same method as relief creation. The attract and smudge tools were also used to build up plate areas and blend them onto the framework sections.

Finishing: Smooth, attract, and smudge tools were used to blend the components together. The buck cast was removed, acting as a Boolean cutting tool to leave just the clay framework.

Active Clasps: The clasps were designed in the same manner as the non-flexible parts of the framework, but using the buck cast with undercuts. The construction lines were joined to the termination point previously marked in the undercut measurement stage.

The buck cast was removed, leaving the clasps. These were joined to the main framework and blended in. Figure 5.169 shows the final, virtual design. The entire framework was exported as an STL file.

5.23.2.8 *Pattern manufacture*

Four RP methods were compared: stereolithography (SL) (3D Systems, Inc., 26,081 Avenue Hall, Valencia, CA, 91355, USA), ThermoJet® (obsolete, 3D Systems, USA), Solidscape®

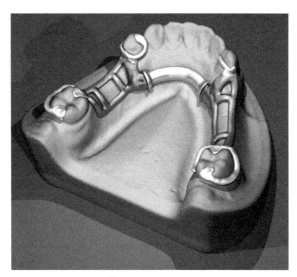

Figure 5.169 The complete FreeForm® design.

T66 (Solidscape, Inc., 316 Daniel Webster Highway, Merrimack, NH 03054-4115, USA), and Perfactory® (EnvisionTEC GmbH, Elbestraβe 10, D-45,768 Marl, Germany). Two STL resins were compared: DSM Somos® 10110 (Waterclear™) (2 Penn's Way, Suite 401, New Castle, DE 19720, USA) and Accura™ Amethyst® (3D Systems, Inc.). Both of the SL patterns were an epoxy-based polymer; the ThermoJet® was a TJ88-grade wax polymer, the Solidscape® was a soft thermoplastic, and Perfactory® was an acrylate-based polymer. The Waterclear and ThermoJet® patterns were manufactured at PDR; the others were prepared and built by external suppliers. The Amethyst, Solidscape®, and Perfactory materials are used by the jewellery industry to produce sacrificial patterns.

SLA-250 in Waterclear™ example:

The STL framework design was prepared using 3D Lightyear™ (3D Systems, Inc.) with a fine point support structure (Figure 5.170). The framework was oriented with the fitting surfaces facing upward to avoid the rough finish created by the support structures affecting fit.

Two build styles were compared: standard 0.1000-mm thick layers and high-resolution 0.0625-mm layers. Once completed, the patterns were carefully removed from the machine platform and cleaned in isopropanol. They were then post-cured in ultraviolet (UV) light to ensure full polymerization. The other patterns were produced according to supplier specifications.

5.23.2.9   Pattern comparison

Of the four RP processes compared in this study, the SL processes provided the most suitable patterns. The SL patterns were accurate and robust and had an acceptable surface finish, but required relatively lengthy cleaning and finishing when removing support structures. The ThermoJet® build preparation was simpler and faster than SL and both the ThermoJet® and Solidscape® processes produced accurate patterns with a

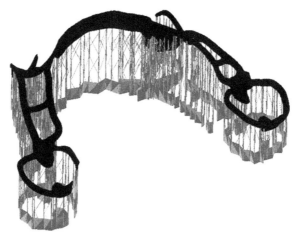

Figure 5.170 The support structure in 3D lightyear™.

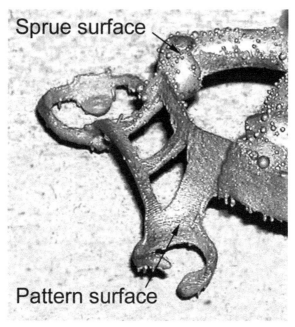

Figure 5.171 Surfaces of the unfinished Amethyst® pattern cast.

good surface finish that required minimal finishing. However, these wax patterns were extremely fragile and could not be cast. The Perfactory®-produced pattern showed a smooth surface finish, but was also extremely flexible and easily distorted when handled.

5.23.2.10 Casting

The SL and Perfactory® patterns were casted in chrome-cobalt without using a refractory cast. A slow mould heating cycle was used to avoid cracking. Figure 5.171 shows the

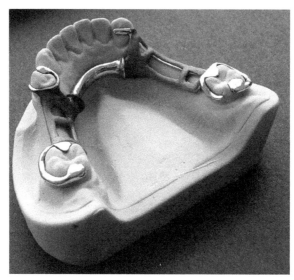

Figure 5.172 The definitive framework.

unfinished cast from the SL, Amethyst® pattern. This shows that air inclusions from the casting process did not adhere to the pattern surface.

Although casts were obtained from the SL and Perfactory® patterns, it proved difficult to add sprues owing to the thin framework sections. To improve casting, the design was thickened in FreeForm® and revised SL patterns were produced and cast. This improved the pattern's strength and the casting reliability.

5.23.2.11   Finishing

The casts produced from the original, thin Amethyst® and thicker Waterclear™ patterns were polished and test fitted to the original, physical cast. These were all judged satisfactory. Figure 5.172 shows the finished RPD framework that was cast from the high-resolution Waterclear™, SLA-250 pattern.

5.23.3   Conclusions

The design stages of this technique rely on having an accurate 3D scan of a patient cast and an understanding of both RPD framework design and CAD techniques. This meant that the time taken to produce castable patterns using the technology described is considerable. However, this would be significantly reduced with familiarity and practice.

The most suitable choice of RP process was determined primarily by accuracy and part strength. The ThermoJet® and Solidscape® patterns, although accurate, were too fragile and were therefore not suitable for the tasks associated with spruing and casting. Although the Perfactory® pattern cast well, the accuracy was poor as a result of distortion inflicted on the flexible pattern during handling. The stiffer patterns produced by SL were easy to handle and accurate and produced satisfactory results. The layer effect exhibited by all RP processes was not evident after finishing and the difference between the high-resolution and standard SLA-250 patterns was negligible.

The techniques undertaken and described here outline a stage in the development of machine-produced RPD frameworks and point to many possible advances that can be achieved. The application of CAD would allow access to new RP technologies that build parts directly in metal alloys, including chromium-cobalt and stainless steel. Sacrificial pattern manufacture and casting may be eliminated altogether. This will be explored in future studies.

The introduction of digital design and RP production into current practices would present a significant change in the field of dentistry; this is unlikely to happen quickly. Studies so far have shown how CAD and RP may be applied and some principles have been developed and established. Possible future benefits and the potential shortfalls have also been discussed.

References

Bibb, R., & Brown, R. (2000). The application of computer aided product development techniques in medical modeling. *Biomedical Sciences Instrumentation*, *36*, 319–324.

Bibb, R., Freeman, P., Brown, R., Sugar, A., Evans, P., & Bocca, A. (2000). An investigation of three-dimensional scanning of human body surfaces and its use in the design and manufacture of prostheses. *Proceedings of the Institute of Mechanical Engineers Part H, Journal of Engineering in Medicine*, *214*(6), 589–594.

Budtz-Jorgensen, E., & Bocet, G. (1998). Alternate framework designs for removable partial dentures. *Journal of Prosthetic Dentistry*, *80*, 58–66.

Hughes, C. W., Page, K., Bibb, R., Taylor, J., & Revington, P. (2003). The custom-made titanium orbital floor prosthesis in reconstruction for orbital floor fractures. *British Journal of Oral and Maxillofacial Surgery*, *41*, 50–53.

Van der Zel, J., Vlaar, S., de Ruiter, W., & Davidson, C. (2001). The CICERO system for CAD/CAM fabrication of full ceramic crowns. *Journal of Prosthetic Dentistry*, *85*, 261–267.

Williams, R., Bibb, R., & Rafik, T. (2004). A technique for fabricating patterns for removable partial denture frameworks using digitized casts and electronic surveying. *Journal of Prosthetic Dentistry*, *91*(1), 85–88.

Williams, R., Eggbeer, D., & Bibb, R. (2004). CAD/CAM in the fabrication of removable partial denture frameworks: a virtual method of surveying 3-dimensionally scanned dental casts. *Quintessence Journal of Dental Technology*, *2*(3), 268–276.

5.24   Dental applications case study 2: trial fitting of an RDP framework made using CAD and RP techniques

Acknowledgments

The work described in this case study was first reported in the references below and is reproduced here in part or in full with the permission of Sage Publishing.

Bibb R, Eggbeer D, Williams RJ, Woodward A, "Trial fitting of a removable partial denture framework made using computer-aided design and rapid proto-typing techniques", Proceedings of the Institute of Mechanical Engineers Part H: Journal of Engineering in Medicine 2006; 220(7): 793–797, ISSN: 0954-4119, http://dx.doi.org/10.1243/09544119JEIM62.

5.24.1   Introduction

The CAD and CAM techniques have been adopted as a method of fabrication for fixed partial denture restorations (Mormann & Bindl, 2000; Van der Zel, Vlaar, de Ruiter, & Davidson, 2001), and CAD/CAM and RP have been extensively used in maxillofacial technology and surgery (Bibb & Brown, 2000; Bibb, Bocca, & Evans, 2002; Hughes, Page, Bibb, Taylor, & Revington, 2003). The application of the principles of CAD and RP to the fabrication of RPDs is in the early stage of development, but the potential advantages are already clear and have been discussed (Eggbeer, Williams, & Bibb, 2004; Williams, Bibb, & Rafik, 2004; Williams, Eggbeer, & Bibb, 2004). Developments achieved so far include electronic surveying of a 3D scanned dental cast (Williams, Eggbeer, et al., 2004) and the production of successful castings from plastic patterns produced by RP technologies (Eggbeer et al., 2004; Williams, Bibb, et al., 2004). The potential future benefits include a rapid and semi-automated method of digital survey-ing, a drag and drop system of virtual patterning using on-screen icons of RPD compo-nents, which could be dragged onto the computer model of a scanned cast of a patient.

However, although the castings fabricated so far by CAD/CAM-produced patterns have been judged to be acceptable for clinical presentation, to date none has been trial-fitted to a patient. This case report follows from previously published work by providing details of the first fitting to a patient of an RPD framework produced by CAD and RP technologies.

5.24.2   Methods

5.24.2.1   The case

A female patient presented to the University Hospital of Wales Dental School with all lower anterior teeth present along with the lower left premolar (referred to as a

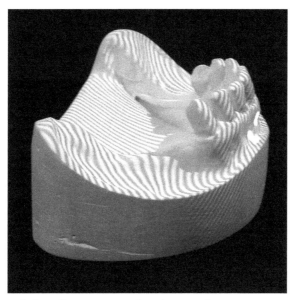

Figure 5.173 Lateral view of master cast undergoing scanning.

Kennedy class 1 case). There was some retroclination, especially of the second incisors, the canines, and the premolar. The most viable form of treatment was considered to be the provision of a cobalt-chromium metal alloy framework RPD. A design based on the 'rest, proximal plate, I clasp' (RPI) (Kratchovil, 1963) principle was formulated.

5.24.2.2   Data capture and digital RPD design

The master cast of the patient's mandibular dental structures was 3D scanned using a structured white light digitizer. The device used is an optical system that uses a projected fringe pattern of light and digital camera technology to capture approximately 140,000 points in 3D on the surface of the object (Comet 250, Steinbichler Optotechnik GmbH, Am Bauhof 4, D-83115 Neubeuern, Germany). The data points collected are referred to as a point cloud. The fringe pattern can be seen in Figure 5.173.

Software (PolyWorks, InnovMetric Software, Inc., 2014 Jean-Talon Blvd. North, Suite 310, Sainte-Foy, Quebec G1N 4N6, Canada) was used to automatically combine multiple scans (which overcomes the problems caused by line of sight) by aligning overlapping areas of scan data. A further software package (Spider, Alias-Wavefront, Inc., 210 King Street East, Toronto, Ontario M5A 1J7, Canada) was used to produce a triangular faceted surface model from the point cloud data, shown in Figure 5.174.

The surface of the computer model created from the point cloud was produced using triangular polygons in the form of an STL file, which is a suitable format for importing into the virtual sculpting environment, FreeForm. This facility was used

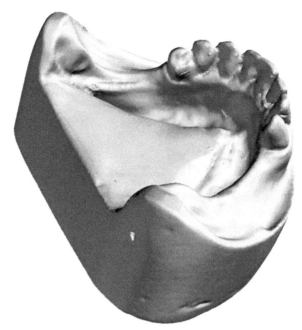

Figure 5.174 The solid computer model of the patient's cast.

to electronically survey the scanned model according to the principles outlined by Williams, Eggbeer, et al. (2004). Although no undercuts were present in the areas of clasp engagement on the abutment teeth, FreeForm is able to measure undercuts, as discussed by Williams, Eggbeer, et al. (2004).

Once surveying was completed, the model was saved in a protected manner so that it could not be altered inadvertently during the next stage of virtual patterning. A pattern was designed onscreen to the design discussed earlier on the digitally scanned model. The design followed the principles described more fully in previous work (Eggbeer et al., 2004; Eggbeer, Bibb, & Williams, 2005). Again, the package FreeForm provided excellent facilities for this process. Accurately defined, semicircular profiles such as the lingual bar were built up using construction curves, and then a groove tool was used to create a raised section. Smudge and smooth tools were used to merge the components together. The process is illustrated in Figure 5.175(a–c).

5.24.2.3   Rapid prototyping and investment casting

Previous work had indicated that STL was capable of producing suitable sacrificial patterns of RPDs (Eggbeer et al., 2004, 2005). Compared with other RP techniques, STL patterns were found to possess a good balance of properties; they were rigid enough to hold their shape, thus maintaining accuracy while being tough enough to allow handling and investment casting without inadvertent damage. Therefore, in this case once the digital design had been finalized, an STL RP machine

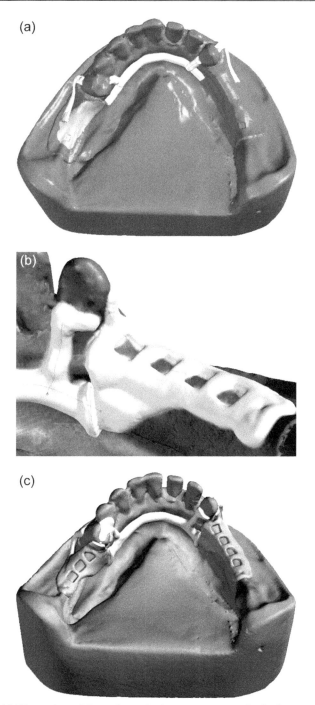

Figure 5.175 (a) Illustration of the major and minor connectors, gingivally approaching clasps, and area of retention for acrylic defined. (b) Right side of framework at a later stage of development with components joined. (c) The final digital pattern.

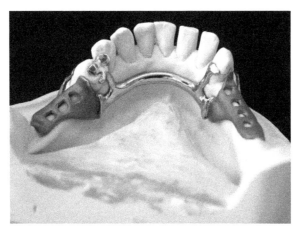

Figure 5.176 Metal alloy framework ready for clinical presentation.

(SLA 250/40; 3D Systems Inc, 26,081 Avenue Hall, Valencia, CA 91355, USA) was used to build a physical pattern in epoxy-based resin (WaterClear 10,110; DSM Somos, New Castle, DE, USA).

The fine-point supporting structures of the sacrificial pattern were thin and easily removed with a scalpel. The pattern then had wax sprues attached and was investment cast according to the procedures typically used in dental technology with the exception of the use of a refractory model. The pattern was cast directly in cobalt-chromium alloy and finished by grit blasting and polishing in the normal manner. The framework was then test fit to the master cast shown in Figure 5.176. Test fitting indicated a good fit and the framework was forwarded to the dental clinic.

This project was undertaken as an elective and was submitted for approval to the internal and external supervisors at the Welsh National School of Medicine. In this case, informed consent was required and subsequently obtained from the patient to undertake a single trial fitting of the experimental framework. The patient was not inconvenienced further and was provided with a partial denture produced using standard techniques.

5.24.3   Results

The framework was prepared for test fitting to the patient in the usual manner. The framework was test fit on the patient by a dentist (clinical supervisor). When fitted to the patient, on initial insertion there were some discrepancies, but with some adjustment with reference to the patient in the dental clinic the framework fitted satisfactorily, as shown in Figure 5.177. It is usual for a cast RPD framework to require minor adjustment to fit the patient perfectly because there may be slight differences between the patient anatomy and the cast. The clinical supervisor also confirmed that the alloy

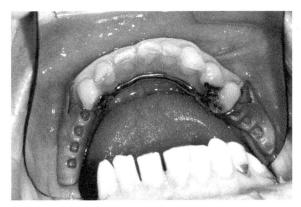

Figure 5.177 The framework in situ.

framework was satisfactory and that it could proceed to the next stage of construction, that of adding acrylic bases and artificial teeth.

5.24.4 Discussion

The concept of introducing CAD/CAM into the fabrication of RPD frameworks is taken a step further than previous work (Eggbeer et al., 2004, 2005). The successful trial fitting of the cast cobalt-chromium framework to a patient in the clinic indicated that the accuracy, tolerances, and overall fitness for purpose suggested in previous research (Eggbeer et al., 2005) can be achieved in an actual clinical case. Although considerable time was involved in producing a framework by this method, it is foreseeable that with more research, the time could be greatly reduced, making the computer-aided method competitive with existing conventional techniques.

That the framework was judged to be acceptable by an independent dentist (clinical supervisor) indicates that the application of CAD/RP techniques to this field stands comparison with existing casting techniques and suggests that there is potential for further investigation. Although the resulting framework required some minor adjustment to achieve a perfect fit, the nature and extent of this adjustment were comparable to those normally undertaken with cast frameworks produced by existing techniques. This suggests that the STL casting pattern was of comparable accuracy to the wax patterns produced by existing techniques.

Because this trial case has proved promising, further cases will be undertaken to verify that the technique is repeatable and consistent and test fitting will also be assessed by physical measurement.

Future research will explore multiple clinical cases of finished prostheses, as well as the application of direct RP technologies such as selective laser melting (SLM) that could be used to build frameworks in appropriate metal alloys, thus eliminating the casting stage altogether.

5.24.5   Conclusions

This chapter demonstrates that a clinically satisfactory RPD framework can be designed and produced by CAD and an RP-built sacrificial pattern. A framework fabricated by the methods described was clinically verified and found to be acceptable and suitable to proceed to the next stage of RPD construction and clinical use.

Acknowledgments

The authors are grateful to Mrs. Rowena Bevan for invaluable support.

References

Bibb, R., Bocca, A., & Evans, P. (2002). An appropriate approach to computer aided design and manufacture of cranioplasty plates. *The Journal of Maxillofacial Prosthetics & Technology*, *5*, 28–31.

Eggbeer, D., Bibb, R., & Williams, R. (2005). The computer-aided design and rapid prototyping fabrication of removable partial denture frameworks. *Proceedings of the Institution of Mechanical Engineers, Part H: Journal of Engineering in Medicine*, *219*, 195–202.

Eggbeer, D., Williams, R., & Bibb, R. (2004). A digital method of design and manufacture of sacrificial patterns for removable partial denture metal frameworks. *Quintessence Journal of Dental Technology*, *2*, 490–499.

Kratchovil, F. J. (1963). Influence of occlusal rest position and clasp design on movement of abutment teeth. *Journal of Prosthetic Dentistry*, *13*, 114–124.

Mormann, W. H., & Bindl, A. (2000). The CEREC 3–a quantum leap for computer-aided restorations: initial clinical results. *Quintessence International*, *31*, 699–712.

Williams, R., Bibb, R., & Rafik, T. (2004). A technique for fabricating patterns for removable partial denture frameworks using digitized casts and electronic surveying. *Journal of Prosthetic Dentistry*, *91*, 85–88.

Williams, R., Eggbeer, D., & Bibb, R. (2004). CAD/CAM in the fabrication of removable partial denture frameworks: a virtual method of surveying 3-dimensionally scanned dental casts. *Quintessence Journal of Dental Technology*, *2*, 242–267.

5.25 Dental applications case study 3: direct additive manufacture of RPD frameworks

Acknowledgments

The work described in this case study was first reported in the reference below and is reproduced here in part or in full with the permission of the copyright holders.

Bibb R, Eggbeer D, Williams R, "Rapid manufacture of removable partial denture frameworks", Rapid Prototyping Journal 2006; 12(2): 95–9, ISSN: 1355-2546, http://dx.doi.org/10.1108/13552540610652438.

5.25.1 Introduction

Over the past decade, CAD, CAM, and RP techniques have been employed in dentistry, but predominantly in the manufacture of crowns and bridges (Duret, Preston, & Duret, 1996; Van der Zel, Vlaar, de Ruiter, & Davidson, 2001; Willer, Rossbach, & Weber, 1998). However, there has been little research into the use of such methods in the field of RPD framework fabrication. Whereas RP and rapid manufacturing (RM) techniques have proved successful in other dental applications, the lack of suitable design software has restricted their application in producing RPD frameworks. Recent studies have established a valid approach to the computer-aided surveying of digital casts, framework design, and the subsequent production of sacrificial patterns using RP technologies (Eggbeer, Williams, & Bibb, 2004; Eggbeer, Bibb, & Williams, 2005; Williams, Bibb, & Rafik, 2004; Williams, Eggbeer, & Bibb, 2004).

Potential advantages offered by the introduction of CAD in the field of RPD framework design include automatic determination of a suggested path of insertion, instant elimination of unwanted undercuts, and the equally rapid identification of useful undercuts, which are all crucial in dental technology. The potential advantages of an RM approach are reduced manufacture time, inherent repeatability, and elimination of inter-operator variation.

5.25.2 Methodology

5.25.2.1 Step 1: Three-dimensional scanning

A 3D scan of a partially dentate patient's dental cast was obtained using a structured white light digitizer (Comet 250; Steinbichler Optotechnik GmbH, AM Bauhof 4, D-83115 Neubeuern, Germany, www.steinbichler.de). Multiple overlapping scans were used to collect point cloud data that were aligned using Polyworks software (InnovMetric Software, Inc., 2014 Jean-Talon Blvd. North, Suite 310, Sainte-Foy,

Quebec G1N 4N6, Canada, www.innovmetric.com). Spider software (Alias-Wavefront, Inc., 210 King Street East, Toronto, Ontario M5A 1J7, Canada, www.alias.com) was used to produce a polygon surface in the STL file format (Manners, 1993).

5.25.2.2 Step 2: Design of the RPD framework

The CAD package used in this study, called FreeForm, was selected for its capability in the design of complex, arbitrary but well-defined shapes that are required when designing custom appliances and devices that must fit human anatomy. The software has tools analogous to those used in physical sculpting and enables a manner of working that mimics that of the dental technician working in the laboratory (Geomagic FreeForm Plus, Geomagic, 3D Systems, USA). The software uses a haptic interface (Phantom Desktop) that incorporates positioning in 3D space and allows rotation and translation on all axes, transferring hand movements into the virtual environment. It also allows the operator to feel the object being worked on in the software. The combination of tools and force feedback sensations mimic working on a physical object and allows shapes to be designed and modified in a natural manner. The software also allows the import of scan data to create reference objects or bucks onto which fitting objects may be designed. The RPD metal frameworks used in this study were designed according to established principles in dental technology using this CAD software and based on a 3D scan of a patient's cast (Budtz-Jorgensen & Bochet, 1998). The CAD of RPD frameworks using this software has been described previously (Eggbeer et al., 2004, 2005; Williams, Eggbeer, & Bibb, 2004). The finished design used in this case is shown in the screen capture shown in Figure 5.178.

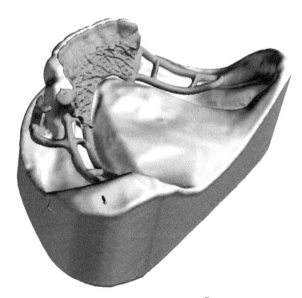

Figure 5.178 The RPD framework designed in FreeForm® CAD.

5.25.2.3 Step 3: Rapid manufacture

In a previous study, the application of RP methods was investigated for the production of sacrificial patterns that were used to investment cast RPD frameworks in cobalt-chrome alloy (Eggbeer et al., 2005). Four RP methods were compared: STL (3D Systems, Inc.), ThermoJet (obsolete, 3D Systems, USA), Solidscape T66 (Solidscape, Inc., 316 Daniel Webster Highway, Merrimack, NH 03054-4115, USA, www.solid-scape.com), and Perfactory (EnvisionTEC GmbH, Elbestraße 10, D-45768 Marl, Germany, www.envisiontec.de). These various RP processes are described more fully in Chapter 4.

In this study, direct manufacture was attempted with the aim was of eliminating the time- and material-consuming investment-casting process. The development of SLM technology showed potential application for dental technologies owing to the ability to produce complex-shaped objects in hard-wearing and corrosion-resistant metals and alloys directly from CAD data. Selective laser melting is described in Chapter 4, Section 4.5.1.

To build the RPD framework successfully using the SLM Realizer machine (MCP Tooling Technologies Ltd., now Renishaw, UK) adequate supports had to be created using Magics software (Version 9.5, Materialise NV, Technologielaan 15, 3001 Leuven, Belgium, www.materialise.com). The purpose of the supports is to provide a firm base for the part to be built onto while separating the part from the substrate plate. In addition, the supports conduct heat away from the material as it melts and solidifies during the build process. Inadequate support results in incomplete parts or heat-induced curl, which leads to build failure. Because the supports need to be removed with tools, the part was oriented such that the supports avoided the fitting surface of the RPD, as shown in Figure 5.179. This meant that the most important surfaces of the resultant part would not be affected or damaged by the supports or their removal.

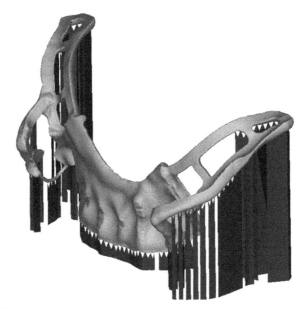

Figure 5.179 Removable partial denture oriented and supported to avoid the fitting surfaces.

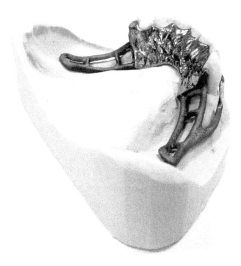

Figure 5.180 316L stainless-steel RPD framework fitted to patient cast.

First experiment

316L stainless steel was selected for the first experiment for its excellent corrosion resistance, which makes it suitable for dental applications. In addition, the SLM machine manufacturers showed that the material is well-suited to processing by SLM. The part and support files were sliced and hatched using the SLM Realizer software with a layer thickness of 0.050 mm. The material used was 316L stainless-steel spherical powder with a maximum particle size of 0.045 mm (particle size range, 0.005–0.045 mm) and a mean particle size of approximately 0.025 mm (Sandvik Osprey Ltd., Red Jacket Works, Milland Road, Neath SA11 1NJ, UK, www.smt.sandvik.com/osprey). The laser had a maximum scan speed of 300 mm/s and a beam diameter of 0.150–0.200 mm. The first two parts attempted were partially successful owing to insufficient support and erroneous slice data. These errors resulted in incomplete RPDs. The third attempt was prepared with more support and the data were sliced using different software (VisCAM RP, Marcam Engineering GmbH, Fahrenheitstrasse 1, D-28359 Bremen, Germany, www.marcam.de). This proved successful and produced a complete stainless-steel RPD framework, shown in Figure 5.180.

Second experiment

The same RPD framework design was manufactured using cobalt-chrome alloy using a layer thickness of 0.075 mm (Sandvik Osprey Ltd.). The principal reason for attempting the design in cobalt-chrome was for direct comparison with traditionally made RPD frameworks, which are typically cast from the same material. Like the previous material, the SLM machine manufacturers showed cobalt-chrome to be

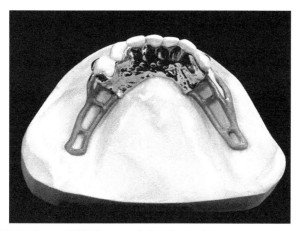

Figure 5.181 Cobalt-chrome RPD framework fitted to patient cast.

suitable for processing by SLM. As before, the laser had a maximum scan speed of 300 mm/s with a beam diameter of 0.150–0.200 mm. The material used was cobalt-chrome spherical powder with a maximum particle size of 0.045 mm (particle size range, 0.005–0.045 mm) and a mean particle size of approximately 0.030 mm. The part proved successful and produced a complete cobalt-chrome RPD framework, shown in Figure 5.181.

5.25.2.4 *Step 4: Finishing*

Supporting structures were removed using a Dremel handheld power tool using a reinforced cutting wheel (Dremel, Reinforced Cutting Disc, Ref. Number 426). The frameworks in their initial form were well formed but showed a fine surface roughness. This roughness was easily removed by bead blasting. This resulted in a framework that showed similar physical appearance and surface qualities as the investment cast items typically used in dental technology. Therefore, the treatment and finishing of the framework from that point onwards was conducted in the same manner as any other RPD framework, using normal dental laboratory techniques and equipment.

5.25.3 **Results**

The successful 316L Stainless Steel RPD framework was assessed for the quality of fit by fitting it to the plaster cast of the patient's oral anatomy. The quality of the fit was assessed according to normal dental practice by an experienced dental technician and found to show excellent fit. The frameworks showed a quality of fit that was comparable with investment cast frameworks. However, repeated insertion and removal from the patient cast resulted in small but permanent deformation of the clasp components. The clasp components are the functional parts of the framework and are designed to grip the teeth to provide a firm location of the denture (the clasps are the elements shown in the close-up photographs in Figure 5.182). Therefore, the permanent deformation reduces the ability of the framework to grip the teeth and the denture becomes loose. This meant that after several operations the clasps no longer held the framework as securely to the existing teeth as deemed necessary by the dental technician.

The cobalt-chrome RPD framework was complete, polished, and finished well with the normal dental technology procedures. The framework proved to be an excellent fit, possessing good clasping when test fitted to the patient's cast (Figure 5.183). The framework was test fit to the patient in the clinic and found to be a precise and comfortable fit with good retention, shown in Figure 5.184. The framework was therefore fitted with the artificial teeth and given to the patient to use in exactly the same manner as a traditionally manufactured item. Unlike the previous stainless-steel framework, the clasping forces did not result in permanent deformation of the clasps and the framework withstood repeated insertion and removal cycles.

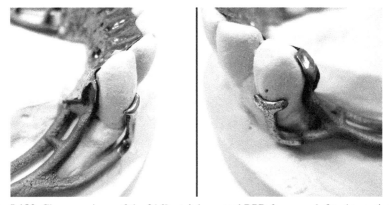

Figure 5.182 Close-up views of the 316L stainless-steel RPD framework fitted to patient cast.

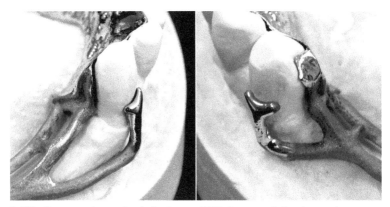

Figure 5.183 Close-up views of cobalt-chrome RPD framework fitted to patient cast.

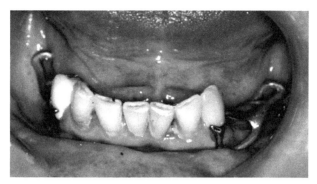

Figure 5.184 Framework fitted to the patient in clinic.

5.25.4 Discussion

5.25.4.1 Sources of error

Various studies have aimed to assess error in cobalt-chrome partial denture frameworks made using traditional investment casting techniques (Barsby & Schwarz, 1989; Murray & Dyson, 1988; Stern, Brudvik, & Frank, 1985). However, in the absence of an appropriate intra-oral scanning technology, the application of CAD/CAM in dental technology depends on the dental model, which is a plaster cast taken from an impression of the dental anatomy taken by a dentist. Clearly, this chapter cannot address issues relating to the quality of the original dental impression or the casting of the dental model from this impression. In addition, human error in the interpretation of the dentist's instructions or in the dental technician's chosen design for the framework is not addressed in here. However, the adoption of CAD/CAM/RP technologies may incur several process steps that may contribute to error between the theoretical designs produced using CAD and the final manufactured item. The effect of these processes will be an accumulation of tolerances at each technology stage. However, certain levels of care and skill may still affect the accuracy of these computer-controlled techniques.

Table 5.17 shows the steps in the process investigated here and indicates nominal tolerances associated with the technologies used. The accumulation of tolerances leads to the maximum error that could be expected to result from the technologies alone, assuming no human error is encountered. Because human skill level and error cannot be attributed a numerical value and might range from zero to complete failure, discussion is not included here. However, because this study aims to investigate the implications of adopting CAD/CAM/RP technologies in dental technology, it is appropriate to attempt to illustrate their potential contribution to error in the final RPD framework. The tolerances used in this table indicate typical or nominal figures, which are quoted by manufacturers or set as parameters in software.

From the processes listed in Table 5.17, it is reasonable to expect a tolerance of approximately 0.2 mm for these parts. Cumulative negative and positive tolerances from the various steps might also partially cancel each other, resulting in a lower overall

Table 5.17 **Process steps and associated tolerances**

Process step	Source of error	Tolerance
Impression taking	Human/skill level	No value
Casting study model	Human/skill level	No value
Optical scanning of study model	Scanner	±0.050 mm
Creating polygon computer model from point cloud data	Software setting	±0.050 mm
Import into CAD software	Software	0.000 mm
Design in CAD software	Software setting	±0.001 mm
Export of CAD data in STL file format	Software setting	±0.010 mm
Physical manufacture using SLM	RP machine	±0.100 mm
Removal of RP pattern from machine, cleaning, and support removal	Human/skill level	No value
Surface preparation and polishing	Human/skill level	No value
	Total	±0.211 mm

tolerance. The contribution of each individual step would be difficult to demonstrate without a statistically significant number of cases. The closeness of the fit and effective clasping observed when fitting the frameworks to the patient cast, as shown in Figures 5.182 and 5.183, suggest that SLM RPD frameworks are in fact within this tolerance.

5.25.4.2   Error analysis

Removable partial denture frameworks are by definition one-off custom-made appliances specifically designed and made to fit a single individual patient. In addition, the anatomically fitting nature of RPD frameworks means that they are complex in form and do not provide convenient datum or reference surfaces. This makes it difficult to achieve an investigation that provides a detailed quantitative analysis of error. Therefore, it is not practical to perform the type of repeated statistical analysis that would be commonly encountered in series production or mass manufacture. Instead, it is normal dental practice to assess the accuracy of an RPD by test fitting the device to the patient cast and subsequently, in clinic, to the patient. In this study, the RPD frameworks created were deemed by a qualified and experienced dental technician to be a satisfactory fit and comparable to those produced by expert technicians (Figure 5.184). This suggests that the approach and technologies used are fit for purpose in this application, although further experiments with a range of patients with differing RPD designs will be required to ensure that this is in fact the general case.

5.25.5   Conclusions

Selective laser melting has been shown to be a viable RM method for the direct manufacture of RPD metal alloy frameworks. Parts produced using the SLM process in conjunction with cobalt-chrome alloy result in RPD frameworks that are comparable in terms of accuracy, quality of fit, and function to existing methods typically used in the dental technology laboratory. The CAD and CAM approaches offer potential advantages in terms of reduced inter-operator variability, repeatability, speed, and economy over traditional handcrafting and investment casting techniques.

References

Barsby, M. J., & Schwarz, W. D. (1989). The qualitative assessment of cobalt-chromium casting for partial dentures. *British Dental Journal*, *166*(6), 211–216.

Budtz-Jorgensen, E., & Bochet, G. (1998). Alternate framework designs for removable partial dentures. *Journal of Prosthetic Dentistry*, *80*(1), 58–66.

Duret, F., Preston, J., & Duret, B. (1996). Performance of CAD/CAM crown restorations. *Journal of the California Dental Association*, *9*(9), 64–71.

Eggbeer, D., Bibb, R., & Williams, R. (2005). The computer aided design and rapid prototyping of removable partial denture frameworks. *Journal of Engineering in Medicine*, *219*, 195–202.

Eggbeer, D., Williams, R. J., & Bibb, R. (2004). A digital method of design and manufacture of sacrificial patterns for removable partial denture metal frameworks. *Quintessence Journal of Dental Technology*, *2*(6), 490–499.

Manners, C. R. (1993). *STL file format*. 26081 Avenue Hall, Valencia, CA, 91355, USA: available on request from 3D Systems Inc. www.3dsystems.com.

Murray, M. D., & Dyson, J. E. (1988). A study of the clinical fit of cast cobalt-chromium clasps. *Journal of Dentistry*, *16*(3), 135–139.

Stern, M. A., Brudvik, J. S., & Frank, R. P. (1985). Clinical evaluation of removable partial denture rest seat adaptation. *Journal of Prosthetic Dentistry*, *53*(5), 658–662.

Willer, J., Rossbach, A., & Weber, H. P. (1998). Computer-assisted milling of dental restorations using a new CAD/CAM data acquisition system. *Journal of Prosthetic Dentistry*, *80*(3), 346–353.

5.26   Dental applications case study 4: a comparison of plaster, digital and reconstructed study model accuracy

Acknowledgments

The work described in this case study was first reported in the reference below and is reproduced here in part or in full with the permission of the copyright holders.

Keating AP, Knox J, Bibb R, Zhurov A, "A comparison of plaster, digital and reconstructed study model accuracy", Journal of Orthodontics 2008; 35: 191–201, http://dx.doi.org/10.1179/146531207225022626

5.26.1   Introduction

Orthodontic treatment outcome and treatment change have traditionally been recorded with gypsum-based study models that are heavy and bulky, pose storage and retrieval problems, are liable to damage, and can be difficult and time-consuming to measure (Ayoub, Wray, & Moos, 1997; Hunter & Priest, 1960; McGuinness & Stephens, 1992; Quimby, Vig, Rashid, & Firestone, 2004; Santoro, Galkin, Teredesai, Nicolay, & Cangialosi, 2003). Legislation relating to the retention of patient records after the completion of treatment (McGuinness & Stephens, 1993) has led to huge demands on space for storage that has prompted the development of alternative methods of recording occlusal relationships (Table 5.18) and electronic storage of records (Delong, Heinzen, Hodges, Ko, & Douglas, 2003; Foong et al., 1999; Hirogaki, Sohmura, Takahashi, Noro, & Takada, 1998; Hirogaki, Sohmura, Satoh, Takahashi, & Takada, 2001; Kuroda, Motohashi, Tominaga, & Iwata, 1996; Kusnoto & Evans, 2002; Yamamoto et al., 1988).

The replacement of plaster study models with virtual images has several advantages, including ease of access, storage, and transfer (Stevens et al., 2006), and the accuracy of image capture techniques has been reported (Alcaniz et al., 1998; Garino & Garino, 2002; Motohashi & Kuroda, 1999; Quimby et al., 2004; Santoro et al., 2003; Sohmura et al., 2000; Tomassetti et al., 2001; Zilberman et al., 2003). However, if the physical restoration of a digital occlusal record is needed, possibly for medicolegal reasons, an accurate method of 3D reconstruction is required.

Rapid prototyping systems such as Stereolithography (SL) generate 3D models from a digital file through incremental layering of photo-curable polymers (Chua, Chou, Lin, Lee, & Saw, 2000). The dimensional accuracy of physical replicas reproduced using the STL technique has been evaluated by a number of authors. Barker, Earwaker, and Lisle (1994) found a mean difference of 0.85 mm between measurements made on actual dry bone skulls, and physical replicas of the skulls, produced by SL from 3D computed tomography (CT) scans of the original dry bone skulls. They concluded that RP models could be confidently used as accurate 3D replicas of complex anatomic structures. Using similar techniques, Kragskov, Sindet-Pedersen,

Table 5.18 **Alternative methods of recording occlusal relationships**

2D techniques	
Conventional photography	Cookson (1970); Burstone (1979); Dervin, Gore, and Kilshaw (1976); McKeown, Robinson, Elcock, Al-Sharood, and Brook (2002)
Photocopying	Singh and Savara (1964); Huddart, Clarke, and Thacker (1971); Mazaheri, Harding, Cooper, Meier, and Jones (1971); Champagne (1992); Schirmer and Wiltshire (1997); McCance, Perera, and Woods (1991); Yen (1991)
Flatbed scanner	Tran, Rugh, Chacon, and Hatch (2003)
3D techniques	
Optocom	Van der Linden, Boersma, Zelders, Peters, and Raaben (1972)
Reflex plotters	Suzuki (1980); Foong et al. (1999)
Reflex metrograph	Scott (1981, 1984); Takada, Lowe, and DeCou (1983); Speculand, Butcher, and Stephens (1988b)
Reflex microscope	Scott (1981); Speculand, Butcher, and Stephens (1988a, 1988b); Johal and Battagel (1997)
Travelling microscope	Bhatia and Harrison (1987)
Moiré topography	Takasaki (1970); Kanazawa, Sekikawa, and Ozaki (1984); Mayhall and Kageyama (1997)
Stereophotogrammetry	Halazonetis (2001); Jones (1979); Richmond (1984); Ayoub et al. (1997); Bell, Ayoub, and Siebert (2003)
Telecentric lens photography	Kennedy (1979)
Holography	Schwaninger, Schmidt, and Hurst (1977); Burstone, Pryputniewicz, and Bowley (1978); Ryden, Bjelkhagen, and Martensson (1982); Keating, Parker, Keane, and Wright (1984); Harradine, Suominen, Stephens, Hathorn, and Brown (1990); Buschang, Ceen, and Schroede (1990); Martensson and Ryden (1992); Romeo (1995)
Optical profilometer	Berkowitz, Gonzalez, and Nghiem-phu (1982)
Image analysis system	Brook, Pitts, and Renson (1983); Brook, Pitts, Yau, and Sandar (1986)
3D computed tomography	Quintero, Trosien, Hatcher, and Kapila (1999); Mah and Baumann (2001); Darvann et al. (2001); Kuo and Miller (2003)
Structured light scanning methods	Harada, Yamamoto, Ohnuma, Mikami, and Nakamura (1985); Yamamoto et al. (1988); Laurendeau, Guimond, and Poussart (1991); Wakabayashi et al. (1997); Hirogaki et al. (1998, 2001); Kojima et al. (1999); Sohmura et al. (2000); Nagao et al. (2001); Hayashi, Araki, Uechi, Ohno, and Mizoguchi (2002); Foong et al. (1999); Lu, Li, Wang, Chen, and Zhao in (2000); Alcaniz et al. (1998); Alcaniz, Grau, Monserrat, Juan, and Albalat (1999); Brosky, Pesun, Lowder, Delong, and Hodges (2002); Brosky, Major, Delong, and Hodges (2003); Delong, Ko, Anderson, Hodges, and Douglas (2002); Delong et al. (2003); Schelb, Kaiser, and Brukl (1985); Braumann, Keilig, Bourauel, Niederhagen, and Jager (1999)
Intra-oral scanning devices	Delong et al. (2003); Commer, Bourauel, Maier, and Jager (2000)
Commercially available 3D digital study models	Zilberman, Huggare, and Parikakis (2003); Redmond (2001); Tomassetti, Taloumis, Denny, and Fischer (2001); Santoro et al. (2003); Baumrind (2001); Kuo and Miller (2003); Freshwater and Mah (2003); Hans et al. (2001); Baumrind et al. (2003); Mah and Sachdeva (2001)

Gyldensted, and Jensen (1996) and Bill et al. (1995) found mean differences of −0.3–
0.8 mm and ±0.5 mm between measurements on 3D-CT images and STL models.

The objectives of this study were to:

- Assess the reproducibility of a conventional method of using a handheld Vernier caliper to measure plaster study models;
- Develop an efficient and reproducible method of capturing a 3D study model image, in a digital format, using the Minolta VIVID 900 noncontact surface laser-scanning device (Konica-Minolta, Inc., Tokyo, Japan);
- Assess the reproducibility of measurements made on the on-screen 3D digital surface models captured using the scanning system setup developed;
- Compare the accuracy of measurements made on the 3D digital surface models and plaster models of the same dentitions; and
- Evaluate the feasibility of fabricating accurate 3D hardcopies of dental models from the laser scan data, by RP (stereolithography).

5.26.1.1 Null hypotheses

- There is no difference in the dimensional accuracy of 3D digital surface models, captured with the surface laser-scanning technique described, and plaster study models.
- There is no difference in the dimensional accuracy of physical model replicas fabricated from the laser scan data by RP, and plaster study models.

5.26.2   Materials and methods

5.26.2.1 Manual measurements

The local Regional Ethics Committee chair confirmed that no ethical approval was required for this study. A minimum of 7–10 models per group were calculated to be required to allow a 90% chance of detecting a 0.3 mm difference in related sample means (standard deviation = 0.2) at the 5% level of significance ($\alpha = 0.05$ and power = 0.90) (Altman, 1991). Thirty randomly selected plaster study models, held in the Orthodontic Unit of University Dental Hospital, Cardiff, were used in the study. Each study model was cast in matte white Crystal R plaster (South Western Industrial Plasters, Wilts, England) and conventionally trimmed (Adams, 1976). To be included in the study the plaster study models had to completely reproduce the arch; show no surface marks, loss of tooth material, voids, or fractures; and demonstrate varying degrees of contact point and buccolingual tooth displacements.

A handheld digital calliper (Series 500 Digimatic ABSolute caliper Mitutoyo Corporation, Japan) was used to manually measure the plaster models. This calliper had a measurement resolution of 0.01 mm, was accurate to ±0.02 mm in the 0–200 mm range, and automatically downloaded data, eliminating measurement transfer and calculation errors.

All plaster models were measured in a bright room without magnification. The plaster models were not prepared in any way before measuring, and the anatomical dental landmarks used in the measurements were not pre-marked. A single examiner conducted all the measurements after an initial training period.

Twenty linear dimensions were measured on each model, in each of the three planes (x, y, and z), with all measurements being recorded to the nearest 0.01 mm. The following dimensions were selected for measurement:

- x plane
  - Intercanine distance – measured as the distance between:
    - The occlusal tips of the upper canines
    - The occlusal tips of the lower canines
  - Interpremolar distances – measured as the distance between:
    - The buccal cusp tips of the upper and lower first and second premolars
    - The palatal cusp tips of the upper first and second premolars
    - The lingual cusp tips of the lower first premolars
    - The mesiolingual cusp tips of the lower second premolars
  - Intermolar distances – measured as the distance between:
    - The mesiopalatal cusp tips of upper first and second molars
    - The mesiobuccal cusp tips of the upper and lower first and second molars
    - The mesiolingual cusp tips of lower first and second molars
    - The distobuccal cusp tips of the upper and lower first molars
- y plane
  - In the upper arch, the distance from the mesiopalatal cusp tip of the upper second molar to:
    - The mesiopalatal cusp tip of the upper first molar
    - The palatal cusp tip of the upper first and second premolar
    - The cusp tip of the upper canine
    - The mesio-incisal corner of the upper lateral incisor

    These dimensions were measured on both sides of the upper arch.
  - In the lower arch, the distance from the mesiolingual cusp tip of the lower second molar to:
    - The mesiolingual cusp tip of the lower first molar and second premolar
    - The lingual cusp tip of the lower first premolar
    - The cusp tip of the lower canine
    - The mesio-incisal corner of the lower lateral incisor

    These dimensions were measured on both sides of the lower arch.
- z plane
  - The clinical crown height of all the teeth, in both upper and lower arches, from the second premolar to second premolar inclusive, measured as the distance between the cusp tip and the maximum point of concavity of the gingival margin on the labial surface.

Measurements were made on two occasions separated by at least 1 week.

5.26.2.2   Virtual measurements

A noncontact laser-scanning device (Minolta VIVID 900, Konica Minolta, Inc., Japan) was used to record the surface detail of each of the 30 study models using a telescopic light-receiving lens (focal distance $f = 25$ mm) and rotary stage (ISEL-RF1, Konica Minolta, Inc., Japan). The rotary stage facilitated the acquisition of multiple range maps by moving the plaster study models in sequence by a controlled rotation as they were being scanned, thus ensuring the entire visible surface of each plaster model was captured. The stage was controlled by a computer software program (Easy3DScan Tower Graphics, Italy) and integrated controller box (IT116G Minolta, Inc., Osaka, Japan).

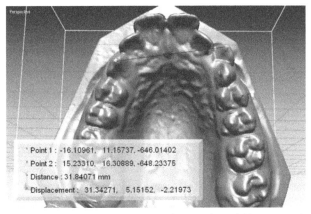

Figure 5.185 On-screen 3D virtual model image: intercanine width measured.

Easy3DScan was used to align, merge and simplify the range maps acquired at different angles to produce a composite surface dataset that was then imported into the RapidForm 2004 software program (INUS Technology, Inc., Seoul, Korea) as a triangulated 3D mesh (Figure 5.185). An automated measuring tool was used to record the same measurements that had been conducted manually on the plaster study models. The 3D digital surface models were magnified and rotated on screen to aid identification of the anatomical landmarks as necessary. Linear distances between landmarks were calculated automatically to five decimal places (Figure 5.185). Replicate measurements were made on all digital model images with a time interval of at least 1 week.

5.26.2.3   Measurement of reconstructed models

One pair of upper and lower plaster models was scanned individually using an identical protocol, adhering to the inclusion criteria listed previously. Only one set of models was evaluated owing to the current cost of SL. The scanned data for both upper and lower plaster models were saved as binary STL files and imported into the Magics RP software (Materialise, Inc., Technologielaan 15, 3001 Leuven, Belgium).

A 3D Systems (SLA-250/40) SL machine (3D Systems, USA) containing a hybrid-epoxy-based resin (10110 Waterclear, DSM Somos, 2 Penn's Way, Suite 401, New Castle, DE 19720, USA) was used to construct RP study models from the digital files using a build layer thickness of 0.15 mm (Figures 5.186 and 5.187).

Measurements in x, y, and z planes were made on the reconstructed models identical to those recorded on the original plaster study models and virtual models. Replicate measurements were made 1 week later.

5.26.2.4   Statistical analysis

Bland Altman analysis (Bland & Altman, 1999) was undertaken to determine agreement between repeat model measurements. Intra-rater reliability was assessed by visually comparing the difference in repeat measurements and performing nonparametric, Wilcoxon signed-rank hypothesis tests. This is described in the Results section.

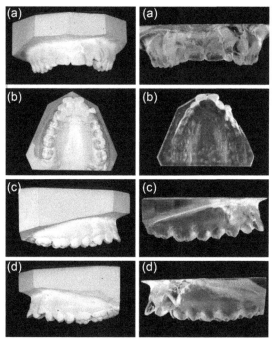

Figure 5.186 Original upper plaster model (left) and SL model (right) as seen from (a) frontal, (b) occlusal, (c) right buccal and (d) left buccal directions.

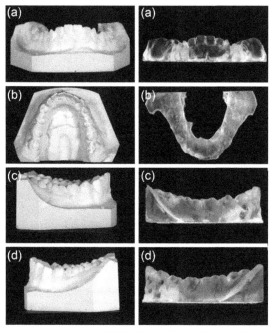

Figure 5.187 Original lower plaster model (left) and SL model (right) as seen from (a) frontal, (b) occlusal, (c) right buccal and (d) left buccal directions.

5.26.3   Results

Data analysis demonstrated a nonnormal distribution of results and non-parametric tests (Wilcoxon signed-rank test) were therefore employed in the statistical analysis.

No significant difference ($p > 0.2$) was demonstrated in measurements at initial time (T1) and 1 week later (T2) for the manual measurement of plaster study models (Table 5.19; Figure 5.188), 3D digital surface model measurement (Table 5.20; Figure 5.189), or manual measurement of the SL, reconstructed models (Table 5.21; Figure 5.190). Almost all points were clustered around the mean difference of 0, within 2 SDs of the mean difference (Figures 5.188–5.190) indicating good intra-rater reliability.

A comparison of linear measurements made on the plaster study models and 3D digital surface (virtual) models is presented in Table 5.22 and Figure 5.191. The mean difference in all planes was 0.14 mm (SD, 0.10 mm) and was not statistically significant ($p > 0.2$).

Measurements made in x and y planes were not significantly different for reconstructed models and plaster models ($p > 0.3$) or 3D digital surface models ($p > 0.5$). However, in the z plane, measurement differences were significantly different ($p < 0.001$) (Tables 5.23 and 5.24; Figures 5.192 and 5.193). All z plane reconstructed model measurements were significantly smaller than the corresponding plaster and 3D digital surface model measurements.

5.26.4   Discussion

This study has demonstrated a simple and reproducible method of study model measurement. The excellent reproducibility of plaster, digital, and reconstructed model measurements reported compares favourably with Zilberman et al. (2003) and Bell et al. (2003), who reported mean intra-operator errors of 0.18 and 0.17 mm, respectively, when the same points were measured by the same operator at different times on plaster study models, and Stevens et al. (2006), who reported a concordance correlation

Table 5.19 **Variation in repeat measurements of plaster models: 20 measurements in each plane repeated on 30 models**

Plane	N	Mean difference, mm	SD, mm	p value
x plane	20	0.15	0.09	0.601
y plane	20	0.16	0.09	0.313
z plane	20	0.11	0.07	0.489
x, y, z planes	60	0.14	0.09	0.558

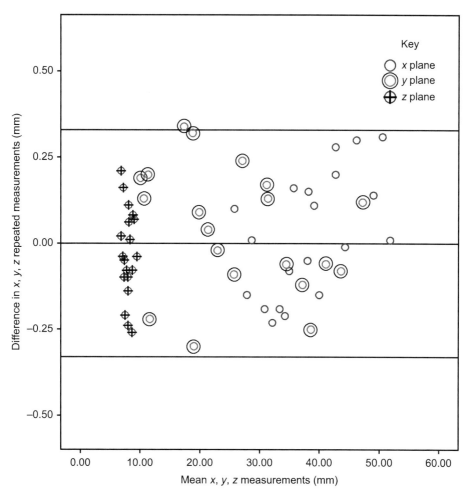

Figure 5.188 Bland Altman plot for repeat measurements plaster model. Twenty measurements in each plane repeated on 30 models (reference lines showing 2 SD).

Table 5.20 Variation in repeat measurements of virtual models: 20 measurements in each plane repeated on 30 models

Plane	N	Mean difference, mm	SD, mm	p value
x plane	20	0.15	0.09	0.823
y plane	20	0.12	0.08	0.549
z plane	20	0.14	0.11	0.501
x, y, z planes	60	0.14	0.09	0.965

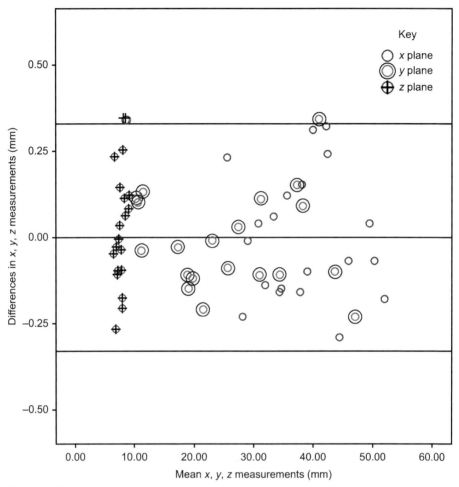

Figure 5.189 Bland Altman plot for repeat virtual model measurements. Twenty measurements in each plane repeated on 30 models (reference lines showing 2 SD).

Table 5.21 **Variation in repeat measurements of the reconstructed model: 20 measurements in each plane repeated on one model**

Plane	N	Mean difference, mm	SD, mm	p value
x plane	20	0.12	0.06	0.985
y plane	20	0.13	0.11	0.985
z plane	20	0.14	0.09	0.550
x, y, z planes	60	0.13	0.09	0.938

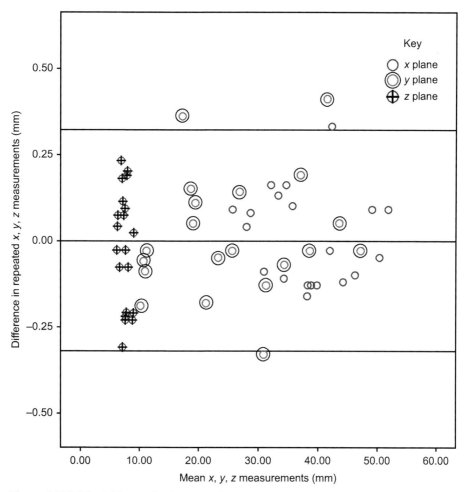

Figure 5.190 Bland Altman plot for repeat SL model measurements. Twenty measurements in each plane repeated on one pair of models (reference lines showing 2 SD).

Table 5.22 Difference between plaster and virtual model measurements (means of 20 measurements in each plane compared)

Plane	N	Mean difference, mm	SD, mm	p value
x plane	20	0.19	0.12	0.765
y plane	20	0.14	0.09	0.501
z plane	20	0.10	0.07	0.218
x, y, z planes	60	0.14	0.10	0.237

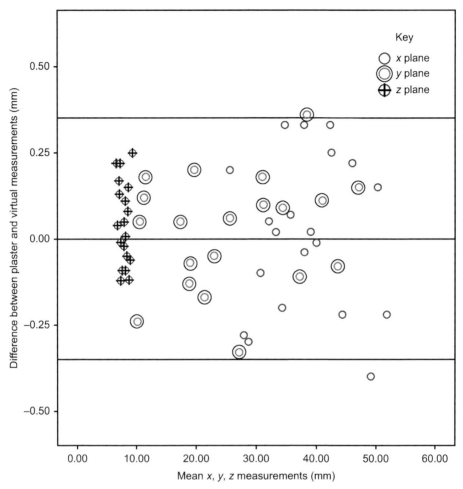

Figure 5.191 Bland Altman plot for differences in plaster and virtual model measurements (reference lines showing 2 SD).

Table 5.23 Difference between plaster and reconstructed model measurements (means of 20 measurements in each plane compared)

Plane	N	Mean difference, mm	SD, mm	p value
x plane	20	0.15	0.16	0.645
y plane	20	0.19	0.15	0.360
z plane	20	0.42	0.23	0.000**
x, y, z planes	60	0.26	0.22	0.000**

** = significant difference (p<0.001).

Table 5.24 **Difference between virtual and reconstructed model measurements (means of 20 measurements in each plane compared)**

Plane	N	Mean difference, mm	SD, mm	p value
x plane	20	0.18	0.12	0.550
y plane	20	0.22	0.16	0.513
z plane	20	0.38	0.21	0.000**
x, y, z planes	60	0.25	0.21	0.000**

** = significant difference ($p<0.001$).

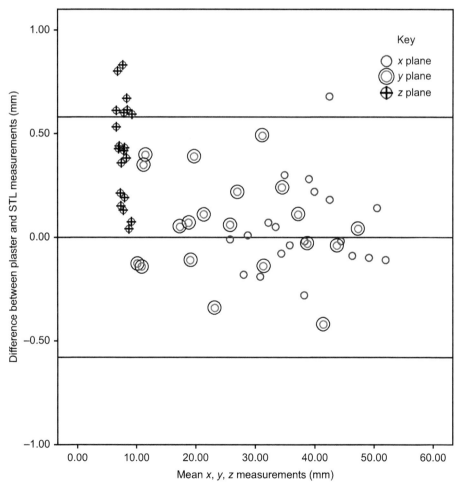

Figure 5.192 Bland Altman plot for differences in plaster and SL model measurements (reference lines showing 2 SD).

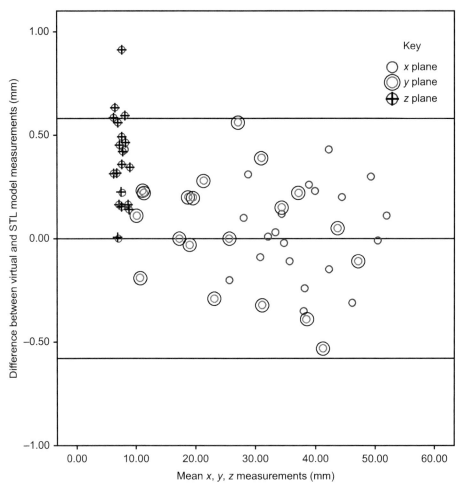

Figure 5.193 Bland Altman plot for differences in virtual and SL model measurements (reference lines showing 2 SD).

coefficient of 0.88 for the measurement of digital models using *e*-model software (GeoDigm, Chanhassen, MN, USA) (Halazonetis, 2001; Stevens et al., 2006; Zilberman et al., 2003). The reconstructed, SL model measurements in this study demonstrated a greater range in repeated absolute measurement differences (0.02–0.41 mm) compared with those for the other two methods (0.01–0.34 mm), which reflects the greater difficulty in measuring these models.

This study has also demonstrated the validity of digital (virtual) models derived from the laser-scanning process described. The problems of trying to acquire dimensionally accurate images using structured light scanning methods have been reported by Bibb et al. (2000) and Mah and Hatcher (2003). The light beam from structured light scanners travels in straight lines, so any object surfaces that are obscured or are at too great an angle to the line of sight of the light source will not be scanned. This results in voids or holes in the scanned surface data. To overcome this problem, the object or scanner needs to be moved to different angulations and the scan process repeated at each angle. For irregular objects, multiple scans of the same object from different angles may need to be acquired. The data from each of these scans can then be stitched (registered and merged) together using special software programs to produce a single composite surface model of the object (Bibb et al., 2000; Halazonetis, 2001; Mah & Baumann, 2001). Compounding these difficulties are the errors introduced during computer processing of the acquired data that is necessary to reduce artefacts and yet retain detail, whereas errors can also be introduced during the merging of the multiple perspectives to form the single composite surface model of the object being scanned (Mah & Hatcher, 2003).

A number of authors who evaluated alternative ways of measuring study models suggested what they considered to be a clinically significant measurement difference. Schirmer and Wiltshire (Schirmer & Wiltshire, 1997) regarded a measurement difference between alternative measurement methods of less than 0.20 mm as clinically acceptable. Hirogaki et al. (2001) suggested the accuracy required with orthodontic study models to be about 0.30 mm, whereas Halazonetis (2001) reported that an accuracy of 0.50 mm was sufficient for head and face laser scanning but would be inadequate for scanning study models. Bell et al. (2003), investigating the accuracy of the stereophotogrammetry technique for archiving study models, decided a mean difference of 0.27 mm (SD, 0.06 mm) between this technique and measurements made by hand on plaster models was unlikely to have a significant clinical impact.

The accuracy of the on-screen virtual models as reported in this study compares favourably with some studies but less favourably than with others. These studies varied greatly in their 3D capture techniques and software analysis systems (Table 5.25).

The statistically significant difference between measurements made directly on the plaster models and those made on the reconstructed models was largely the result of errors in the z plane. The SL models were built in 0.15-mm layers from a clear resin. Model translucency made landmark identification difficult and layering resulted in some loss of surface detail, particularly at the cervical margin (Figures 5.186 and 5.187). In addition, errors in data conversion and data manipulation generated while converting digital surface models to STL file format can result in some distortion (Cheah, Chua, Tan, & Teo, 2003; Mah & Hatcher, 2003) and the rapid prototyping technique can also introduce errors owing to model shrinkage during building and post-curing (Barker et al., 1994). However, the clinical significance of these errors will depend on the intended purpose of the reconstructed model. The models may not be sufficiently accurate for appliance construction but may be sufficient to demonstrate pre- or post-treatment occlusal relationships. Unfortunately, the current prohibitive cost of SL limited this study to the evaluation of only one pair of reconstructed models.

This study has presented a novel method of digitally recording study model data, offering the profession a valid alternative to the use of conventional plaster models and the potential to significantly reduce the burden of model storage. In addition, the potential for physical reconstruction of a model from the digital archive has been demonstrated, which may go towards addressing medicolegal concerns.

Table 5.25 **Accuracy of various 3D capture techniques (average measurement error is the mean difference between measurements made on virtual models and the original plaster models)**

Authors	Type	Device	Average measurement	Landmark/ identification
Wakabayashi et al. (1997)	Point laser	Mitsubishi, MD1211-40	<0.072 mm	Landmarks not identified
Kojima et al. (1999)	Line laser	Hitachi, HL6712G laser	0.03 mm	Not specified
Motohashi and Kuroda (1999)	Line laser	UNISN, 3D VMS-250R	0.00–0.20 mm	Not specified
Lu et al. (2000)	Laser	Not indicated	<0.10 mm	Not specified
Commer et al. (2000)	Line laser	POS-PLD-50, laser 2000	Not more than +0.2 mm	Landmarks identified by marking
Sohmura et al. (2000)	Line laser	Minolta VIVID700	0.08–0.35 mm	Landmarks not identified
Hirogaki et al. (2001)	Line laser	Cubesper laser	<0.30 mm	Landmarks not identified
Tomassetti et al. (2001)	Destructive laser scanner	OrthoCAD	1.2 mm	Overall Bolton discrepancy
Kusnoto and Evans (2002)	Line laser	Minolta VIVID700	$x = 0.20$ mm $z = 0.70$ mm	Landmarks identified by marking
Santoro et al. (2003)	Destructive laser scanner	OrthoCAD	0.16–0.49 mm	Landmarks not identified
Bell et al. (2003)	Stereophoto-grammetry	Not indicated	0.27 mm	Landmarks identified by marking
Quimby et al. (2004)	Destructive laser scanner	OrthoCAD	0.15–0.66 mm	Landmarks not identified

5.26.5   Conclusions

- The use of using a handheld Vernier calliper to measure plaster study models was reliable and reproducible.
- The Minolta VIVID 900 is a reliable device for capturing the surface detail of plaster study models 3D in a digital format using the protocol described.
- Measurement of the on-screen 3D digital surface models captured was reproducible.
- Measurement of 3D digital surface models and plaster models of the same dentitions showed good agreement.
- The detail and accuracy of physical models, reconstructed from digital data, may not be sufficient for certain applications, using the standard SL techniques described.
- Improved RP techniques may offer a more accurate method of model reconstruction from digital archives.

5.26.6   Future work

A SL process employing thinner layers or other RP technologies that use significantly thinner build layers may address the deficiencies of the reconstructed models used in this study. Techniques that should be investigated include digital light processing-based machines (EnvisionTEC GmbH, Brüsseler Straße 51, 45968 Gladbeck, Germany) and the various printing based processes such as poly-jet modelling (Objet Geometries, Ltd., Headquarters, 2 Holzman Street, Science Park, PO Box 2496, Rehovot 76124, Israel), multi-jet modelling (3D Systems, Inc., Headquarters, 333 Three D Systems Circle, Rock Hill, SC 29730, USA), and single-head jetting (Solidscape, Inc., 316 Daniel Webster Highway, Merrimack, NH 03054-4115, USA). These processes are all capable of producing physical models with a layer thickness of up to 10 times thinner than the STL models described in this chapter (layer thicknesses range from 0.013 to 0.150 mm).

5.26.7   Contributors

Andrew Keating, Richard Bibb, and Jeremy Knox were responsible for study design. Richard Bibb contributed to this work while at the National Centre for Product Design and Development Research, Cardiff Metropolitan University, Cardiff, which supplied the SL models. Andrew Keating was responsible for data collection. Andrew Keating and Alexei Zhurov were responsible for data manipulation and processing. All authors were responsible for data analysis and preparation of manuscript. Jeremy Knox is the guarantor.

References

Adams, P. C. (1976). *The design and construction of removable orthodontic appliances* (4th ed.). Bristol: John Wright and Sons Ltd.

Alcaniz, M., Grau, V., Monserrat, C., Juan, C., & Albalat, S. (1999). A system for the simulation and planning of orthodontic treatment using a low cost 3D laser scanner for dental anatomy capturing. *Studies in Health Technology and Information, 62*, 8–14.

Alcaniz, M., Montserrat, C., Grau, V., Chinesta, F., Ramon, A., & Albalat, S. (1998). An advanced system for the simulation and planning of orthodontic treatment. *Medical Image Analysis, 2*, 61–77.

Altman, D. G. (1991). *Practical statistics for medical research.* London: Chapman and Hall, 455–58.

Ayoub, A. F., Wray, D., Moos, K. F., Jin, J., Niblett, T. B., Urquhart, C., et al. (1997). A three-dimensional imaging system for archiving dental study casts: a preliminary report. *International Journal of Adult Orthodontics & Orthognathic Surgery, 12*, 79–84.

Barker, T. M., Earwaker, W. J. S., & Lisle, D. A. (1994). Accuracy of stereolithographic models of human anatomy. *Aus Rad Journal, 38*, 106–111.

Baumrind, S. (2001). Integrated three-dimensional craniofacial mapping: background, principles, and perspectives. *Seminars in Orthodontics, 7*, 223–232.

Baumrind, S., Carlson, S., Beers, A., Curry, S., Norris, K., & Boyd, R. L. (2003). Using three-dimensional imaging to assess treatment outcomes in orthodontics: a progress report from the University of the Pacific. *Orthodontics and Craniofacial Research, 6*(1), 132–142.

Bell, A., Ayoub, A. F., & Siebert, P. (2003). Assessment of the accuracy of a three-dimensional imaging system for archiving dental study models. *Journal of Orthodontics, 30*, 219–223.

Berkowitz, S., Gonzalez, G., & Nghiem-Phu, L. (1982). An optical profilometer – a new instrument for the three dimensional measurement of cleft palate casts. *Cleft Palate Journal, 19*, 129–138.

Bhatia, S. N., & Harrison, V. E. (1987). Operational performance of the travelling microscope in the measurement of dental casts. *British Journal of Orthodontics, 14*, 147–153.

Bibb, R., Freeman, P., Brown, R., Sugar, A., Evans, P., & Bocca, A. (2000). An investigation of three-dimensional scanning of human body surfaces and its use in the design and manufacture of prostheses. *Proceedings of the Institution of Mechanical Engineers, Part H, Journal of Engineering in Medicine, 214*, 589–594.

Bill, J. S., Reuther, J. F., Dittmann, W., Kübler, N., Meier, J. L., Pistner, H., et al. (1995). Stereolithography in oral and maxillofacial operation planning. *International Journal of Oral and Maxillofacial Surgery, 24*, 98–103.

Bland, J. M., & Altman, D. G. (1999). Measuring agreement in method comparison studies. *Statistical Methods in Medical Research, 8*, 135–160.

Braumann, B., Keilig, L., Bourauel, C., Niederhagen, B., & Jager, A. (1999). Three-dimensional analysis of cleft palate casts. *Annals of Anatomy, 181*, 95–98.

Brook, A. H., Pitts, N. B., & Renson, C. E. (1983). Determination of tooth dimensions from study casts using an Image Analysis System. *Journal of the International Association of Dentistry for Children, 14*, 55–60.

Brook, A. H., Pitts, N. B., Yau, F. S., & Sandar, P. K. (1986). An image analysis system for the determination of tooth dimensions from study casts: comparison with manual measurements of mesio-distal diameter. *Journal of Dental Research, 65*, 428–431.

Brosky, M. E., Major, R. J., Delong, R., & Hodges, J. S. (2003). Evaluation of dental arch reproduction using three-dimensional optical digitisation. *The International Journal of Prosthetic Dentistry, 90*, 434–440.

Brosky, M. E., Pesun, I. J., Lowder, P. D., Delong, R., & Hodges, J. S. (2002). Laser digitization of casts to determine the effect of tray selection and cast formation technique on accuracy. *Journal of Prosthetic Dentistry, 87*, 204–209.

Burstone, C. J. (1979). Uses of the computer in orthodontic practice. *Journal of Clinical Orthodontics, 13*, 442–453, 539–551.

Burstone, C. J., Pryputniewicz, R. J., & Bowley, W. W. (1978). Holographic measurement of tooth mobility in three dimensions. *Journal of Periodontal Research, 13*, 283–294.

Buschang, P. H., Ceen, R. F., & Schroeder, J. N. (1990). Holographic storage of dental casts. *Journal of Clinical Orthodontics, 24*(5), 308–311.

Champagne, M. (1992). Reliability of measurements from photocopies of study models. *Journal of Clinical Orthodontics, 26*, 648–650.

Commer, P., Bourauel, C., Maier, K., & Jager, A. (2000). Construction and testing of a computer-based intraoral laser scanner for determining tooth positions. *Medical Engineering and Physics, 22*, 625–635.

Cookson, A. M. (1970). Space closure following loss of lower first premolars. *Dental Practitioner, 21*, 411–416.

Cheah, C. M., Chua, C. K., Tan, K. H., & Teo, C. K. (2003). Integration of laser surface digitizing with CAD/CAM techniques for developing facial prostheses. Part 1: design and fabrication of prosthesis replicas. *International Journal of Prosthodontics, 16*, 435–441.

Chua, C. K., Chou, S. M., Lin, S. C., Lee, S. T., & Saw, C. A. (2000). Facial prosthetic model fabrication using rapid prototyping tools. *International Journal of Manufacturing Technology and Management*, *11*, 42–53.

Darvann, T. A., Hermann, N. V., Huebner, D. V., Nissen, R. J., Kane, A. A., Schlesinger, J. K., et al. (2001). *The CT-scan method of 3D form description of the maxillary arch. Validation and an application. Transactions of the 9th International Con on Cleft Palate and Related Craniofac Anomalies.* Goteborg: Erlanders Novum, pp. 223–233.

Delong, R., Heinzen, M., Hodges, J. S., Ko, C. C., & Douglas, W. H. (2003). Accuracy of a system for creating 3D computer models of dental arches. *Journal of Dental Research*, *82*, 438–442.

Delong, R., Ko, C., Anderson, G. C., Hodges, J. S., & Douglas, W. H. (2002). Comparing maximum intercuspation contacts of virtual dental patients and mounted dental casts. *Journal of Prosthetic Dentistry*, *88*, 622–630.

Dervin, E., Gore, R., & Kilshaw, J. (1976). The photographic measurement of dental models. *Medical and Biological Illustration*, *26*, 219–222.

Foong, K. W. C., Sandham, A., Ong, S. H., Wong, C. W., Wang, Y., & Kassim, A. (1999). Surface laser-scanning of the cleft palate deformity-validation of the method. *Annals of the Academy of Medicine of Singapore*, *28*, 642–649.

Freshwater, M., & Mah, J. (2003). The cutting edge. *Journal of Clinical Orthodontics*, *37*, 101–103.

Garino, F., & Garino, G. B. (2002). Comparison of dental arch measurements between stone and digital casts. *World Journal of Orthodontics*, *3*, 250–254.

Halazonetis, D. J. (2001). Acquisition of 3-dimensional shapes from images. *American Journal of Orthodontics and Dentofacial Orthopedics*, *119*, 556–560.

Hans, M. G., Palomo, J. M., Dean, D., Cakirer, B., Min, K. J., Han, S., et al. (2001). Three-dimensional imaging: the Case Western Reserve University method. *Seminars in Orthodontics*, *7*, 233–243.

Harada, R., Yamamoto, K., Ohnuma, H., Mikami, T., & Nakamura, S. (1985). Three-dimensional measurement of dental cast using laser and image sensor. (In Japanese) *Japanese Journal of Medical, Electronic and Biological Engineering*, *23*, 166–171.

Harradine, N., Suominen, R., Stephens, C., Hathorn, I., & Brown, I. (1990). Holograms as substitutes for orthodontic study casts: a pilot clinical trial. *American Journal of Orthodontics and Dentofacial Orthopaedics*, *98*, 110–116.

Hayashi, K., Araki, Y., Uechi, J., Ohno, H., & Mizoguchi, I. (2002). A novel method for the three-dimensional (3-D) analysis of orthodontic tooth movement – calculation of rotation about and translation along the finite helical axis. *Journal of Biomechanics*, *35*, 45–51.

Hirogaki, Y., Sohmura, T., Satoh, H., Takahashi, J., & Takada, K. (2001). Complete 3-D reconstruction of dental cast shape using perceptual grouping. *IEEE Transactions on Medical Imaging*, *20*, 1093–1101.

Hirogaki, Y., Sohmura, T., Takahashi, J., Noro, T., & Takada, K. (1998). Construction of 3-D shape of orthodontic dental casts measured from two directions. *Dental Materials-Journal*, *17*, 115–124.

Huddart, A. G., Clarke, J., & Thacker, T. (1971). The application of computers to the study of maxillary arch dimensions. *British Dental Journal*, *130*(9), 397–404.

Hunter, W. S., & Priest, W. R. (1960). Errors and discrepancies in measurement of tooth size. *Journal of Dental Research*, *39*, 405–414.

Johal, A. S., & Battagel, J. M. (1997). Dental crowding: a comparison of three methods of assessment. *European Journal of Orthodontics*, *19*, 543–551.

Jones, M. L. (1979). *An investigation into stereophotogrammetric measurement of routine study casts and its use in relating palatal cortica adaption to incisor movement. MScD dissertation*, University of London.

Kanazawa, E., Sekikawa, M., & Ozaki, T. (1984). Three-dimensional measurements of the occlusal surfaces of upper molars in a Dutch population. *Journal of Dental Research, 63,* 1298–1301.

Keating, P. J., Parker, R. A., Keane, D., & Wright, L. (1984). The holographic storage of study models. *British Journal of Orthodontics, 11,* 119–125.

Kennedy, D. (1979). *The photography of orthodontic study casts. MScD dissertation.* Cardiff: University of Wales.

Kojima, T., Sohmura, T., Wakabayashi, K., Nagao, M., Nakamura, T., Takasima, F., et al. (1999). Development of a new high-speed measuring system to analyze the dental cast form. *Dental Materials Journal, 18,* 354–365.

Kragskov, J., Sindet-Pedersen, S., Gyldensted, C., & Jensen, K. L. (1996). A comparison of three-dimensional computed Tomography scans and stereolithographic models for evaluation of craniofacial anomalies. *International Journal of Oral and Maxillofacial Surgery, 54,* 402–411.

Kuo, E., & Miller, R. J. (2003). Automated custom-manufacturing technology in orthodontics. *American Journal of Orthodontics and Dentofacial Orthopedics, 123,* 578–581.

Kuroda, T., Motohashi, N., Tominaga, R., & Iwata, K. (1996). Three-dimensional dental cast analyzing system using laser-scanning. *American Journal of Orthodontics and Dentofacial Orthopedics, 110,* 365–369.

Kusnoto, B., & Evans, C. A. (2002). Reliability of a 3D surface laser scanner for orthodontic applications. *American Journal of Orthodontics and Dentofacial Orthopedics, 122,* 342–348.

Laurendeau, D., Guimond, L., & Poussart, D. (1991). A computer-vision technique for the acquisition and processing of 3-D profiles of dental imprints: an application in orthodontics. *IEEE Transactions on Medical Imaging, 10,* 453–461.

Lu, P., Li, Z., Wang, Y., Chen, J., & Zhao, J. (2000). The research and development of a non contact 3-D laser dental model measuring and analysing system. *The Chinese Journal of Dental Research, 3,* 386–387.

Mah, J., & Baumann, A. (2001). Technology to create the three-dimensional patient record. *Seminars in Orthodontics, 7,* 251–257.

Mah, J., & Hatcher, D. (2003). Current status and future needs in craniofacial imaging. *Orthodontics & Craniofacial Research, 6*(Suppl. 1), 10–16.

Mah, J., & Sachdeva, R. (2001). Computer-assisted orthodontic treatment: the SureSmile process. *American Journal of Orthodontics and Dentofacial Orthopaedics, 120,* 85–87.

Martensson, B., & Ryden, H. (1992). The Holodent system, a new technique for measurement and storage of dental casts. *American Journal of Orthodontics and Dentofacial Orthopaedics, 102,* 113–119.

Mayhall, J. T., & Kageyama, I. (1997). A new three-dimensional method for determining tooth wear. *American Journal of Physical Anthropology, 103,* 463–469.

Mazaheri, M., Harding, R. L., Cooper, J. A., Meier, J. A., & Jones, T. S. (1971). Changes in arch form and dimensions of cleft patients. *American Journal of Orthodontics, 60,* 19–32.

McCance, A., Perera, S., & Woods, S. J. W. (1991). Trimming study models for photocopying. *Journal of Clinical Orthodontics, 25,* 445–447.

McGuinness, N. J., & Stephens, C. D. (1992). Storage of orthodontic study models in hospital units in the UK. *British Journal of Orthodontics, 19,* 227–232.

McGuinness, N. J., & Stephens, C. D. (1993). Holograms and study models assessed by the PAR (peer assessment rating) index of malocclusion - a pilot study. *British Journal of Orthodontics, 20,* 123–129.

McKeown, H. F., Robinson, D. L., Elcock, C., Al-Sharood, M., & Brook, A. H. (2002). Tooth dimensions in hypodontia patients, their unaffected relatives and a control group measured by a new image analysis system. *European Journal of Orthodontics*, *24*, 131–141.

Motohashi, N., & Kuroda, T. A. (1999). A 3D computer-aided design system applied to diagnosis and treatment planning in orthodontics and orthognathic surgery. *European Journal of Orthodontics*, *21*, 263–274.

Nagao, M., Sohmura, T., Kinuta, S., Wakabayashi, K., Nakamura, T., & Takahashi, J. (2001). Integration of 3-D shapes of dentition and facial morphology using a high-speed laser scanner. *The International Journal of Prosthodontics*, *14*, 497–503.

Quimby, M. L., Vig, K. W. L., Rashid, R. G., & Firestone, A. R. (2004). The accuracy and reliability of measurements made on computer based digital models. *Angle Orthodontist*, *74*, 298–303.

Quintero, J. C., Trosien, A., Hatcher, D., & Kapila, S. (1999). Craniofacial imaging in orthodontics: historical perspective, current status, and future developments. *Angle Orthodontist*, *69*, 491–506.

Redmond, R. W. (2001). The digital orthodontic office: 2001. *Seminars in Orthodontics*, *7*, 266–273.

Richmond, S. (1984). *The feasibility of categorising orthodontic treatment difficulty: the use of three-dimensional plotting. MScD dissertation*. University of Wales.

Romeo, A. (1995). Holograms in orthodontics: a universal system for the production, development, and illumination of holograms for the storage and analysis of dental casts. *American Journal of Orthodontics and Dentofacial Orthopaedics*, *108*, 443–447.

Ryden, H., Bjelkhagen, H., & Martensson, B. (1982). Tooth position measurements on dental casts using holographic images. *American Journal of Orthodontics*, *81*, 310–313.

Santoro, M., Galkin, S., Teredesai, M., Nicolay, O. F., & Cangialosi, T. J. (2003). Comparison of measurements made on digital and plaster models. *American Journal of Orthodontics and Dentofacial Orthopedics*, *124*, 101–105.

Schelb, E., Kaiser, D. A., & Brukl, C. E. (1985). Thickness and marking characteristics of occlusal registration strips. *Journal of Prosthetic Dentistry*, *54*, 122–126.

Schirmer, U. R., & Wiltshire, W. A. (1997). Manual and computer-aided space analysis: a comparative study. *American Journal of Orthodontics and Dentofacial Orthopedics*, *112*, 676–680.

Schwaninger, B., Schmidt, R. L., & Hurst, R. V. (1977). Holography in dentistry. *Journal of American Dental Association*, *95*, 814–817.

Scott, P. J. (1981). The reflex plotters: measurements without photographs. *Photogrammetric Record*, *10*, 435–446.

Scott, P. J. (1984). Pepper's ghosts observed in Cape Town. South African Journal of Photogrammetry. *Remote Sensing and Cartography*, *40*, 89–95.

Singh, I. J., & Savara, B. S. (1964). A method for making tooth and dental arch measurements. *The Journal of the American Dental Association*, *69*, 719–721.

Sohmura, T., Kojima, T., Wakabayashi, K., & Takahashi, J. (2000). Use of an ultrahigh speed laser scanner for constructing three-dimensional shapes of dentition and occlusion. *Journal of Prosthetic Dentistry*, *84*, 345–352.

Speculand, B., Butcher, G. W., & Stephens, C. D. (1988a). Three-dimensional measurement: the accuracy and precision of the reflex metrograph. *British Journal of Oral and Maxillofacial Surgery*, *26*, 265–275.

Speculand, B., Butcher, G. W., & Stephens, C. D. (1988b). Three-dimensional measurement: the accuracy and precision of the reflex metrograph. *British Journal of Oral and Maxillofacial Surgery*, *26*, 276–283.

Stevens, D. R., Flores-Mir, C., Nebbe, B., Raboud, D. W., Heo, G., & Major, P. W. (2006). Validity, reliability and reproducibility of plaster v's digital study models: comparison of PAR and Bolton analysis and their constituent measurements. *American Journal of Orthodontics and Dentofacial Orthopedics*, *129*, 794–803.

Suzuki, S. (1980). An application of the computer system for three-dimensional cast analysis. *Journal of the Japanese Orthodontic Society*, *39*, 208–228.

Takada, K., Lowe, A. A., & DeCou, R. (1983). Operational performance of the Reflex Metrograph and it's applicability to the three-dimensional analysis of dental casts. *American Journal of Orthodontics*, *83*, 195–199.

Takasaki, H. (1970). Moire topography. *Applied Optics*, *9*(6), 1467–1472.

Tomassetti, J. J., Taloumis, L. J., Denny, J. M., & Fischer, J. R. (2001). A comparison of 3 computerized Bolton tooth-size analyses with a commonly used method. *Angle Orthodontist*, *71*, 351–357.

Tran, A. M., Rugh, J. D., Chacon, J. A., & Hatch, J. P. (2003). Reliability and validity of a computer-based Little irregularity index. *American Journal of Orthodontics and Dentofacial Orthopaedics*, *123*, 349–351.

Van der Linden, F. P., Boersma, H., Zelders, T., Peters, K. A., & Raaben, J. H. (1972). Three-dimensional analysis of dental casts by means of the Optocom. *Journal of Dental Research*, *51*, 1100.

Wakabayashi, K., Sohmura, T., Takahashi, J., Kojima, T., Akao, T., Nakamura, T., et al. (1997). Development of the computerised dental cast form analyzing system – three dimensional diagnosis of dental arch form and the investigation of measuring condition. *Dental Materials Journal*, *16*, 180–190.

Yamamoto, K., Toshimitsu, A., Mikami, T., Hayashi, S., Harada, R., & Nakamura, S. (1988). Optical measurement of dental cast profile and application to analysis of three-dimensional tooth movement in orthodontics. *Frontiers of Medical & Biological Engineering*, *1*, 119–130.

Yen, C. H. (1991). Computer-aided space analysis. *Journal of Clinical Orthodontics*, *25*, 236–238.

Zilberman, O., Huggare, J. A. V., & Parikakis, K. A. (2003). Evaluation of the validity of tooth size and arch width measurements using conventional and three-dimensional virtual orthodontic models. *Angle Orthodontist*, *73*, 301–306.

5.27 Dental applications case study 5: design and fabrication of a sleep apnoea device using CAD/AM technologies

Acknowledgments

The work described here was first reported in the reference below and is reproduced here with kind permission of Sage Publishing.

Al Mortadi N, Eggbeer D, Lewis J, Williams RJ, Design and fabrication of a sleep apnea device using CAD/AM technologies. Proceedings of the Institution of Mechanical Engineers, Part H, Journal of Engineering in Medicine 2013; 227(4): 350–355. http://dx.doi.org/10.1177/0954411912474741

5.27.1 Introduction

In dentistry, a three-stage process is commonly used to produce patient-specific devices: a scanner/digitization tool for data acquisition (either intra-oral scanning, scanning of a poured cast obtained through a traditional impression, or direct scan of a conventional impression), CAD software for data processing, and a subtractive CAM production technology for manufacturing (Bever & Brown, 2011; Van Noort, 2012). Systems can be categorized into in-office (chair side) systems, laboratory-based systems, and milling (production) centre systems (Liu & Essig, 2008), depending on the locations of their components (Beuer, Schweiger, & Edelhoff, 2008). Additive manufacture (AM) building centres should also now be included in this list. The term appropriate for this study is CAD/AM.

The main principle of AM is that the final model consists of processed layers joined together. The layers are organized and bonded together beginning from the base to the uppermost aspect of the build. AM has previously been employed to build complex 3D models in medicine, but its application to orthodontics is comparatively limited. The technique is adept at producing the complex medical and/or dental models that have fine details, such as cavities, sinuses, undercuts, and thin bars (Azari & Nikzad, 2009; Bibb, Eggbeer, & Williams, 2006; Van Noort, 2012). AM technology is also suitable for building thin parts because it is difficult or impossible for these parts to be created by milling owing to the potential flexure of the parts during production (Williams, Bibb, & Eggbeer, 2008). The technique also has the advantage of potentially allowing translucent structures and internal anatomy to be easily seen (Liu, Leu, & Schmitt, 2006). Additive manufacturing has been used in areas such as the production of occlusal splints (Lauren & McIntyre, 2008) and maxillofacial prosthetics (Bibb, Eggbeer, & Evans, 2010; Eggbeer, Bibb, & Evans, 2007; Hughes, Page, Bibb, Taylor, & Revington, 2003). The technology has also been used in orthodontic treatments (Abe & Maki, 2005) for the invisible orthodontic appliance fabricated without metal wire or brackets (Align Technology, Inc., Santa Clara, CA) and in producing customized lingual brackets (Wiechmann, Rummel, Thalheim, Simon, & Wiechmann, 2003).

The current report highlights a new application of the fabrication of a sleep apnoea device with rotating parts produced in a single AM build.

Snoring and obstructive sleep apnoea is a medical condition that can result in abnormal pauses in breathing owing to the soft palate obstructing the oropharynx. Disturbances in normal sleep patterns can leave individuals with daytime fatigue, slower reaction times, difficulties with vision, and moodiness, among other major problems (Dieltjens, Vanderveken, van de Heyning, & Braem, 2012; Malhotra & White, 2002). It is estimated that about 80,000 people in Britain suffer from obstructive sleep apnoea. They are mostly (but not all) men, mostly (but not all) overweight, especially around the neck, and they all snore. These conditions are treatable with dental sleep apnoea appliances that are often designed to posture the mandible forward. Current sleep apnoea devices are designed using laboratory-based techniques and require component assembly (Cistulli, Gotsopoulos, Marklund, & Lowe, 2004; Stradling and Dookun, 2011).

The problem this study addressed was to apply and evaluate novel CAD/AM design solutions that take advantage of AM to produce a hinged, patient-specific, and sleep apnoea device that did not require component assembly.

5.27.2   Methods and materials

Opposing dental casts used for technique development available at a university were scanned by a handheld laser sensor (Creaform, HandyScan 3D, 5825, rue St-Georges Lévis, Quebec, Canada) (Figure 5.194). The manufacturer had calibrated this equipment to ±0.075 mm. Laser lines were projected from the scanner onto the surface of each cast to produce a 3D image that appeared on the computer screen as a faceted surface (Figure 5.195).

The scanner was moved around the casts by hand to produce a clear and complete 3D surface in the STL file format (Figure 5.196) with minimal missing data. Some manufacturers use a specific data format that is not compatible with other construction programs (Beuer et al., 2008). The STL file format is defined as a network of triangular facets that describes the 3D model design mathematically using a network of triangular facets (Gronet, Waskewicz, & Richardson, 2003).

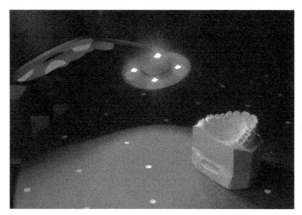

Figure 5.194 Scanning the cast with the handheld scanner.

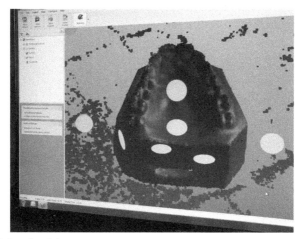

Figure 5.195 Point cloud data.

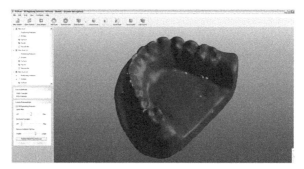

Figure 5.196 Completed 3D model.

The obscured areas appeared on the screen as holes. These mesh holes could be filled by software (Creaform, VX elements version 2.1) together with a specific tool. Data were processed and STL files generated using the proprietary scanner software (VX elements, version 2.1). Computer software (FreeForm Modeling Plus, version 11, Geomagic, 3D Systems, USA) that enables the design of complex, freeform-shaped objects and is ideal for designing anatomically based devices (Bibb et al., 2006, 2010; Eggbeer et al., 2007; Williams et al., 2008) was used and is the one developed in this study. This specialist CAD software was use, in conjunction with a Phantom Desktop (Geomagic, 3D Systems, USA) arm (Figure 5.197) to build the sleep apnoea device in a virtual environment.

The software has different tools to accomplish different tasks and ranges from sculpting to trimming and from contraction to stretching of the clays applied to the virtual cast. The virtual material, in this case the sleep apnoea device, so formed is termed 'clay' and is manipulated and shaped in a similar way to wax in conventional methods. An experienced CAD/CAM dental technician can choose the appropriate tool for the required steps easily and quickly. The Phantom device works as

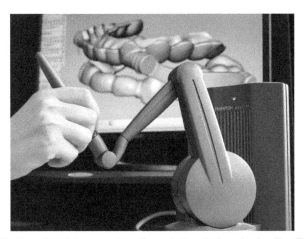

Figure 5.197 Phantom (haptic) arm and appliance being designed using FreeForm.

a 3D mouse mimicking the hand of a dental technician, enabling the creation of appropriate appliances in a way similar to the conventional laboratory method.

Before the virtual build, it was necessary to save the scanned dental casts as buck files by altering the settings appearing at the foot of the screen. This ensures the models are protected because any change to this saved form is not permitted.

The sleep apnoea device was developed in the virtual environment by initially orienting the scanned casts and creating a normal (class I) molar relationship or as close to it as possible. This was achieved within the virtual environment by dragging the mandible forward and by opening the vertical dimension between the upper and lower casts from a closed position until there was a space of 4–5 mm between the opposing premolars, which is considered enough to open the palatal blockage and therefore help prevent snoring. Spaces were measured using the ruler function of the software.

The inclination of upper central incisors towards the lip was selected as a reference for the path of insertion of the upper cast, and the lower cast followed the upper according to the corrected occlusal relationship. The midlines between upper and lower incisors were aligned until coincident. Re-entry points or undercuts on both virtual models that would prevent the seating of the appliance were detected automatically using the set pull direction option from the view tool. These areas were blocked out using the layer tool to add clay, with the thickness depending on the size of the undercut. The blocked-out layers were smoothed using the smooth tool and the files were then saved as buck files, so that the models were protected and could not be affected by the software tools. It is essential that undercut areas are identified and eliminated to allow the creation of parallel guide planes so that the appliance will fit onto a physical model, and ultimately the patient.

In the virtual environment, a 2-mm layer was applied to cover the tooth surfaces, as shown in Figure 5.198, using the layer tool with an offset thickness of 2 mm. The layers covered the lingual tooth surfaces and further extended to 3 mm below the gingival margins. The layers kept 1 mm above the gingival margins of the molars and premolars buccally to avoid soft tissue irritation. One-millimetre capping was applied on the

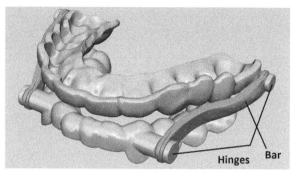

Figure 5.198 Sleep apnoea device design.

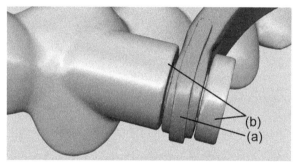

Figure 5.199 The hinge. (a) A rod with its stops and (b) cylinder.

four incisors of opposing casts to minimize the inclination towards the lips or to prevent the over-eruption.

The hinges on both sides connected the upper molars to lower canines (Figure 5.198). The role of these hinges is to move the mandible forward into the corrected position. Four hinges were built for each single appliance: two on the lower canines and two on the upper first molars. A bar on either side connected the hinges. The four hinges were positioned parallel to each other to allow free rotation.

Each hinge consisted of two parts (Figure 5.199). The first part was a circular rod (2.5 mm in length, 3 mm in diameter), with two wide, circular stops (6 mm in diameter) at the ends (Figure 5.199(a)). The second part was a hollow cylinder with an internal diameter of 3.5 mm and external diameter of 6 mm (Figure 5.199(b)). The two parts were separated in all areas to permit rotational freedom of the cylinder around the rod. There was a gap clearance between sides of the cylinder and the first part. The inner gap was 0.2 mm and the outer gap was 0.1 mm. The gap between the inner surface of the cylinder and the rod was 1 mm. The internal stop connected the teeth continuously.

All hinges were built in the same way by use of the wire cut tool. This tool can build a 3D model from a 2D sketch by producing an additional prismatic plane, which can either cut or fill the profile area between it and the original plane. It is critical to position the original plane in the correct place and to determine how far from the additional prismatic plane by moving (dragging).

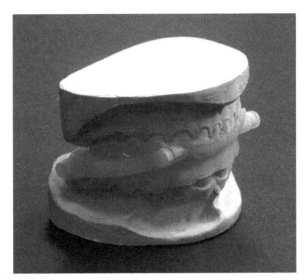

Figure 5.200 Sleep apnoea prototype fitted to cast.

To build a hinge virtually, a new plane was created on the frontal view of the models. Three combined circles to make up the hinges were drawn on the plane. The diameter of the outer circle was 6 mm, the middle circle was 3 mm, and the inner was 2.5 mm. The outer and inner circles were used to produce the stops labelled as (b) in Figure 5.199.

All the hinges were drawn on the same plane. It was only possible to modify the plane by moving it using snap axis, which moves the plane in and out without rotation in any axis. This ensures the parts of each hinge are parallel and guarantees the four hinges are parallel to each other. A profile of the bar to connect the two hinges was drawn on the same plane, as well. The thickness of the bar was 2.9 mm, similar to the cylinder's thickness. However, the transverse dimension of the bar was 4 mm.

The bars were constructed with a bend to avoid interference with the teeth and device during movement, but at the same time allowing positioning to be as close as possible to them (Figure 5.198).

Finally, the design was exported as an STL file. The ProJet 3000 Plus HD 3D printing system (3D Systems, USA) was chosen for fabrication.

The process is capable of building in extremely fine detail (656 × 656 × 1600 dots per inch in x–y–z axes) necessary to define the intricate features. It uses an acrylate-based polymer (EX200, 3D Systems) for the component and, crucially, uses a wax support structure. This enables the production of high-definition hinge mechanisms that can be built without assembly. Once the build is complete, the wax that supports and separates can be melted away to free the hinge mechanisms. This also provides the freedom to incorporate other features such as slides and screws that do not require assembly to be fabricated in a single build. Even with the ability to melt the support structure, it is necessary to ensure a sufficient gap between hinge components to enable trapped wax to be removed. Parallelism of hinges is also important for hinge function.

The device was transferred into proprietary software (Client Manager) and oriented in the same plane in which it would be worn (i.e. the lowest z height) to obtain the optimum build speed. The build took 8 h to complete. The completed build was post-processed by using an oven set at 72 °C to melt the wax support structure and an oil bath to remove residual wax followed by degreasing to clean the device. This was then trial fitted to the stone casts (Figure 5.200).

5.27.3   Results

The device was fitted on the physical casts and its features checked and compared with the expectations of conventionally formed appliances. The device had smooth surfaces with no sharp edges, which could not be easily dealt with by normal finishing procedures.

Because of the virtual preparation of the models, no areas were in undercuts. The device was inserted and removed to and from the casts easily without interference. The sleep apnoea device displayed an excellent fit on the teeth and palatal and gingival tissues.

The appliance was considered functional and the hinges worked well to posture the mandible forward. The hinges allowed free forward and reverse movements. The hinges also permitted some lateral movement. Further experimentation could reduce this, but there may be an argument in favour of maintaining such movement to allow the device to be myodynamic – that is, allowing muscular exercise.

5.27.4   Discussion

Currently, most sleep apnoea devices are built relying on fairly intensive handcrafting methods. The design described in this research for a sleep apnoea appliance highlighted the technical achievements necessary to use CAD/AM technologies for its fabrication. With the AM technique, it is possible to build all parts of a sleep apnoea device in one build, layer by layer, and parts of hinges can be manufactured without assembly. In future, there may also be opportunities for greater automation.

Another benefit of using CAD/AM technologies as described earlier is the possibility of linking it to intra-oral scanning. This may allow an appliance to be made without the use of a dental cast. Hence, the difficulties of tray selection, impression material selection, and inadequate impression tray adhesive, dimensional changes in impression material, and bubble entrapment during impression pouring could be avoided. There is evidence that misunderstanding between laboratory technicians and dentists (Polido, 2010) and patient discomfort (Christensen, 2008) may also be avoided. The technique offers the negation of repeated disinfection necessary as items pass between the laboratory and clinic, which adversely affects the dimensional accuracy of impression materials (Al Mortadi & Chadwick, 2010). Other advantages include ease of identification of undercuts and the parallel positioning of hinges, which is much less precise when accomplished by hand/eye coordination.

It is suggested that AM techniques are advantageous because there is little waste of material with this technology, and production is potentially fast as any number of objects can be produced at once as long as they fit within the build envelop of the machine. Other interesting applications with potential include the following (Van Noort, 2012). A recent study concluded that AM techniques could be used in the fabrication of RPDs (Williams et al., 2008).

Although the techniques described demonstrate the potential benefits in workflow and design freedom, technical and material challenges currently prevent the chosen technologies from being suitable for clinical application. The ProJet EX200 material used in this study has not undergone toxicity testing to the necessary standards to enable it to be used in intraoral device production. There are currently no AM polymers approved for long-term intraoral use, which remains a challenge for material developers. However, there is significant pace in the development of AM processes and materials, particularly in the medical device sector, that may shortly overcome these issues.

It is the considered opinion of the authors that the appliance described earlier is worthy of presentation to a dental clinic.

5.27.5 Conclusion

A method of applying CAD/AM to patient-specific devices requiring rotating parts has been described through employing handheld scanners, a virtual environment, and AM technologies. The benefits, difficulties, and barriers have also been discussed. However, the built example shown earlier demonstrates the enormous potential for the application of CAD/AM technologies to the treatment of sleep apnoea and the benefits of accuracy and control that become available with these technologies.

References

Abe, Y., & Maki, K. (2005). Possibilities and limitation of CAD/CAM based orthodontic treatment. *International Conference Series, 1281*, 1416.

Al Mortadi, N., & Chadwick, R. G. (2010). Disinfection of dental impressions – compliance to accepted standards. *British Dental Journal, 209*(12), 607–611.

Azari, A., & Nikzad, S. (2009). The evolution of rapid prototyping in dentistry: a review. *Rapid Prototyping Journal, 15*(3), 216–222.

Beuer, F., Schweiger, J., & Edelhoff, D. (2008). Digital dentistry: an overview of recent developments for CAD/CAM generated restorations. *British Dental Journal, 204*(9), 505–511.

Bever, P., & Brown, C. (2011). The CAD in CAD/CAM: CAD design software's powerful tools continue to expand. Inside Dental Technology: published by AEGIS communications, 2(9), 52.

Bibb, R., Eggbeer, D., & Evans, P. (2010). Rapid prototyping technologies in soft tissue facial prosthetics: current state of the art. *Rapid Prototyping Journal, 16*(2), 130–137.

Bibb, R., Eggbeer, D., & Williams, R. (2006). Rapid manufacture of removable partial denture frameworks. *Rapid Prototyping Journal, 12*(2), 95–99.

Christensen, G. J. (2008). The challenge to conventional impressions. *Journal of American Dental Association*, *139*(3), 347–348.

Cistulli, P. A., Gotsopoulos, H., Marklund, M., & Lowe, A. A. (2004). Treatment of snoring and obstructive sleep apnea with mandibular repositioning appliances. *Sleep Medicine Reviews*, *8*, 443–457.

Dieltjens, M., Vanderveken, O. M., van de Heyning, P. H., & Braem, M. J. (2012). Current opinions and clinical practice in the titration of oral appliances in the treatment of sleep-disordered breathing. *Sleep Medicine Reviews*, *16*(2), 177–185.

Eggbeer, D., Bibb, R., & Evan, S. P. (2007). Digital technologies in extra-oral, soft tissue facial prosthetics: current state of the art. *The Institute of Maxillofac Prosthetits Technologists (JMPT) – Winter*, *10* 9–8.

Gronet, P. M., Waskewicz, G. A., & Richardson, C. (2003). Preformed acrylic cranial implants using fused deposition modeling: a clinical report. *Journal of Prosthetic Dentistry*, *90*(5), 429–433.

Hughes, C. W., Page, K., Bibb, R., Taylor, J., & Revington, P. (2003). The custom-made titanium orbital floor prosthesis is reconstruction for orbital floor fractures. *British Journal of Oral and Maxillofacial Surgery*, *41*(1), 50–53.

Lauren, M., & McIntyre, F. (2008). A New computer-assisted method for design and fabrication of occlusal splints. *American Journal of Orthodontics and Dentofacial Orthopedics*, *133*, S130–S135.

Liu, P.-R., & Essig, M. E. (2008). A panorama of dental CAD/CAM restorative systems. *CAD/CAM System Update*, *29*(8), 482–493.

Liu, Q., Leu, M. C., & Schmitt, S. M. (2006). Rapid prototyping in dentistry: technology and application. *International Journal of Advanced Manufacturing Technology*, *29*, 317–335.

Malhotra, A., & White, D. P. (2002). Obstructive sleep apnoea. *Lancet*, *360*, 237–245.

Polido, W. D. (2010). Digital impressions and handling of digital models: the future of Dentistry. *Dental Press Journal of Orthodontics*, *15*(5), 18–22.

Stradling, J., & Dookun, R. (March 2011). The role of the dentist in sleep disorders. *Dental Update*, *38*(2), 136.

Van Noort, R. (2012). The future of dental devices is digital. *Dental Materials*, *28*(1), 3–12.

Wiechmann, D., Rummel, V., Thalheim, A., Simon, J.-S., & Wiechmann, L. (2003). Customized brackets and archwires for lingual orthodontic treatment. *American Journal of Orthodontics and Dentofacial Orthopedics*, *124*(5), 593–599.

Williams, R. J., Bibb, R., & Eggbeer, D. (2008). CAD/CAM- fabricated removable partial denture alloy frameworks. *Practical Procedures and Aesthetic Dentistry (PPAD)*, *20*(6), 349–351.

5.28 Dental applications case study 6: computer-aided design, CAM and AM applications in the manufacture of dental appliances

Acknowledgments

The work described here was first reported in the reference below and is reproduced here with kind permission of the copyright holders.

Al Mortadi N, Eggbeer D, Lewis J, Williams R J., "CAD/CAM/AM applications in the manufacture of dental appliances", American Journal of Orthodontics & Dentofacial Orthopedics 2012; 142(5): 727–733. http://dx.doi.org/10.1016/j.ajodo.2012.04.023

5.28.1 Introduction

The CAD, CAM, and AM and scanning systems have been successfully introduced in dentistry (Stevens, Yang, Mohandas, Stucker, & Nguyen, 2007; Sun, Lǔ, & Wang, 2009; Williams, Bibb, & Eggbeer, 2008). Researchers have used this technology as a tool for orthodontic diagnosis and treatment planning to determine the position of impacted maxillary canines (Faber, Berto, & Quaresma, 2006) and for the fabrication of occlusal splints (Lauren & McIntyre, 2008). However, CAD/CAM/AM innovations have not been used successfully on a wide scale for removable orthodontic appliances.

Sassani and Roberts (Sassani & Roberts, 1996) reported that it was possible to use CAD/CAM/AM techniques to build base plates of orthodontic appliances, but they stated that it was not possible to incorporate wires into the built parts. However, this study discusses a method to accomplish such incorporation. The technology mentioned earlier has been used in orthodontic treatments such as the invisible orthodontic appliance fabricated without metal wire or brackets (Align Technology, Inc., Santa Clara, CA) (Abe & Maki, 2005) and in producing customized lingual brackets (Wiechmann, Rummel, Thalheim, Simon, & Wiechmann, 2003). Recently, Wiechmann, Schwestka-Polly, and Hohoff (2008) used this computer-based technology to make connectors for Herbst appliance hinges to custom lingual brackets, and a more recent study used this technique for a digital titanium Herbst appliance (Farronato, Santamaria, Cressoni, Falzone, & Colombo, 2011).

Because of this lack of application in general to removable appliances, the aim of this study was to apply new methods of producing dental appliances with wires and hinges using CAD/CAM/AM based techniques and to evaluate these. Andresen and sleep apnoea appliances were chosen. Although an Andresen appliance is rarely used, it provides a clear illustration for the inclusion of wire into a build, which is straightforward. A sleep apnoea device provides a case in which four coordinating hinges can be studied. Both devices were produced in a single build.

5.28.2   Material and methods

Opposing Class II case models suitable to be treated with Andresen and sleep apnoea removable devices were scanned by a handheld laser sensor (Creaform, HandyScan 3D, 5825, rue St-Georges Lévis, Quebec, Canada) (Figure 5.201 in the previous case study). Laser lines were projected from the scanner onto the surface of each model to produce a 3D image from a point cloud. The software automatically transformed the point cloud into an STL facet file, which is shown as it appears on a screen in Figure 5.202 in the previous case study. The scanner was moved around the casts by hand to produce a clear and complete 3D surface in the STL file format (Figure 5.203 in the previous case study) with minimal missing data.

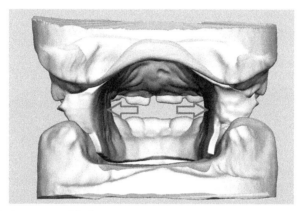

Figure 5.201 Andresen monoblock.

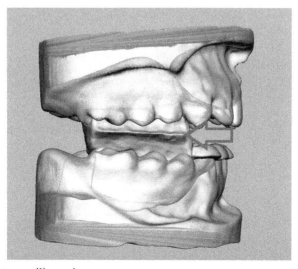

Figure 5.202 Intermaxillary edge.

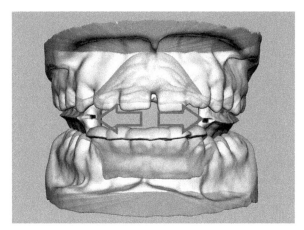

Figure 5.203 The tubes seen from the outside.

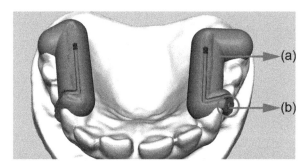

Figure 5.204 Guiding jigs.

Data were processed and STL files were generated using the proprietary scanner software (Creaform, VX elements, version 2.1). The appliances were designed using specialist CAD software (FreeForm Modeling Plus, version 11, Geomagic, 3D Systems, USA), in conjunction with a Phantom Desktop (Geomagic, 3D Systems, USA) arm (Figure 5.204 in the previous case study). This apparatus allowed appliances to be built on scans of a patient's cast in a virtual environment.

The software enables the design of complex, nongeometric forms and is ideal for designing anatomical-based devices. It has different tools to accomplish specific functions. Some tools are used for building, whereas others are used for trimming and shaping the constructed parts. The Phantom device works as a 3D mouse as it mimics the work of a dental technician, enabling the design and construction of the appropriate appliances in a way similar to the conventional method. The operator is able to feel force feedback sensations during the process.

Before the virtual build, it was necessary to create a Class I molar relationship or achieve this as closely as possible. This relationship was accomplished by posturing the mandible forward, until a Class 1 M relationship existed and the vertical opening between opposing premolars was 4–5 mm at the lingual cusps.

The labial inclination of upper central incisors was selected as a reference for the path of insertion of the appliance onto the upper model. Undercuts on both virtual models were detected and eliminated virtually automatically using the extrude-to-plane tool. The files were then saved as buck files, which meant that the models were protected and could not be changed.

The following is a brief description of the design process for each appliance.

5.28.2.1   Andresen

A 2-mm layer was applied to the upper and lower models with capping on the lower incisors. Palatal coverage was made by selecting the required area using a cursor and offsetting a 2-mm thickness to form the plate. Upper and lower plates extended to the first molars. Capping was 1 mm thick on the anterior lower teeth to prevent proclination and over-eruption. The thickness of the appliance was set at 2 mm to withstand occlusal forces and patient handling. The virtual material so formed is termed 'clay' and is added and shaped in a way similar to physical wax. It was necessary to add clay in some areas of the appliance to make the surfaces smooth and to form junctions of two surfaces continuous to each other. Borders were smoothed.

Finally, the occlusal surfaces of the opposing offset clay areas were connected together to produce a one piece monoblock at the lingual surfaces of the teeth extending to the middle of the occlusal surfaces (Figure 5.201).

These blocks were attached to the plates and the junctions were smoothed using the sculpting and smooth tools. The mid distances of the blocks were stretched buccally using the smudge tool up to the outline of the buccal surfaces to form a sharp intermaxillary edge. The blocks were cleared from the mandibular teeth to permit upward and forward eruption (Figure 5.202).

To allow wire to be inserted within the physical build, a square section tube (1.25 × 1.25 mm) on either side of the appliance was created. Each tube consisted of two parts. The first was horizontal, 15 mm in length and 1 mm from the lingual walls of the intermaxillary blocks, and was located in the maxillary part with its lower border at the level of the intermaxillary edge. The second was vertical projecting into the maxillary part and 5 mm in length. The first part was at right angles to the second part, and these were connected at a point between the second premolar and first molar. These tubes were made so that the retentive part of a 0.9-mm hard stainless-steel (S–S) wire labial bow could be inserted, as shown in Figure 5.203.

For the wire to be inserted in the tubes within the appliance, it was necessary to build guiding jigs. Fitting the wire in the guiding jigs before the build ensured that the wire tagging would fit into the appliance during the build.

5.28.2.2   The guiding jigs

The guiding jigs (i.e. small plastic components around which wire could be bent before the main build) were built from RP material for the 3D Systems Project 3000 Plus HD Machine, which was acrylate based. The jigs were produced using the copy tool to duplicate the upper virtual model in conjunction with the outline of the tubes. Clay was

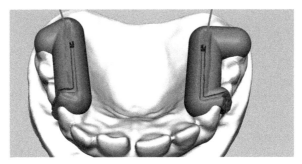

Figure 5.205 Outline of the tubes in guiding jigs.

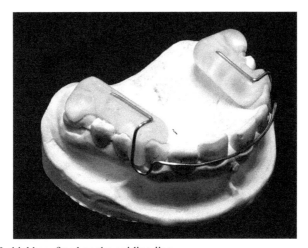

Figure 5.206 Labial bow fitted on the guiding jigs.

constructed in the space between the model and the level of the tubes (Figure 5.204(a)).
Also, a small amount of clay was added under the tubes to make sure the wire would
not be in contact with the model (Figure 5.204(b)). This was accomplished by offsetting a 1-mm layer underneath the tubes using a sculpting clay tool.

The tubes formed spaces along which wire would be bent before insertion into
the final device during the build process. Figure 5.205 shows 1.25-mm deep grooves.
Sharp angles were smoothed to facilitate positioning of the wire in the tubes.

Next, the guiding jigs were exported as an STL file to an AM machine (3D Systems
Project 3000 Plus HD Machine) for building. VisiJet EX200 (3D-Systems, USA), a
3DP build material that is acrylate-based and near colourless, was used. The wax supporting material was removed by melting using an oven set at 72 °C and an oil bath
was used to remove residual wax. A citrus-based degreaser was then used to clean the
jigs before they were fitted to the stone models.

The labial bow was bent in the conventional way and the taggings were fitted to
the jigs positioned on the gypsum cast (Figure 5.206). The labial bow was constructed
from 0.9-mm hard stainless-steel wire. The bow extended from canine to canine,

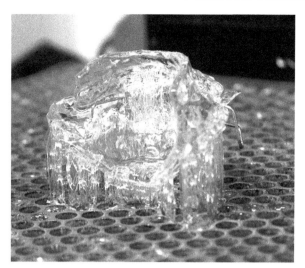

Figure 5.207 The final prototype before cleaning.

touching the labial surfaces of the upper incisors and the canines at the mesial third area. From that point on each side, U loops were formed 1–2 mm below the gingiva on both sides. The wire was then passed between the canines and first premolars to enter the plastic between the occlusal surfaces. The wire tagging followed the tubes formed in the jigs. The horizontal parts of the wire tags were made straight and passively fitted on the jigs.

The next stage was to export the Andresen design to an AM machine as an STL file ready for fabrication in WaterShed XC 11,122 (DSM Somos). Stereolithography was chosen for fabrication because it enables builds to be paused and prefabricated pieces to be inserted and built around. This was necessary for embedding wires.

Preparation software (Magics, Materialise, Leuven, Belgium) was used to orientate the Andresen design with the recessed section to accommodate the wire set horizontally. The same software was also used to generate support structures that are necessary to attach the part to the build platform. Lightyear (version 1.1, 3D Systems) was used to prepare and slice the build in 0.1-mm layers before transferring the build to the SLA machine (SLA 250-50, 3D Systems, USA). The machine was paused before it reached the top of the tubes to allow insertion of the wire in the correct place, and then restarted. The zephyr blade recoating mechanism, which normally sweeps unprocessed liquid from the build surface of the machine, was stopped from sweeping for the three layers after insertion of the wire so that the wire was not disturbed during the process. The procedure was continued until building was completed (Figure 5.207).

Support structures were removed from the device and it was cleaned in fresh isopropanol solvent (99% minimum) for 20 min, and then dried using compressed air and ventilated for a further 6 h. Next, the device was post-cured for 30 min in ultraviolet light to ensure full polymerization. The final design is shown in Figure 5.208 and the built Andresen fitted to the master model is in Figure 5.209.

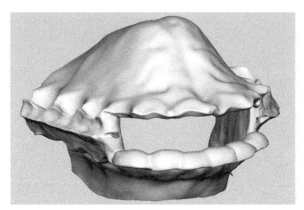

Figure 5.208 Final Andresen in CAD.

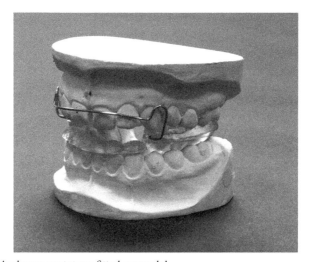

Figure 5.209 Andresen prototype fitted on model.

5.28.2.3 Sleep apnoea device

The material used was VisiJet EX200. In the virtual environment, a 2-mm layer was applied to the upper and lower models and extended along the occlusal and buccal surfaces to approximately 1 mm short of the gingival margins facially to avoid irritation to the gingivae. One-millimetre capping was applied on the four incisors to prevent labial inclination and extended 2–3 mm below the palatal and lingual gingival margins. The hinges on both sides connected the upper molars to lower canines and were parallel to allow free rotation. The role of these hinges is to move the mandible forward into the corrected position (see Case study 6 in Section 5.5.6 for more detail and figures).

Each bar connected two hinges located on the opposing arches. One bar was built on each side to connect the upper first molar hinge to the lower canine hinge. The four hinges were parallel to each other. Each hinge consisted of two parts. The first was a circular rod (2.4 mm in length, 3 mm in diameter) and a wide circular stop (6 mm in diameter). The second was a hollow cylinder with an internal diameter of 3.5 mm and external diameter of 6 mm. The two parts were separated in all areas to permit rotational freedom of the cylinder around the rod. There was a gap clearance between sides of the cylinder and the first part. The inner gap was 0.2 mm and the outer gap was 0.1 mm. The gap between the inner surface of the cylinder and the rod was 0.5 mm.

Finally, the design was exported as an STL file. A ProJet 3000 Plus HD 3D printing machine (3D Systems, USA) was chosen to manufacture the sleep apnoea device. The process fabricates using an acrylate-based polymer with a wax support structure, which effectively allows parts to be separated after building. Fundamentally, the process allows features such as hinges to be fabricated as one piece, with the wax melted away to free the mechanism once produced. The ProJet process is also capable of building extremely fine detail ($656 \times 656 \times 1600$ dots per inch in x–y–z).

The device was transferred into proprietary software and orientated in the same plane in which it would be worn (i.e. the lowest z height) to obtain the optimum build speed. The build took 8 h to complete. The completed build was post-processed by using an oven set at 72 °C to melt the wax support structure and an oil bath to remove residual wax followed by degreasing to clean the device. This was then trial fitted to the stone models.

5.28.3 Results

The appliances were fitted on the physical models and their features were checked and compared with the expectations of conventional appliances. Both appliances built by the AM method had smooth surfaces with no sharp edges that could not be easily dealt with by normal finishing procedures.

With regard to the Andresen, the fit palatally and lingually was satisfactory. Because of the virtual preparation of the models, no areas were positioned in undercuts. The labial bow fitted firmly in the tubes, was well connected to the appliance, and was deemed functional. The acrylic guiding channels could be trimmed according to conventional practice to enhance the anterior movement of the lower teeth and the posterior movement of the upper teeth and to encourage expansion of the upper arch and upward eruption of the lower teeth into the corrected occlusion.

The sleep apnoea device displayed an excellent fit on the teeth and palatal and gingival tissues. The hinges allowed free anterior and posterior movements. Some sideways movement was also permitted by the hinges. Further experimentation could reduce this, but there may be an argument in favour of maintaining such movement to allow the device to be myodynamic.

5.28.4 Discussion

The appliances described earlier could be presented to a clinic for fitting to a patient. The two appliances illustrated in this research highlight the technical possibilities of using CAD/CAM/AM technologies in dental device design. Although the techniques are currently slow and expensive, the benefits of introducing digital technologies into other areas of dentistry are well known. The ability to fabricate layer by layer enables features such as inserting wire, producing hinges, and building threads to be made in one build without assembly. In orthodontics, this has potential application in fabricating devices incorporating built expansion screws and twin blocks, which is part of ongoing research. Other benefits of using CAD are accurate control of the thickness of the material, producing an even shape when required, and exact positioning of the periphery of the design. Precise measurements can be easily taken as the virtual build progresses or programmed in as the software produces layers. A huge potential advantage is that the techniques could be linked to intraoral scanning, which would help to streamline the process of producing appliances by removing the impression and dental cast production stages.

5.28.5 Conclusion

This report indicates that digital technologies may be applied to the fabrication of appliance types not previously considered possible.

References

Abe, Y., & Maki, K. (2005). Possibilities and limitation of CAD/CAM based orthodontic treatment. *International Congress Series*, *1281*, 1416.

Bibb, R., Eggbeer, D., & Williams, R. (2006). Rapid manufacture of removable partial denture frameworks. *Rapid Prototyping Journal*, *12*, 95–99.

Faber, J., Berto, P. M., & Quaresma, M. (2006). Rapid prototyping as a tool for diagnosis and treatment planning for maxillary canine impaction. *American Journal of Orthodontics and Dentofacial Orthopedics*, *129*, 583–589.

Farronato, G., Santamaria, G., Cressoni, P., Falzone, D., & Colombo, M. (2011). The cutting edge. *Journal of Clinical Orthodontics*, *XLV* (5), 263–267.

Lauren, M., & McLntyre, F. A. (2008). New computer-assisted method for design and fabrication of occlusal splints. *American Journal of Orthodontics and Dentofacial Orthopedics*, *133*, S130–S135.

Sassani, F., & Roberts, S. (1996). Computer-assisted fabrication of removable appliances. *Computers in Industry*, *29*, 179–195.

Stevens, B., Yang, Y., Mohandas, A., Stucker, B., & Nguyen, K. T. (2007). A review of materials, fabrication methods, and strategies used to enhance bone regeneration in engineered bone tissues. *Journal of Biomedical Materials Research Part B: Applied Biomaterials*, *85*, 573–582.

Sun, Y., Lǚ, P., & Wang, Y. (2009). Study on CAD & RP for removable complete denture. *Computers Methods and Programs in Biomedicins*, *93*, 266–271.

Wiechmann, D., Rummel, V., Thalheim, A., Simon, J.-S., & Wiechmann, L. (2003). Customized brackets and archwires for lingual orthodontic treatment. *American Journal of Orthodontics and Dentofacial Orthopedics*, *124*, 593–599.

Wiechmann, D., Schwestka-Polly, R., & Hohoff, A. (2008). Herbst appliance in lingual orthodontics. *American Journal of Orthodontics and Dentofacial Orthopedics*, *134*, 429–446.

Williams, R. J., Bibb, R., & Eggbeer, D. (2008). CAD/CAM-fabricated removable partial denture alloy frameworks. *Practical Procedures and Aesthetic Dentistry*, *20*, 349–351.

Research applications

5.29 Research applications case study 1: bone structure models using stereolithography

Acknowledgements

The work described in this case study was first reported in the reference below and is reproduced here in part or in full with the permission of MCP UP Ltd. now Emerald Publishing

Bibb R, Sisias G, Bone Structure Models using Stereolithography: A technical note, Rapid Prototyping Journal, 2002, vol. 8, (1), 25–9. http://dx.doi.org/10.1108/13552540210413275

5.29.1 Introduction

To further the understanding of osteoporosis and its dependence upon the material and structural properties of cancellous bone, many experimental studies continue to be undertaken on natural tissue samples obtained from human subjects and animals. A significant difficulty, however, in analysing in vitro bone samples is that the structural parameters have to be elucidated by destructive means. In addition, the human in vitro samples most readily available tend to be from an elderly population and therefore may be of limited structural variation compared with the full population age range. The development of a physical model of cancellous bone whose structure could be controlled would provide significant advantages over the study of in vitro samples. This would also enable the relationship between the mechanical integrity and hence fracture risk of cancellous bone and its structural properties to be more exactly defined.

This paper describes how these complex three-dimensional (3D) structures can be physically reproduced using rapid prototyping (RP) techniques. As the paper will show, although use of RP techniques allows the generation of physical objects that would have previously been impossible to manufacture problems are still encountered. The data, generated from microcomputed tomography (µCT), were used to perform finite element analysis (FEA) on the structure of various human and animal bones. The physical models were required to validate the results of the FEA.

The difficulties encountered in creating physical models of these structures arose from the nature of the structure. Not only does the highly complex porous structure result in extremely large computer files, but also presents problems of support during the build process.

5.29.2   Human sample data

The techniques of serial sectioning and μCT reconstruction were used to obtain 3D reconstructions of in vitro samples of natural cancellous bone tissue. The physical size of each sample is approximately 4 mm × 4 mm × 4 mm and their relative densities are 9–25%. The models were to be scaled up by an approximate factor of 10 and physically produced using RP. The considerable effort that went in to generating the data from which these models would be made are described elsewhere (Sisias, Phillips, Dobson, Fagan, & Langton, 2002).

5.29.3   The use of stereolithography in the study of cancellous bone

For this project, stereolithography was the preferred method for several essential reasons. First, it is capable of building models at an exact layer thickness of 0.1 mm. This was desirable because the FEA was generated from voxel data with a voxel size of 0.1 mm. Therefore, the stereolithography model would replicate the FEA mesh exactly, i.e. with no smoothing between the layers or within the plane of each slice. Second, it is the most accurate method available (except Solidscape machines, but this would have been extremely slow and the models would have proved far too delicate to mechanically test). Third, although selective laser sintering (SLS®) and 3D printing had the advantage of not requiring support structures the finished models are not completely dense or sufficiently accurate. Fused deposition modelling (FDM™) parts can also show small degree of porosity and are unable to match the accuracy desired in this case. However, since these models were built, the water-soluble supports that are now available for FDM™ would prove extremely useful for structures such as these. Finally, and most importantly, stereolithography could be used to generate models from slice data rather than triangular facetted data. The intention was to use the SLC file format as it resulted in dramatically smaller files than the same data generated in the Standard Triangulation Language (STL) file format.

A general description of all of these RP technologies can be found in Chapter 4. For a general description of the SLC, STL and SLI data formats (see Chapter 3).

5.29.4   Single human bone sample (approximate 45-mm cube)

Because of initial problems with the SLC data files, the first single model was built from an STL file (see Figure 5.210). This was attempted to test the general capability of stereolithography to manufacture these forms. Although the STL file was much larger than the equivalent SLC file, as the model was small it was still a feasible option. However, the standard procedures for generating supports were simply not suitable as the software automatically attempts to support all of the down-facing

Figure 5.210 STL file of the first sample.

areas of the model. This resulted in overlong processing times and a vast number of supports with a correspondingly large support STL file. In addition, the nature of the structure would make it extremely difficult to remove supports from the innermost areas of the model. (Note: This model was constructed before Fine-point supports became available. Although they would have been an improvement in this respect, they would have still resulted in an excessive number of supports with an even larger support STL file.)

To avoid these problems, a novel strategy for producing the necessary support was attempted. The approach was based on two fundamental assumptions. First, as bone is a naturally occurring, load-bearing structure it is made up of self-supporting arches. Therefore, the structure should in theory, support itself except for the sides of the model where the structure has been sliced through. The second assumption was that the open spaces would all be intercommunicating and therefore there would be no 'trapped volumes' (a recognised problem in stereolithography).

As automatically generated supports were impractical, to support the base and sides of the model, a very thin crate was designed using computer-aided design (CAD) that would support the sides and base yet still allow good draining (see Figure 5.211). This is necessary to avoid the 'trapped volume' effect that adversely affects the stereolithography process. The crate was then exported as an STL file. Curtain support structures were then automatically generated (using 3D Lightyear™) for the crate only in order to separate the whole build from the platform. This combination can be seen in Figure 5.212. The objects were then prepared for stereolithography in the usual manner and built on an SLA-250 using SL5220 resin.

Figure 5.211 The crate structure.

Figure 5.212 Sample, supports and crate.

The model was built successfully suggesting that the fundamental assumptions regarding the self-supporting and self-draining nature of the object were indeed sound. The crate structure was carefully broken away along with the supports using a scalpel.

5.29.5   Multiple human samples (approximate 50-mm cube)

For the second batch, five copies each of the five types of model were required. The STL approach was not feasible because of the incredibly excessive size of the files. Instead, contour files were generated from the original data in the SLC file format.

However, the problem of how to support the sides and base was encountered again. Although automatic support software is available to generate supports for SLC files, it presented exactly the same problems as the previous attempt. To create minimal supports a combination of two approaches was used. The first involved the use of C-Sup to generate supports automatically that would separate the bottom of the model from the build platform. These supports were automatically generated to end just above the bottom of the part as it was again assumed that the internal structure of the model would be self-supporting. To support the sides, the crate structure was used again (minus the bottom). Automatic support generation software was then used to create supports that would separate the crate from the build platform. The STL files of these structures were then converted into the SLC format (using Magics). This resulted in four SLC files that could be prepared for the stereolithography process. These are not shown here, as the files are only contours and therefore cannot be rendered and viewed from an angle.

However, this was not a simple task to achieve. The first problem was encountered when arranging the items in the correct positions for part building because of the use of the SLC format. The SLC file is essentially the contours of the model at the layer thickness intended to build the model. Therefore, it has to be generated in the correct position and orientation relative to the z-axis. Once generated the files cannot be repositioned or rotated relative to the z-axis. Initially, this was overlooked resulting in corrupt SLC files after repositioning. The SLC generation code was therefore rewritten to create the first contour and 8 mm in z height to allow room for support structures.

The second, more fundamental and difficult, problem was discovered with the software that generated the bone structure SLC files. The SLC files created were invalid and were not recognised by current stereolithography software (3D Lightyear™). However, no error files are generated by 3D Lightyear™ and therefore it was impossible to ascertain the nature of the problem. However, when using the obsolete Maestro software, the SLC files were again found invalid but error message files were generated. Reading these error messages showed us that the orientation of the contours was incorrect. Crucially when attempting to prepare a build using these SLC files Maestro was able to reorient them, resulting in valid slice files in the SLI format. These error messages also highlighted the fact that the final layer had zero thickness. Although Maestro was unable to correct this error automatically, once it had been discovered it was a relatively simple matter to correct the code.

Figure 5.213 Model of one of the samples supported by the crate.

The reason that the SLC files were invalid in 3D Lightyear™ is that they were generated according to an obsolete specification from 3D Systems. This means that they are not recognised by current 3D Systems software releases. A similar issue can be found when attempting to use SLI files generated by CTM (a module of Mimics), which are also not recognised by Lightyear. This effectively renders the SLI export option of CTM redundant unless the user is prepared to maintain and use Maestro. This is an important point as the ability to move from computed tomography (CT) or magnetic resonance data directly to the SLI format represents by far the most efficient route to a stereolithography build.

Given that Maestro is old software running on obsolete UNIX hardware, the SLC files took an incredibly long time to prepare (approximately 3 days). To reduce the file size requirement, four types of model were prepared in a line that would fit across the width of the SLA-250 build platform. Once generated, this vector file was then copied four times at the SLA machine. This method does not increase the size of the vector file, but offsets the whole set and repeats it. This enabled us to complete 16 models without creating an unnecessarily large vector file. Even after taking these steps, the vector file was large and the build took approximately 65 h to build using SL5220 resin. A second build was implemented along the same lines to produce the remaining models of the 25 required.

An example of one of the models with its supporting crate is shown in Figure 5.213. Because of the extremely delicate nature of the models, they were painstakingly hand finished using scalpels to remove the supporting structures without causing damage.

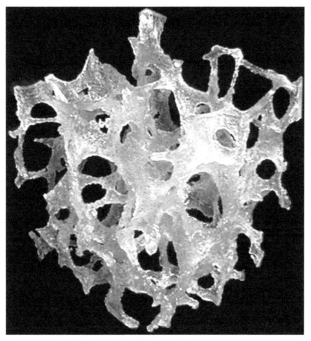

Figure 5.214 Final cleaned model of one of the samples.

This was complicated by the inability to familiarise the technicians with the models because it is not possible generate 3D rendered views of SLC data. One of the finished models is shown in Figure 5.214.

5.29.6 Conclusion

Issues caused by RP software made these items a particular challenge to the application of stereolithography. The machine operators' complete dependence on the preparation software that is supplied with the machine presents two main reasons behind the difficulties.

First, as RP develops and improves, the software is increasingly designed to automate as many functions as possible. This is intended to improve the ease and speed of use in the commercial environment. However, this increasing level of automation reduces accessibility to variables and parameters removing many options from the expert user, particularly those in the research community. In this case, typical semiautomated methods for support generation would have proved utterly impractical for the nature of these objects.

The second issue is the industry's concentration on the STL file format and complete neglect of 2½D alternatives. As this case in particular shows, contour-based formats can be dramatically more efficient. However, current stereolithography users have no functions within the preparation software (3D Lightyear™) for the verification, translation and orientation or crucially support generation of contour formats, specifically the SLC. This is despite the fact that the necessary code for each of these functions exists elsewhere (C-Sup, VerSLC).

These problems meant that to successfully build these models necessitated the utilisation and combination of several different pieces of software where appropriate and a thorough understanding of the stereolithography process. Objects that appear unfeasible for the standard practises of STL file and automatic support generation may in fact be perfectly possible if they are considered carefully (there is always more than one way around a problem). For example, these models were made possible because (1) the nature of the objects suggested that they would form self-supporting and self-draining structures in the inner volume and (2) the SLC file format was used.

As building these models has shown, seemingly obsolete software may still possess a level of accessibility that is extremely useful to the expert user, especially in research. Consequently, for many researchers in this field it is advisable to maintain copies of superseded RP software.

5.29.7   Software

C-Sup: a module of Mimics Version 7.0, Materialise NV, Technologielaan 15, 3001 Leuven, Belgium, 2001.

3D Lightyear™ version 1.1: Advanced User Licence, 3D Systems Inc., 26,081 Avenue Hall, Valencia, CA 91355, USA, 2000.

VerSLC: 3D Systems Inc., 26081 Avenue Hall, Valencia, CA 91355, USA, 1993.

Maestro version 1.9.1: 3D Systems Inc., 26081 Avenue Hall, Valencia, CA 91355, USA, 1996.

Slicer: a module of Magics version 6.3, Materialise NV, Technologielaan 15, 3001 Leuven, Belgium, 2000.

Reference

Sisias, G., Phillips, R., Dobson, C. A., Fagan, M. J., & Langton, C. M. (2002). Algorithms for accurate rapid prototyping replication of cancellous bone voxel maps. *Rapid Prototyping Journal*, 8(1), 6–24.

5.30 Research applications case study 2: recreating skin texture relief using computer-aided design and rapid prototyping

Acknowledgements

The work described in this case study was first reported in the reference below and reproduced here with kind permission of Sage Publishing.

Eggbeer D, Evans P, Bibb R, "A pilot study in the application of texture relief for digitally designed facial prostheses", Proceedings of the institute of mechanical engineers part H: Journal of Engineering in Medicine 2006; 220(6): 705-714, ISSN: 0954-4119, http://dx.doi.org/10.1243/09544119JEIM38.

5.30.1 Introduction

Maxillofacial prosthetics and technologists (MPTs) seek to meet the needs of patients with various degrees of facial deformity by restoring aesthetic and functional portions of missing tissue using artificial materials. Prosthetic restoration of lost tissue precedes surgical reconstruction and despite recent advances in surgery, many cases remain where prosthetic rehabilitation is a more appropriate treatment (Thomas, 1994). Patients typically suffer from conditions resulting from traumatic injury (such as road traffic accidents), congenital deformity or diseases that cause significant tissue damage such as cancer.

Factors that contribute to the aesthetic success of prostheses include skin colour match, appropriate contours and realistic texture (Ansgar, Cheng, Wee, Li, & Archibald, 2002). The MPTs who create the prostheses attempt to address these factors, which are conventionally assessed by eye and carved by hand in wax on a plaster replica of the patient's defect. First, the external and fitting surfaces are shaped so that the contours of the prosthesis are established. This is often done with the patient present for test fittings. The detail is gradually refined using sculpting tools to define features, creases, folds and smaller skin details that recreate missing anatomy (perhaps using old photographs as a guide) and that match the topography of the surrounding anatomy.

To create a more realistic appearance for the prosthesis skin texture may be added. This can be achieved in a variety of techniques, such stippling with a stiff brush or taking an impression from orange peel or gauze. Conversely, a flame torch may be used to locally melt or soften the wax to selectively smooth areas or decrease the prominence of textures and bumps. When the wax sculpting is completed, a plaster mould is made from it. The mould surface picks up the texture and detail of the wax carving. Once the plaster mould is set, the wax is melted out and the mould is packed with silicone elastomer that has been matched to the patient's skin colour. Other details may be added at this stage such as the use of red rayon fibres that replicate superficial capillaries and veins.

The silicone is heated under pressure to produce the final solid but flexible prosthesis. Depending on the complexity and size of the prosthesis, this process may take 2–3 days.

Improved surgical and medical techniques have led to improved survival rates from accidents and cancer treatments, which has in turn led to increased workload on the MPTs. This has driven growing interest in the application of advanced CAD and manufacturing technologies. Technologies such as CAD and RP have shown benefits in reducing time and labour in product design and development and initial research suggests that similar benefits may be realised in the production of facial prosthetics. However, although some RP technologies have been successfully exploited in maxillofacial surgery for many years, their application in facial prosthetics remains relatively unexplored (Wolfaardt, Sugar, & Wilkes, 2003). Recent technological advances have increased opportunities for MPTs to benefit from these technologies, and this can be seen in recent research (Bibb et al., 2000; Cheah, Chua, & Tan, 2003; Cheah, Chua, Tan, & Teo, 2003; Coward, Watson, & Wilkinson, 1999; Eggbeer, Evans, & Bibb, 2004; Evans, Eggbeer, & Bibb, 2004; Reitemeier et al., 2004; Tsuji et al., 2004; Verdonck, Poukens, Overveld, & Riediger, 2003). Despite this interest and some promising results, most of this research has focused on the creation of the overall shape of the prosthesis and has not considered the importance of the smaller details that make a prosthesis visually convincing. Given the importance of details such as texture and wrinkles in achieving a natural and realistic result, it was important to explore the problem through the study described here.

This research aimed to identify and assess suitable technologies that may be used to create and produce fine textures and wrinkles that may be conveniently incorporated into prosthesis design and production techniques.

5.30.2   Definition of skin texture

Visible skin texture may be classified according to the orientation and depth of the lines (Piérard, Uhoda, & Piérard-Franchimont, 2003). Primary and secondary lines form a pattern on the skin surface and are only noticeable on close observation. They typically form a polygon pattern ranging from 20 to 200 μm in depth (Hashimoto, 1974). The back of the hand often shows a good example. It has been suggested that the term 'wrinkle' should apply when an extension of the skin perpendicular to the axis of the skin surface change leaves a marked line representing the bottom of the wrinkle (Piérard et al., 2003). Further, an assessment scale that was subsequently used to assess and quantify deep facial wrinkles has been developed by Lemperle, Holmes, Cohen, & Lemperle, (2001). Wrinkles from various facial locations were subjectively graded from 0 to 5 by dermatologists, with 0 describing no wrinkles and 5 describing very deep wrinkles, redundant folds. Following the visual grading, the wrinkles were then measured using profilometry and the results correlated. This produced a graded wrinkle scale table with associated depth of wrinkle values for the various facial locations. Using this scale, a nasolabial wrinkle (side of the nose) with a grading of 1 would correlate to a wrinkle depth of less than 0.2 mm and a grading 5 would be greater than 0.81 mm in depth. This varied with other facial locations with the minimum measured depth being 0.06 mm and maximum 0.94 mm. The proposed margins on the scale ranged from >0.1 mm to <0.81 mm.

5.30.3   Identification of suitable technologies

5.30.3.1   Specification requirements

Based upon the rating scale developed by Lemperle et al. (2001), potential digital technologies must be capable of creating and reproducing wrinkle and texture details with a minimum depth of 0.1 mm. The CAD software used must be capable of creating and manipulating complex anatomical forms and the RP process capable of producing parts or patterns to this resolution in a material compatible with current prosthetic methods (Eggbeer et al., 2004; Evans et al., 2004; Verdonck et al., 2003).

5.30.3.2   Computer representation and manipulation
of skin textures

3D CAD packages have traditionally been developed for two main markets: engineering design and computer gaming/animation. Engineering CAD has been developed to define exact shapes using mathematical geometry (e.g. lines, arcs, circles, squares). Relatively complex surfaces may be generated, for example the surfaces of cars, but the mathematical geometry used is aimed towards smooth flowing surfaces and limits the ability to define the levels of contouring, such as creases, folds and sharp radii required to represent anatomical forms and textures. In fact, in applications such as automotive and aerospace design it is highly desirable to avoid unwanted creases in the surfaces being created. Software aimed towards 3D computer gaming and animation, such as 3D Studio (Discreet – AutoDesk Inc., 10 Duke Street, Montreal, Quebec, H3C 2L7, Canada) exhibits many of the same limitations as engineering CAD, but typically allows a greater freedom for surface manipulation. Because these objects are not actually physically produced, as long as the visual effect on the screen is convincing there is no need to go further in terms of detail. The textures that appear on these animations and games are normally represented by two-dimensional images that are 'wrapped' around the object. This creates an illusion of texture rather than true 3D relief. Therefore, for the purposes of prosthesis manufacture this wrapped texture cannot be used, because it cannot be physically reproduced using RP techniques.

Recently, methods of true 3D texture creation have been explored (Yean, Kai, Ong, & Feng, 1998). The application of textures has been applied in the jewellery industry and software such as ArtCAM (Delcam plc, Small Heath Business Park, Birmingham, B10 OHJ, UK) incorporate tools to map 3D textures around a CAD model. However, ArtCAM and other jewellery design software construct their shapes in the same manner as engineering CAD and are therefore not suited to the representation and manipulation of anatomical forms. Recent developments in CAD software has led to design packages that offer a more intuitive and freehand interface to the design process. Software such as ZBrush (Pixologic Inc., 320 West 31st Street, Los Angeles, CA 90007, USA) and Geomagic® FreeForm® Plus (Geomagic, 3D Systems, USA) may provide a CAD environment that is more analogous to sculpting by hand which clearly is more appropriate to the design and manufacture of a facial prosthesis. Both FreeForm® and ZBrush allow complex 3D forms to be manipulated and given high-resolution textures in a freehand manner. FreeForm® has been shown to be suitable for facial prosthesis design in previous research (Eggbeer et al., 2004; Evans et al., 2004; Verdonck et al., 2003). In addition, the haptic interface between the user and FreeForm® software allows shapes to be manipulated in ways that more closely mimic the hand carving techniques used in conventional prosthesis sculpting. In addition, FreeForm® has a number of functions that can create relief on a model surface derived from a two-dimensional image.

5.30.3.3   RP reproduction of skin textures

RP offers the most suitable solution to the production of a prosthesis or pattern from CAD data (Eggbeer et al., 2004; Evans et al., 2004). Computer numerically controlled machining (CNC) has also been used to create textures (Yean et al., 1998), but is not as well adapted to create fitting and undercut surfaces and is also limited by suitable material choice (machining of soft flexible materials is difficult). CNC becomes very slow when creating intricate or small scale detail such as textures and requires a cutting tool with a very small diameter. A review of the currently available RP technologies highlights a number of technologies that are capable of creating the level of detail required to reproduce realistic skin textures. A critical parameter in order to achieve the level of detail required is the layer thickness that the RP system uses. To achieve the level of detail identified above a layer thickness of below 0.1 mm is necessary. Currently available RP technologies that can achieve a layer thickness of below 0.1 mm include:

- ThermoJet® wax printing (obsolete, 3D Systems, USA)
- Perfactory® digital light processing (EnvisionTEC GmbH, Elbestrasse 10, D-45768 Marl, Germany)
- Solidscape wax printing (Solidscape Inc., 316 Daniel Webster Highway, Merrimack, NH 03054-4115, USA)
- Objet poly-jet modelling (Objet Geometries Ltd, 2 Holzman St., Science Park, P.O. Box 2496, Rehovot 76124, Israel)
- Stereolithography (3D Systems Inc.)

Of these, only the ThermoJet® and Solidscape printing technologies are capable of producing parts in a material directly compatible with current prosthetic construction techniques. Therefore, it was decided that these would provide the focus of the study. The Solidscape process uses a single jetting head to deposit a wax material and another one to deposit a supporting material, which can be dissolved from the finished model using a solvent. This process produces very accurate, high-resolution parts but due to its single jetting head is extremely slow. The process is therefore highly appropriate for small, intricate items such as jewellery or dentures but proves unnecessarily slow for facial prosthetic work. Like the Solidscape process, the ThermoJet® process deposits a wax material through inkjet-style printing heads, building a solid part layer by layer. The object being built requires supports, which are built concurrently as a lattice, which can be manually removed when the part is completed. The ThermoJet® process uses an array of jetting heads to deposit the material and is therefore much faster. The material is also softer than that used by Solidscape, making it more akin to the wax already used by MPTs and is therefore more appropriate for manipulation using conventional sculpting techniques. Although no accuracy specifications are given for

ThermoJet®, it is advertised as having a very high resolution ($300 \times 400 \times 600$ dots per inch in x-, y- and z-axes) and aimed at producing finely detailed parts (a drop of wax approximately every 0.085 mm by 0.064 mm in layers 0.042 mm thick).

5.30.3.4   Suitable technologies identified

From the review of currently available technologies and the required parameters stated above the following technologies were selected. Prosthesis design was to be performed using FreeForm® CAD software, including the addition of texture. The completed prosthesis design would be produced from this design file using the ThermoJet® wax printing process.

5.30.4   Methods

Case studies were conducted that explored utilising the technologies identified. First, the application of two-dimensional texture maps to create 3D relief on anatomical models was explored using FreeForm CAD and ThermoJet® RP. This was followed by a series of test pieces designed and manufactured to test the ability of the selected CAD software to create and export texture relief that would provide a range of realistic skin textures and the capability of the RP machine to reproduce these textures to the required accuracy.

5.30.4.1   Assessment

Subjective analysis of the results was used in this study, because the nature of prosthesis production and texture detailing results in complex forms that do not provide convenient datum or reference surfaces from which to measure accuracy. By definition, prostheses are one-off custom-made appliances made to fit individual patients. Therefore, it is not practical to perform the type of repeated statistical analysis that would be commonly encountered in series production or mass manufacture. The aesthetic outcome and accuracy of prostheses is subjectively assessed by the prosthetist and subsequently the patient.

The RP produced patterns were first assessed against the CAD models and, second, the level of detail in each of the model sets were assessed by a qualified and experienced prosthetist and the ability of the technologies to create and manufacture realistic and convincing skin textures was commented on.

5.30.5   Case studies

5.30.5.1   Pilot study 1

An initial pilot study was used to assess the capability of FreeForm® in creating texture relief and ThermoJet® in reproducing the detail. A series of texture images were

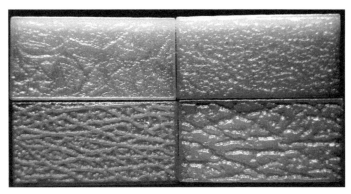

Figure 5.215 Close-up photograph of the four trial pieces.

converted into high-contrast black and white images using photographic manipulation software. These images were used to create relief textures on small rectangular samples (20 mm × 10 mm × 5 mm high). A four-sided patch was drawn on top surface of the blocks and the 'emboss with wrapped image' function used to overlay the texture maps. The scaling tools were used to gauge the approximate depth and density of the required texture relief. The method of creating the textures is described in detail in the experimental section that follows.

ThermoJet® was used to produce the physical models of the test pieces shown in Figure 5.215. The details were produced with good visual effect. However, the texture was not sufficiently detailed or deep enough to prove conclusive.

5.30.5.2   Pilot study 2

Another slightly larger test piece was created with an image that more closely resembled a detailed skin texture. The texture was created on a segment of ana-tomical data to assess the effect on a contoured piece rather than the simple flat area of the previous test pieces. The test contoured test piece was approximately 27 mm × 21 mm × 10 mm at its highest point. The model was built using the Ther-moJet® RP machine, shown in Figure 5.216. Again, the depth was judged by eye and the resulting pattern suggested that a realistic skin texture could be produced over a contoured surface.

5.30.5.3   Experimental series

Step 1: producing the texture image

A sample skin texture image was located in the form of a two-dimensional grey-scale bitmap, shown in Figure 5.217. This image was then manipulated to produce a high-contrast, black and white image shown in Figure 5.218 using Photoshop software (Adobe Systems Inc., 345 Park Avenue, San Jose, CA 95110-2704, USA). Suitable texture Images may be obtained from databases, digital macro-photographs or a pad print of skin produced from an impression.

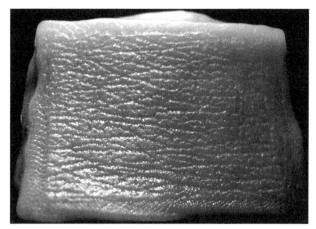

Figure 5.216 A close-up photograph showing the contoured test piece.

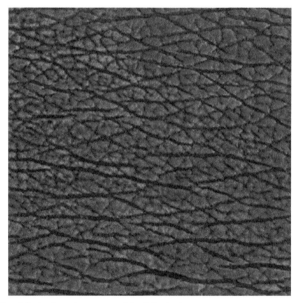

Figure 5.217 The skin texture image used in this study.

Step 2: application of textures in CAD

To assess the effect of creating textures on anatomical shapes associated with max-illofacial prosthetics, a small section of data were taken from a 3D CAD model of a human face derived from a 3D computed tomography (CT) scan. The area was selected to display a variety of compound curved surfaces, whereas the size of the selected area was kept small to minimise build time and cost. The selected region is shown in Figure 5.219. A series of test pieces based on the selected data were created in FreeForm® with a 0.1 mm edge definition.

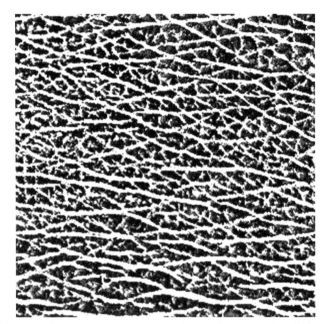

Figure 5.218 The manipulated, high-contrast version used to create the relief.

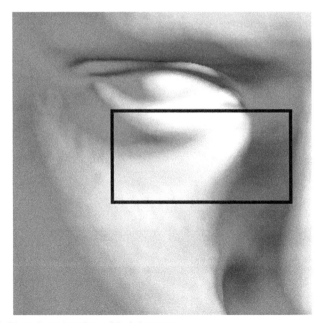

Figure 5.219 The selected region of facial anatomy.

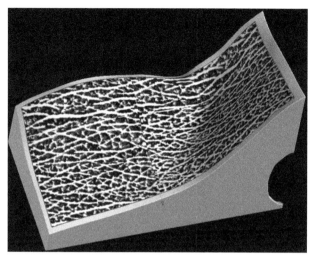

Figure 5.220 Preview of the image for the embossing function.

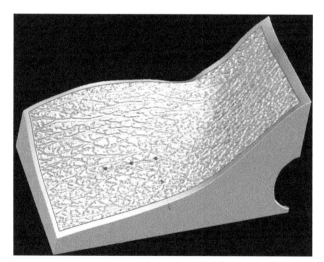

Figure 5.221 The texture relief created by the emboss function.

A rectangular box was drawn on the contoured surface and the 'emboss with wrapped image' function was used to overlay the sample two-dimensional texture image in the box. The 'emboss' function was used to set varying depths of texture relief, at 0.1, 0.15, 0.25, 0.35, 0.5 and 0.8 mm. These depths corresponded to the wrinkle depth scale and associated measurements in various facial areas developed by Lemperle et al. (2001). The actual emboss depth produced in the model is also influenced by the greyscale values in the image. Black areas are embossed to the full depth selected whereas grey areas will be proportionally embossed to lesser degree. The embossing effect may be previewed as an image (Figure 5.220) or as the resultant texture relief (Figure 5.221).

Figure 5.222 The six ThermoJet® produced skin texture sample patterns.

The 'ruler' function in FreeForm® was used to ascertain the depth of the textures by measuring the distance from an original smooth copy of the part, to the deepest part of the grooves in the textured part.

Step 3: RP manufacture

The blocks were manufactured using ThermoJet® printing with the contour surface set facing upwards. The test pieces were 18.5 mm × 35 mm × approximately 27 mm high at the tallest point. All six patterns were built in less than 3 h. The results can be seen in Figure 5.222, with the shallowest relief (0.1 mm) on the left ranging to the deepest relief (0.8 mm) on the right.

5.30.6   Results

All of the texture depths were visible on the ThermoJet® patterns. This indicated that the process was capable of producing patterns with sufficient detail to describe realistic skin textures. The layer stepping effect commonly exhibited by layer manufacture RP processes was not visible on the surface and did not interfere with the texture pattern. Textures would not however be well-defined on down facing surfaces due to the dense support structure that when removed, left a rough finish. This may be a problem if the technique was used in the production of complex prosthetic forms where all surfaces are on show such as hands.

5.30.7   Discussion

The tools available in FreeForm® were well-adapted to creating accurate texture relief from two-dimensional images. One possible limitation is file size. To represent texture faithfully and wrinkle detail at each stage of both methods, large amounts of data are required. A high-resolution model setting that demands a lot of computer processing power must be used in the CAD stage. The output STL (Standard Triangulation Language) file required for the subsequent ThermoJet® production stage must also have good detail definition that translates to a large file size. Although more modern, high-specification computers may be able to handle the large file sizes, it may make the process unmanageably slow for others. Further research will be undertaken to optimise the file size versus quality settings.

The ThermoJet® process was capable of producing all of the texture samples faithfully and did not exhibit the stair stepping effect that some other RP processes display. This ability combined with the suitable material properties demonstrates how the process may be integrated into digital prosthesis design and production techniques that are compatible with conventional handcrafting techniques.

5.30.7.1 Limitations

Although the visual effect is ultimately more important to measure subjectively by a qualified and experienced prosthetist, this study was unable to quantify the accuracy and resolution of the ThermoJet® produced patterns. This study has shown that CAD and RP processes may be used to generate fine texture detailing, but further research is required to quantify the resolution and accuracy requirements of these processes. The authors intend to apply profilometry to assess the surface of the RP-produced patterns and compare these with the CAD models.

5.30.8 Conclusions

This research has identified methods of capturing, creating and reproducing 3D, skin texture like relief using CAD and RP technologies. Furthermore, it has highlighted how these techniques may be integrated into digital prosthesis design and construction processes. Limitations of current technologies have also been highlighted and the authors intend to refine the techniques and evaluate their effectiveness in suitable patient case studies.

References

Ansgar, C., Cheng, A. C., Wee, A. G., Li, J. T. K., & Archibald, D. (2002). A new prosthodontic approach for craniofacial implant-retained maxillofacial prostheses. *Journal of Prosthetic Dentistry*, *88*(2), 224–228.

Bibb, R., Freeman, P., Brown, R., Sugar, A., Evans, P., & Bocca, A. (2000). An investigation of three-dimensional scanning of human body surfaces and its use in the design and manufacture of prostheses. *Proceedings of the Institute of Mechanical Engineers Part H: The Journal of Engineering in Medicine*, *214*(6), 589–594.

Cheah, C. M., Chua, C. K., & Tan, K. H. (2003). Integration of laser surface digitizing with CAD/CAM techniques for developing facial prostheses Part 2: development of molding techniques for casting prosthetic parts. *International Journal of Prosthodontics*, *16*(5), 543–548.

Cheah, C. M., Chua, C. K., Tan, K. H., & Teo, C. K. (2003). Integration of laser surface digitizing with CAD/CAM techniques for developing facial prostheses Part 1: design and fabrication of prosthesis replicas. *International Journal of Prosthodontics*, *16*(4), 435–441.

Coward, T. J., Watson, R. M., & Wilkinson, I. C. (1999). Fabrication of a wax ear by rapid-process modelling using stereolithography. *International Journal of Prosthodontics*, *12*(1), 20–27.

Eggbeer, D., Evans, P., & Bibb, R. (2004). The application of computer aided techniques in facial prosthetics. In *Abstract in the proceedings of the 6th International Congress on Maxillofacial Rehabilitation*, Maastricht, The Netherlands (p. 55).

Evans, P., Eggbeer, D., & Bibb, R. (2004). Orbital prosthesis wax pattern production using computer aided design and rapid prototyping techniques. *The Journal of Maxillofacial Prosthetics and Technology*, *7*, 11–15.

Hashimoto, K. (1974). New methods for surface ultra structure: comparative studies scanning electron microscopy and replica method. *International Journal of Dermatology*, *13*, 357–381.

Lemperle, G., Holmes, R. E., Cohen, S. R., & Lemperle, S. M. (2001). A classification of facial wrinkles. *Plastic Reconstructive Surgery*, *108*, 1735–1750.

Piérard, G. E., Uhoda, I., & Piérard-Franchimont, C. (2003). From skin micro relief to wrinkles: an area ripe for investigation. *Journal of Cosmetic Dermatology*, *2*(1), 21–28.

Reitemeier, B., Notni, G., Heinze, M., Schone, C., Schmidt, A., & Fichtner, D. (2004). Optical modeling of extraoral defects. *Journal of Prosthetic Dentistry*, *91*(1), 80–84.

Thomas, K. (1994). *Maxillofacial prosthetics*. London: Quintessence publishing.

Tsuji, M., Noguchi, N., Ihara, K., Yamashita, Y., Shikimori, M., & Goto, M. (2004). Fabrication of a maxillofacial prosthesis using a computer-aided design and manufacturing system. *Journal of Prosthodontics*, *13*(3), 179–183.

Verdonck, H. W. D., Poukens, J., Overveld, H. V., & Riediger, D. (2003). Computer-assisted maxillofacial prosthodontics: a new treatment protocol. *International Journal of Prosthodontics*, *16*(3), 326–328.

Wolfaardt, J., Sugar, A., & Wilkes, G. (2003). Advanced technology and the future of facial prosthetics in head and neck reconstruction. *International Journal of Oral and Maxillofacial Surgery*, *32*(2), 121–123.

Yean, C. K., Kai, C. C., Ong, T., & Feng, L. (1998). Creating machinable textures for CAD/CAM systems. *International Journal of Advanced Manufacturing Technologies*, *14*, 269–279.

5.31   Research applications case study 3: comparison of additive manufacturing materials and human tissues in computed tomography scanning

Acknowledgements

The work described in this case study was first reported in the references below and is reproduced here with the permission of Elsevier publishing and CRDM/Lancaster University respectively.

Bibb R, Thompson D, Winder J, "Computed tomography characterisation of additive manufacturing materials", Medical Engineering & Physics, 2011; 33(5): 590–596, ISSN: 1350–4533, http://dx.doi.org/10.1016/j.medengphy.2010.12.015.

Winder RJ, Thompson D, Bibb RJ, "Comparison of additive manufacturing materials and human tissues in computed tomography scanning", 12th National conference on rapid design, Prototyping & Manufacture, eds Bocking CE, Rennie AEW, 2011, CRDM Ltd., High Wycombe, ISBN: 978-0-9566643-1-0, pp. 79–86.

5.31.1   Introduction

Additive manufacturing (AM) technologies have rapidly developed and can now produce objects in a wide variety of materials ranging from soft, flexible polymers to high-performance metal alloys (Chua, Leong, & Lim, 2010; Gibson, Rosen, & Stucker, 2009; Hopkinson, Hague, & Dickens, 2005; Noorani, 2005) and they have been successfully employed in medicine since the early 1990s (Arvier et al., 1994). Initially, rapid prototyping (RP) processes, such as stereolithography, were used to make highly accurate models of skeletal anatomy directly from 3D CT data. Typically referred to as medical modelling or biomodelling, this has now become widely accepted as good practise with much literature reporting cases and the benefits achieved, particularly in craniomaxillofacial surgery. Medical models have been typically used to plan and rehearse surgery and in the design and manufacture of custom-fitting prostheses. The use of medical models has become routine and it is not necessary to discuss these applications in detail here. A number of texts and review papers are available that describe a wide range of medical applications and their principal advantages (Azari & Nikzad, 2009; Bibb, 2006; Giannatsis & Dedoussis, 2009; Gibson, 2005; Petzold, Zeilhofer, & Kalender, 1999; Webb, 2000).

Recently, AM technologies have been used to manufacture custom-fitting medical devices; for example, facial prosthetics, removable partial denture frameworks, surgical guides and even implants directly from 3D CAD data (Bibb, Eggbeer, & Evans, 2010; Bibb, Eggbeer, Evans, Bocca, & Sugar, 2009; Bibb, Eggbeer, & Williams, 2006). AM principles are also being exploited in tissue engineering where the advantages of layer additive manufacture are being used to build highly complex porous scaffolds that can support the growth of living cells (Leong, Chua, Sudarmadji, & Yeong, 2008; Peltola, Melchels, Grijpma, & Kellomäki, 2008). Polymer-based AM materials have not been approved for implantation and most of the materials tested in this research are not considered biocompatible. However, these materials may be used externally or to produce working templates from which biocompatible devices may be developed.

Some assessment of the physical properties of AM materials has been carried out. For example, dimensional accuracy, roughness of surface and mechanical properties have been established for ZPrinter 310 Plus and the Objet Eden 330 (Pilipović, Raos, & Šercer, 2009). Also, much research has been conducted on the utilisation of CT data in building objects using AM technologies and some of the materials have been characterised using CT scanning (Bibb, Thompson, & Winder, 2011). A CT scanner measures the spatial distribution of the linear attenuation coefficient or amount of absorption of X-rays. To enable this measure to be compared between scanners the CT number range (also known as the Hounsfield unit), was developed which is based on the linear attenuation to X-rays of water. The CT number range is typically from −1024 to 3072. Table 5.26 shows typical CT number ranges of selected human tissues (Kalender, 2000, p. 101). The characteristics of AM materials under radiological conditions will become important in the future as a variety of medical devices and custom-fitting patient products may be manufactured using AM and subsequently scanned using CT for either design, testing or treatment purposes. Therefore, the aim of this work was to determine the CT number of a wide selection of AM materials and establish their appearance in CT images.

It should be noted that since this work was carried out, there have been consolidation through mergers and acquisitions in the AM industry. Z Corp was acquired by 3D Systems and the technology is now marketed under their ProJet label. Stratasys merged with Objet but the technology remains labelled Objet and Connex. The Huntsman materials business has been acquired by 3D Systems. These developments have not altered the basic materials or process capabilities of the machines used and these results remain valid.

Table 5.26 CT number ranges of selected human tissues

Tissue	CT number
Air	−1005 to −995
Lungs	−950 to −550
Fat	−100 to −80
Water	−4 to 4
Kidney	20 to 40
Pancreas	30 to 50
Blood	50 to 60
Liver	50 to 70
Spongiosus bone	50 to 300
Cortical bone	300+

Adapted from Kalender, 2000, p. 30 Figure 1.9.

5.31.2   Materials and methods

A total of 29 AM samples were constructed from a CAD-generated Standard Triangulation Language (STL) file defining a rectangular block of material with dimensions $40 \times 20 \times 10$ mm. The samples represented a variety of commonly used materials from the most popular AM processes. However, the sample set is not intended to be comprehensive as there are potentially hundreds of processes and material combinations that could have been used. Twenty-five of the blocks were solid and five 'sparse' or quasi-hollow. Quasi-hollow parts are created to reduce material consumption, build time and therefore cost in some AM processes. In this study for each quasi-hollow sample, there was an equivalent solid sample. The CT number ranges were determined only for the solid samples and the quasi-hollow samples were included only to investigate their appearance in CT images.

Objet 'digital materials' (DM) materials are made from combinations of the rigid VeroWhite and soft, flexible Tango + materials. For the range DM9740 to DM9795 the final two digits relate to Shore hardness (i.e. from 40 to 95 Shore). DM8410 and DM8430 approach the rigidity of some common thermoplastics.

The samples underwent a CT scan in contact with a tissue-equivalent head phantom, as shown in Figure 5.223 (phantom supplied by Imaging Equipment Ltd., Bristol, UK) to mimic the situation of the materials adjacent to the body, as they might be in the case of an AM prosthesis or wearable medical device. CT scanning was performed using a Philips Brilliance 10 multislice system (www.medical.philips.com) using a sinus/facial/head CT protocol (exposure of 67 mAs, peak voltage 120 kV, slice thickness 2 mm, rotation time 1 s and convolution kernel type 'D', software version 1.2.0). CT images were stored in DICOM format and imported into image analysis software, AnalyzeAVW Version 9.0 (Lenexa, Kansas, US) for visualisation and CT number measurement. Visual inspection and analysis of the images was also performed using Mimics version 13 (Materialise NV, Leuven, Belgium). The density of each sample (excluding the quasi-hollow samples) was calculated by measuring the sample weight in grammes using a Sartorius precision balance and the sample volume in cubic centimetres using a digital Vernier Calliper (g/cm^3).

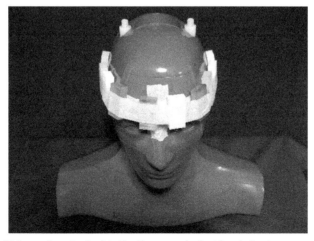

Figure 5.223 AM samples attached to the tissue-equivalent head phantom.

5.31.3 Results

Table 5.27 shows the sample name, material, mean CT number, the standard deviation of pixel values and the density of each solid sample. ABS is acrylonitrile butadiene styrene, PPSF is polyphenylsulphone and PC is polycarbonate. The Objet material Vero White is rigid white acrylate based material and the Tango+ material is a soft rubber like material. The Objet DM are composite materials made from a selective mixture of Vero White and Tango+. This produces a range of physical properties that can replicate the stiffness of a variety of thermoplastics. All materials are proprietary and specific to the relevant AM process.

The table presents the average density for each sample and it should be noted that some samples are not homogeneous and their density varies considerably across their sections (especially the Z Corp samples). As might be expected, the relationship between CT number and average sample density is essentially linear (Figure 5.224). It is well known that the CT number of a material is dependent on a range of properties including density, X-ray beam energy and sample thickness. As X-ray beam energy and section thickness were constant the large variations in CT number can be attributed to the differences in material and is related to their density. It can be seen that there is a cluster of samples around the density of 1.0–1.2 g/cm^3, which is typical for polymers, and the CT numbers are clustered suggesting that the CT number for these polymers is also similar. The two denser materials are from the Z Corp process, which are not polymers but it is interesting to note that their CT number is also proportional to their density.

The mean CT number for the solid samples ranged from a minimum −359 to a maximum of 1146. It is interesting to note that many of the AM sample CT number ranges coincide with or are similar to those of the human tissues as shown in Table 5.26. For example, samples 1 and 2 have CT number ranges that are similar to cortical bone, which may range from 200 to 1200. Samples 4, 6, 7 and 8 are similar to cancellous bone with a CT number range 50–300. Both samples 9 and 10 have mean CT numbers, which are very similar to the range found for fat tissue in the body at approximately −100. The standard deviation of the sample CT numbers range from approximately 5.0 to more than 70.0, the larger deviations measured in samples, which had a higher CT number. This is in keeping with CT scans of human tissue where bone (CT number > 300) has the highest standard deviation because of increased noise present in that tissue type, whereas air (CT number = −1024) had the lowest standard deviation. The standard deviation of the measurements within the AM samples was due to two factors, inherent noise due to the CT imaging system and any material density variation within structure of the AM sample. Samples, numbered 21–29 were all constructed using the Objet Connex 500 system and they demonstrate a limited range of CT numbers, from 70 to 120. This CT number range is similar in CT scanning to contrast enhanced blood.

Although it is known that certain AM processes produce inherently porous parts, the porosity is not apparent in all of the CT images. This is because the porosity is at a very small scale compared with the resolution of the CT scanner (high-contrast objects less than 0.5 mm can be visualised) and appears uniform throughout the parts. If we consider the example of SLS®, the process works by sintering together

Table 5.27 CT number, standard deviation of CT numbers and density of AM samples

No.	Manufacturer and machine	Material	CT number	Standard deviation	Density g/cm³
1	Z Corp 450	Z Bond (cyan-acrylate)	850.17	51.28	1.44
2	Z Corp 450	ZP130 (wax)	1146.41	71.72	1.64
3	EOS P100 Formiga	Nylon 12 (polyamide)	−17.80	29.88	1.00
4	3D Systems 250	ProtoCast AF19120, DSM Somos	168.50	28.57	1.20
5	3D Systems 250	Watershed XC11122, DSM Somos	320.82	27.62	1.17
6	3D Systems 250	9420 EP (White) DSM Somos	251.57	26.35	1.19
7	3D Systems 250	RenShape stereo-olithography Y-C 9300, Huntsman	142.43	28.67	1.22
8	3D Systems InVision	VisiJet SR	126.44	26.05	1.18
9	Dimension 1200 SST	ABS	−115.74	34.43	0.96
10	Fortus 400mc	ABS	−102.86	32.61	0.97
12	Fortus 400mc	ABS+	−358.93	31.08	0.79
14	Fortus 400mc	PPSF	151.60	46.01	1.17
16	Fortus 400mc	PC	−26.37	29.88	1.11
18	Fortus 400mc	PC/ABS	−30.21	26.34	1.05
19	Fortus 400mc	PC/ABS	110.36	8.86	1.17
20	Objet Connex 500	Vero white	99.75	5.06	1.17
21	Objet Connex 500	Tango+	118.28	6.00	1.17
22	Objet Connex 500	DM 9740	111.96	5.99	1.14
23	Objet Connex 500	DM 9750	93.13	5.54	1.13
24	Objet Connex 500	DM 9760	75.09	5.54	1.14
25	Objet Connex 500	DM 9770	72.60	5.70	1.13
26	Objet Connex 500	DM 9785	69.65	6.17	1.14
27	Objet Connex 500	DM 9795	75.96	5.08	1.16
28	Objet Connex 500	DM 8410	71.55	6.16	1.18
29	Objet Connex 500	DM 8430	119.82	6.16	1.17

Data for samples 1–18 have been used with permission from Elsevier (Bibb, Thompson, & Winder, 2011).

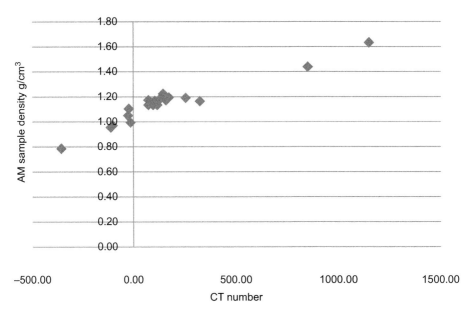

Figure 5.224 Linear relationship between CT number and average sample density.

Figure 5.225 Demonstrating uniform CT appearance of (left) sample 3, EOS P100 Formiga and (right) nonuniform appearance of sample 1, Z Corp Z450 ZBond.

thermoplastic particles with a typical average particle size of around 60 μm (material data sheet for PA2200, EOS GmbH, Munich, Germany). The particles do not fully melt but fuse together to form a sintered, porous structure. Therefore, it is reasonable to assume that this results in a slight lowering of the CT number compared to fully dense nylon produced by injection moulding, extrusion or casting. Further work will be conducted to ascertain whether the difference between SLS® nylon and solid nylon can be detected in CT images.

Figure 5.225 shows the internal structure of two AM samples (3 and 1) using CT scanning. Figure 5.225(a) shows an overall uniform internal structure, although there is some noise present within the image. This is typical of CT scanning where the X-ray photons are detected in a random manner creating small variations in pixel values for a constant material, whereas Figure 5.225(b) demonstrates internal variation in the sample. The variation is caused by the AM method where more material is deposited internally in a structured way to support the sample. This distinct variation of internal density was also visible in the images of the 3D printed samples shown in Figure 5.226(b). A variation in the sample density can be seen in the CT images, which show a higher density around the periphery of the sample. This is a result of the 3D printing process, whereby the manufactured part is initially very fragile. The parts are subjected to infiltration of a liquid hardener, typically a cyano-acrylate resin (as in sample 1)

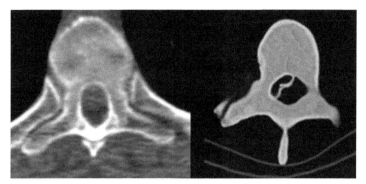

Figure 5.226 (left) A CT of human vertebra (used with permission of National Library of Medicine Visible Human Project®) and (right) a CT scan of a Z Corp lumbar spine model.

or wax (as in sample 2). It is known that the hardeners penetrate into the part through capillary action but that this penetration is limited to a few millimetres. This leads to a higher density 'skin' or 'shell' that is clearly visible in the CT image. We have observed that the internal structure in Figure 5.225(b) simulates variations in human cancellous bone making this type of modelling ideal for bone simulation. Figure 5.226(a) shows a normal CT scan of a human spine vertebra and (b) show a CT scan of a spine model manufactured using a Z Corp 3D Printing system (the same process as used to produce sample one). The CT number range for the cortical bone is 1000–1300 whereas the range within the cancellous bone is 490–815. This mimics the CT number ranges for human cortical and cancellous bone very closely. As described previously this is due to the AM process, which hardens the outer few millimetres of the model, resulting in an elevated CT number at the periphery. This relatively simple example demonstrates the potential to manufacture anatomically correct, sophisticated test objects with a mixture of hard and soft tissue materials, useful in radiation therapy dosimetry experiments where test objects may be created from a combined approach to model creation.

We have demonstrated the potential of AM to create a full size, anatomically accurate test object (phantom) for use in image quality and machine (system) performance testing of a multislice CT scanner. Data from the Digital Human Project was modelled and converted into an STL data file (with permission of National Library of Medicine Visible Human Project®). This file was subsequently used to make a life size physical model of the spine using the Z Corp AM system. Figure 5.227 shows the Z Corp model viewed from an anterior aspect. To determine how well this model mimicked human bone, the model was scanned in a Siemens Sensation 16 CT scanner (www.medical.siemens.com), with the following spine protocol (120 kVp, 134 mAs, slice = 1.0 mm, 125.0 mm field of view and 512 × 512 image acquisition matrix).

Figure 5.228 shows the coronal, sagittal and transverse reformats through the data. Excellent anatomical detail is demonstrated of the individual vertebrae, spinal canal, nerve foramen, and transverse, spinous and superior articular processes. Both cortical and cancellous bone simulates the CT image representation of a human scan with cortical bone indicating higher density that the internal cancellous bone. The anatomy is also clearly visualised on surface rendered images of the posterior and anterior surface of the spine model (Figure 5.229).

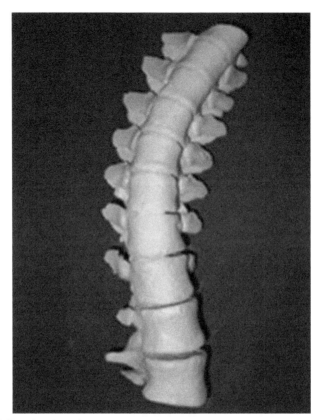

Figure 5.227 Showing the spine model manufactured using the Z Corp rapid prototyping system. The model represents the thoracolumbar region.

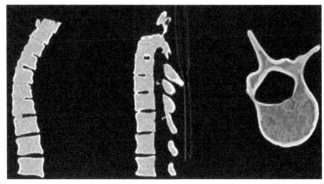

Figure 5.228 Showing coronal, sagittal transverse CT images of the 3D Z Corp spine model. Note the difference in CT number (related to bone density) of cortical bone versus cancellous bone.

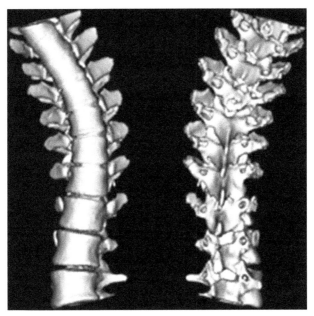

Figure 5.229 Showing surface rendered images of the CT scan, a threshold of CT number = 110 were used to segment the bone anatomy of the model.

5.31.4 Discussion

These CT number and standard deviation findings demonstrate that there is significant potential to use AM materials for sophisticated test objects for use in medical image modality testing, including image quality and radiation dose. Some AM materials have CT numbers very similar to those of human tissues, as summarised in Table 5.27, and therefore may be used to develop anatomically accurate phantoms produced from real patient CT scans using AM. Phantoms designed using these materials may have the added advantage of having CT numbers corresponding to real tissues. Anatomically complex, multitissue phantoms could be developed from existing patient CT scan data using well-established image segmentation techniques providing more accurate phantoms for test purposes. Anthropomorphic phantoms have been developed for use in radiation dose studies for diagnostic radiology and therapeutic radiology. Soft-tissue, lung and bone equivalent tissue substitutes (at diagnostic X-ray energy range 80–120 kVp) were created from urethane-based compounds mixed with other materials (Winslow, Hyer Ryan, Fisher, Tien, & Hintenlang, 2009). This particular phantom suffered from manufacturing difficulties, in that moulds would display variation in depth or suffer from physical distortion. AM has the capability to provide accurate anatomical definition, geometrical shape and the appropriate X-ray attenuation.

5.31.5   Conclusions

This study has revealed several interesting facts relating to AM materials in CT images. First, the images provide an indication of material uniformity of density at the macro scale. This analysis can be used to corroborate other observations from visual analysis and mechanical testing. Second, the actual CT numbers of a number of commonly used AM materials has been established. This may enable the specification of AM materials for specific medical devices that are required to present a specific CT number or characteristic in CT images. Further work is required to analyse a greater variety of AM materials and in particular samples from AM processes that produce mixed, graded and multiple material parts.

Acknowledgements

The authors would like to thank the following for the donation of the samples:

Phil Dixon, Loughborough Design School, and Dr Russ Harris, Wolfson School of Mechanical and Manufacturing Engineering, Loughborough University, UK; Dr Dominic Eggbeer, National centre for product design & development research, University of wales institute cardiff; Jeremy Slater, Technical sales engineer, Design Engineering Group, Laser Lines Ltd., Banbury UK.

Manufacturer contact details

3D Systems Corporation, 333 Three D Systems Circle, Rock Hill, SC 29730, USA
http://www.3dsystems.com.
AnalyzeAVW V9.0, Lenexa, Kansas, USA
http://www.analyzedirect.com
DSM Somos, 1122 St. Charles Street, Elgin, IL 60120, USA
http://www.dsm.com/en_US/html/dsms/home_dsmsomos.htm
EOS GmbH, Robert-Stirling-Ring 1, D-82,152 Krailling, Munich, Germany
http://www.eos.info/en/home.html
Materialise NV, Technologielaan 15, 3001 Leuven, Belgium
www.materialise.com/mimics
Stratasys, Inc., 7655 Commerce Way, Eden Prairie, MN 55344-2020, USA
http://www.dimensionprinting.com
http://www.fortus.com

References

Arvier, J. F., Barker, T. M., Yau, Y. Y., D'Urso, P. S., Atkinson, R. L., & McDermant, G. R. (1994). *British Journal of Oral and Maxillofacial Surgery, 32*(5), 276–283.
Azari, A., & Nikzad, S. (2009). *Rapid Prototyping Journal, 15*(3), 216–225.

Bibb, R. (2006). *Medical modelling: the application of advanced design and development technologies in medicine*. Woodhead Publishing Ltd, ISBN: 1-84569-138-5.

Bibb, R., Eggbeer, D., & Evans, P. (2010). *Rapid Prototyping Journal*, *16*(2), 130–137. http://dx.doi.org/10.1108/13552541011025852.

Bibb, R., Eggbeer, D., Evans, P., Bocca, A., & Sugar, A. W. (2009). *Rapid Prototyping Journal*, *15*(5), 346–354. http://dx.doi.org/10.1108/13552540910993879.

Bibb, R., Eggbeer, D., & Williams, R. (2006). *Rapid Prototyping Journal*, *12*(2), 95–99. http://dx.doi.org/10.1108/13552540610652438.

Bibb, R., Thompson, D., & Winder, J. (2011). *Medical Engineering and Physics*, *33*(5), 590–596. http://dx.doi.org/10.1016/j.medengphy.2010.12.015.

Chua, C. K., Leong, K. F., & Lim, C. S. (2010). *Rapid prototyping: principles and applications* (3rd ed.). WSPC. ISBN-13: 978–9812778987.

Giannatsis, J., & Dedoussis, V. (2009). *International Journal of Advanced Manufacturing Technology*, *40*(1–2), 116–127.

Gibson, I. (Ed.). (2005). *Advanced manufacturing technology for medical Applications: reverse engineering, software conversion, and rapid prototyping*. Wiley Blackwell. ISBN-13: 978–0470016886.

Gibson, I., Rosen, D. W., & Stucker, B. (2009). *Additive manufacturing technologies: rapid prototyping to direct digital manufacturing*. Springer. ISBN-13: 978–1441911193.

Hopkinson, N., Hague, R., & Dickens, P. (Eds.). (2005). *Rapid manufacturing: an industrial revolution for a digital Age*. Wiley Blackwell. ISBN-13: 978–0470016138.

Kalender, W. A. (2000). *Computed tomography* (2nd ed.). Wiley VCH, ISBN: 978-3895780813.

Leong, K. F., Chua, C. K., Sudarmadji, N., & Yeong, W. Y. (2008). *The Journal of the Mechanical Behavior of Biomedical Materials*, *1*(2), 140–152.

Noorani, R. I. (2005). *Rapid prototyping: principles and applications*. John Wiley & Sons. ISBN-13: 978–0471730019.

Peltola, S. M., Melchels, F. P., Grijpma, D. W., & Kellomäki, M. (2008). *Annals of Medicine*, *40*(4), 268–280.

Petzold, R., Zeilhofer, H., & Kalender, W. (1999). *Computerised Medical Imaging & Graphics*, *23*, 277–284.

Pilipović, A., Raos, P., & Šercer, M. (2009). *The International Journal of Advanced Manufacturing Technology*, *40*, 105–115.

Webb, P. A. (2000). *The Journal of Medical Engineering & Technology*, *24*(4), 149–153.

Winslow, J. F., Hyer Ryan, D. E., Fisher, F., Tien, C. J., & Hintenlang, D. E. (2009). *The Journal of Applied Clinical Medical Physics*, *10*(3), 195–204.

5.32   Research applications case study 4: producing physical models from computed tomography scans of ancient Egyptian mummies

Acknowledgements

This project was conducted in collaboration with Dr John Taylor, Assistant Keeper at the Department of Ancient Egypt & Sudan. The 'Tamut' project was performed on computed tomography (CT) data acquired by Clive Baldock, Reg Davies, Ajit Sofat, Stephen Hughes and John Taylor (British Museum), in 1993. The CT data were gratefully obtained from Stephen Hughes via the Internet. The Nespurennub project was conducted on CT scans acquired at the National Hospital for Neurology and Neurosurgery, London. The facial reconstruction work was undertaken by Dr Caroline Wilkinson at the Unit of Art in Medicine, The University of Manchester.

Figure 5.236 is reproduced from Taylor JH, Mummy: the inside story, 2004 with the permission of the Trustees of the British Museum and Dr Caroline Wilkinson, University of Manchester.s

5.32.1   Introduction

The development of computed tomography (CT) has allowed archaeologists to gain access to the internal details of mummies without destroying the cartonnage cases or disturbing the wrappings and remains. This nondestructive investigation has proved very successful and several investigations have been conducted in this way at various locations in the world, improving with advances in CT technology (Dawson & Gray, 1968, p. 8; Harwood-Nash, 1979; Marx & D'Auria, 1998; Pickering, Conces, Braunstein, & Yurco, 1986; Vahey & Brown, 1984). These scans have given archaeologists and forensic experts many insights into the condition of the remains and provided additional evidence relating to Egyptian funerary practise and the health of the individual. Figure 5.230 shows an axial CT slice of a mummy.

PDR, the Department of Ancient Egypt and Sudan at the British Museum and the Unit of Art in Medicine at Manchester University have formed a long-term relationship exploring the noninvasive investigation and reconstruction of ancient Egyptian mummies. The first of two mummies investigated, called Tayesmutengebtiu ('Tamut' for short), was the subject of X-ray investigation in the 1960s (Baldock et al., 1994) and subsequent investigation by CT scan more recently in 1993. The second mummy, called Nesperennub, had also undergone previous X-ray investigation. However, this mummy was recently scanned to capture better data. These CT data, specifically the series of scans of the head was used in these studies. Both mummies belong to the British Museum.

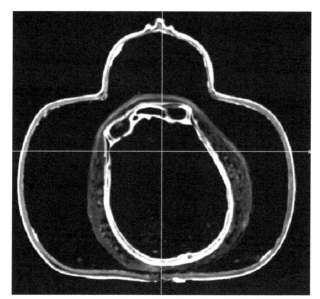

Figure 5.230 Axial CT image of 'Tamut'.

The aim of this work was to go a step beyond viewing 2D images of the mummies and to use the data to manufacture precise physical replicas of the mummies' skulls. This would allow the skulls to be investigated and handled at will without causing any damage to the original priceless remains. Facial reconstructions could also then be performed on the models. The more recent Nesperennub case was used for investigation into the cause of death, reconstruction of the facial features and other artefacts of interest for a major new exhibition at the British Museum called 'Mummy: the inside story'.

5.32.2   Technology

A range of advanced computer software and rapid prototyping (RP) hardware was required to achieve accurate digital and physical models of the two mummies. Mimics software was instrumental in importing, segmenting, cleaning and outputting the required files in order to produce the 3D model files for visualisation and manufacture. The physical models of the skulls were produced by stereolithography (Jacobs, 1996). Stereolithography is increasingly used to produce models of patients with many medical conditions and has subsequently been applied to other archaeological and palaeontological remains (Hjalgrim, Lynnerup, Liversage, & Rosenklint, 1995; Nedden et al., 1994; Seidler et al., 1997). Stereolithography and laminated object manufacture were used to reproduce physical models of some objects that were wrapped up with the mummy.

5.32.3   Case studies

Mimics software is typically used in medicine to segment CT data to isolate the tissue of interest. Frequently the tissue of interest is bone. This is accomplished by using the 'threshold' functions to select the appropriate limits of density and the numerous editing and segmentation tools to make adjustments to ensure the desired bone structures are isolated.

In contrast when attempting the first case, 'Tamut', a number of challenges made accurate reconstruction of the bony anatomy extremely challenging. In living patients, the difference in density between bone and the adjacent soft tissues is quite marked. This allows the segmentation of bone to be performed relatively easily. However, in these ancient Egyptian mummy cases, all of the soft tissues were completely desiccated by the mummification process. This resulted in the soft tissue remains having an artificially high density compared with the remaining bone (see Figures 5.231 and 5.232).

This effect is confounded by the demineralisation of some bone structures also resulting from the mummification process. Therefore, performing the segmentation by density-threshold only results in a poor 3D reconstruction. It loses some data from the skull whereas including unwanted elements of desiccated soft tissue. This effect can be seen in the reconstruction on the left of Figure 5.233.

In previous image reconstructions, higher thresholds had been used to try to eliminate some of the desiccated soft tissue remains. Although this improves matters, it results in the loss of low-density bone structures whereas some soft-tissue structures remain. In addition, the high-density artificial objects also remained present in the

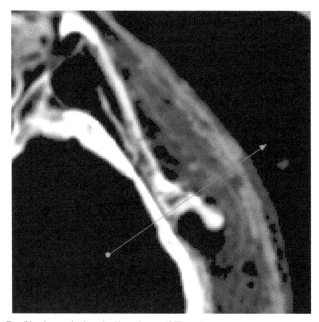

Figure 5.231 Profile through the skull and ear of Tamut.

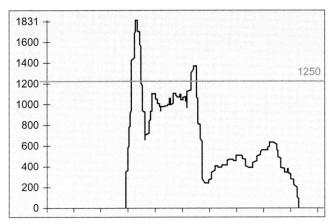

Figure 5.232 Graph of density through the profile.

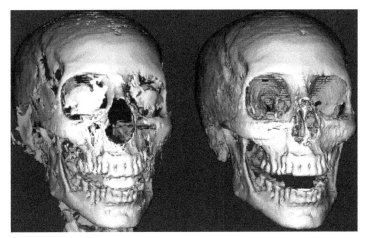

Figure 5.233 3D reconstruction of data segmented by the standard threshold for bone (left) and the manually edited data (right).

eyes, mouth and neck. To produce a model of the skull from these data would have resulted in gaps in the surface of the skull and the absence of some of the more delicate bone structures. For example, gaps could be seen in the temporal bone and zygoma (cheekbones) and the soft tissue of the ears are still present. As facial reconstruction depends on imposing known depths of soft tissue on to the facial bones, these defects would make the whole process more difficult and less reliable.

To improve the data describing the surface of the facial bones required the extensive use of the manual editing tools in Mimics. Some areas were improved by using the local Thresholding tools within the Edit tools in Mimics. This allowed small areas to be selected according to higher or lower densities without affecting the overall segmentation. However, some areas required specific areas to be edited manually using the Edit tools to delete data relating to soft tissue and draw in data relating to bone. The flexibility of these editing tools combined with good anatomical knowledge resulted in accurate segmentation of the skull. Although this was quite time-consuming the improved results are well worth the effort and would ensure that the subsequent facial reconstruction was carried out on the best model possible.

The inner surfaces of the skulls were more easily identified as the brains had been removed as part of the mummification process. However, other areas proved much more difficult to pick out from the surrounding tissue. This was especially the case in the mouth, palate and nasal cavities where desiccated soft tissue remained in place. The nose also posed problems due to the presence of cartilage and the fact that the nasal bones were broken and displaced. Again, this damage resulted from the mummification process.

The 3D reconstruction functions of Mimics were used to assess the quality of the segmentation on screen. Finally, the highest quality 3D reconstructions were created to check the data before building the models. The final reconstruction can be seen on the right of Figure 5.233.

Once we were satisfied that we had a good segmentation of the skull, the data were prepared for model building using stereolithography. The RP Slice module of Mimics was used to generate SLC files of the skulls. The RP Slice module was then also used to create the necessary supports, also in the SLC format. This direct interface to a layer format suitable for stereolithography results in smaller file sizes whilst retaining excellent detail and accuracy. In addition, the single support file generated proves simpler to process and subsequently easier to remove from the model when compared with alternative formats. The skull models produced provided Caroline Wilkinson at the Unit of Art in Medicine at Manchester with a sound basis for the facial reconstructions as shown in Figures 5.234 and 5.235 (Wilkinson, 2004).

With the Nesperennub case, the Standard Triangulation Language (STL) + module of Mimics was also used to export the skull data as an STL file. The availability of a high quality STL file enabled Caroline to attempt a digital facial reconstruction using a sophisticated virtual sculpture system (Wilkinson, 2003).

The Nesperennub CT data also showed a number of other articles of interest that had been wrapped up with the mummy. These included objects found in some other mummies of the period, a snake amulet on the forehead and artificial eyes in the sockets. Uniquely, Nesperennub also appeared to have a bowl on top of his head (see Figure 5.236). The British Museum was very keen to have replicas of these objects for use in the exhibition.

Mimics was used to segment the data describing these objects so that they could be made using RP techniques. The artificial eyes and snake amulet appeared to have high densities and were therefore relatively easy to threshold and segment from their surroundings. The bowl however, proved more difficult as the density was quite high but similar to bone and in close contact with the head in some areas. The bowl was therefore segmented using the same approach used for the skulls. The bowl was of particular interest as it had never been seen in a mummy before and nobody was quite sure what its function was.

The STL + module of Mimics was used to generate STL files of the objects that were made by RP. The snake amulet and eyes were made using stereolithography and the bowl was made using laminated object manufacture (see Figure 5.237). The model of the bowl was used to help museum staff to identify the object and speculate how it came to be in the mummy's wrappings. Handling and inspecting the bowl led

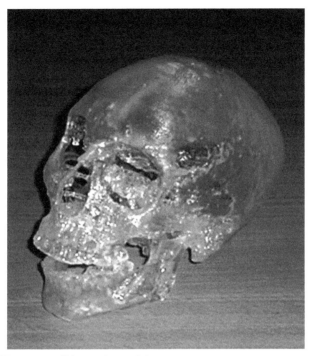

Figure 5.234 Tamut stereolithography model.

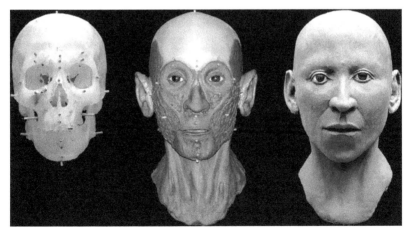

Figure 5.235 The stages of facial reconstruction on the Nesperennub stereolithography model. Image courtesy of Dr. Caroline Wilkinson, The University of Manchester.

them to the conclusion that it was a simple unfired clay bowl that was probably used to hold the resin used in the mummification process. It had probably become glued to Nesperennub's head during the mummification process by accident and was simply covered up to hide the mistake. The bowl model was also used in a film reconstructing

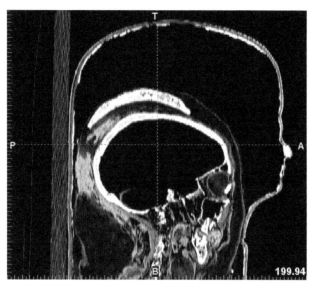

Figure 5.236 A sagittal CT image of Nesperennub.

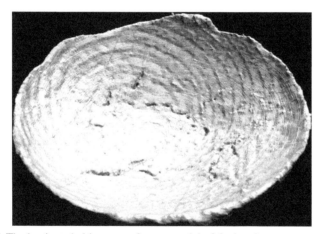

Figure 5.237 The laminated object manufacture model of the bowl.

the mummification process, which was shown as part of the exhibition. The amulet, eyes and bowl were on display in the exhibition.

5.32.4 Conclusions

'The exhibition at the British Museum proved a great success, attracting 388,000 visitors. Since the mummy has never been unwrapped, the models of the skull, bowl and amulets are key elements in the display, providing an essential complement to the noninvasive images obtained from CT scans. The models have also been used successfully at the Museum in handling sessions for visually impaired visitors.'

Quote by Dr John Taylor, Assistant Keeper at the Department of Ancient Egypt & Sudan, The British Museum.

References

The Nesperennub case was shown in the major exhibition 'Mummy: the inside story' at the British Museum, London, and described in the accompanying book of the same name by Dr John Taylor (Taylor, 2004) and featured in a TV documentary shown in the United Kingdom on Channel 5.

Baldock, C., Hughes, S. W., Whittaker, D. K., Taylor, J., Davis, R., Spencer, A. J., et al. (1994). 3-D reconstruction of an ancient Egyptian mummy using X-ray computed tomography. *Journal of the Royal Society of Medicine, 87,* 806–808.

Dawson, W. R., & Gray, P. H. K. (1968). *Catalogue of Egyptian antiquities in the British museum: 1 mummies and human remains.* London: British Museum Press.

Harwood-Nash, D. C. F. (1979). Computed tomography of ancient Egyptian mummies. *Journal of Computer Assisted Tomography, 3,* 768–773.

Hjalgrim, H., Lynnerup, N., Liversage, M., & Rosenklint, A. (1995). Stereolithography: potential applications in anthropological studies. *American Journal of Physical Anthropology, 97,* 329–333.

Jacobs, P. F. (1996). *Stereolithography and other RP&M technologies.* Dearborn, MI: Society of Manufacturing Engineering.

Marx, M., & D'Auria, H. D. (1998). Three-dimensional CT reconstruction of an ancient Egyptian mummy. *American Journal of Radiology, 150,* 147–149.

Nedden, D., Knapp, R., Wicke, K., Judmaier, W., Murphy, W. A., Seidler, H., et al. (1994). Skull of a 5,300 year old mummy: reproduction and investigation with CT guided stereolithography. *Radiology, 193,* 269–272.

Seidler, H., Falk, D., Stringer, C., Wilfing, H., Muller, G. B., zur Nedden, D., et al. (1997). A comparative study of stereolithographically modelled skulls of Petralona and Broken Hill: implications for future studies of middle Pleistocene hominid evolution. *Journal of Human Evolution, 33*(6), 691–703.

Taylor, J. H. (2004). *Mummy: the inside story.* London, UK: The British Museum Press, ISBN: 0-7141-1962-8.

Vahey, T., & Brown, D. (1984). Comely Wenuhotep: computed tomography of an Egyptian mummy. *Journal of Computer Assisted Tomography, 8,* 992–997.

Wilkinson, C. M. (2003). Virtual sculpture as a method of computerised facial reconstruction. In *Proceedings of the 1st International conference on reconstruction of soft facial parts,* Potsdam, Germany (pp. 59–63).

Wilkinson, C. M. (2004). *Forensic facial reconstruction.* Cambridge, UK: Cambridge University Press.

5.33   Research applications case study 5: trauma simulation of massive lower limb/pelvic injury

Acknowledgements

The work described in this case study was undertaken as part of a project led by Professor Ian Pallister (Programme Director for the MSc Trauma Surgery (Civilian and Military)) Course at the College of Medicine, Swansea University, in collaboration with Professor Mark Waters of MBI (Wales) Ltd and Dr Dominic Eggbeer at PDR. The project was funded via the Centre for Defence Enterprise (project number CDE33698) and was in response to the call "THE MEDIC OF THE FUTURE Challenge 1: "SimTraining".

Much of the introduction text in this section was drafted by Prof. Pallister as part of the project proposal and has been edited for inclusion in this book.

5.33.1   Introduction

High-fidelity training is a crucial part of skill development for trauma and orthopaedic surgeons, yet the tools currently available are frequently poor representations of what is experienced in real-life cases. There are currently four primary training tools available: artificial anatomical models, cadavers, animal models and clinical cases. Each of these options is flawed.

Artificial mannequin models are extensively used in both undergraduate and postgraduate medical training for fundamental procedures such as intubation, cannulation and blood gas sampling. These mannequins are characterised by their single function capabilities. The synthetic bones currently available for teaching and training in fracture surgery incorporate very poor quality simulated soft tissue envelopes, which do not reflect human tissue either in terms of anatomical detail or handling characteristics. Although some simulation mannequins are anatomically accurate, they are typically limited in terms of the body area reproduced. However, although their tissue-handling properties are far from perfect, they do offer distinct advantages over alternative means for delivering similar trauma training. Indeed, the alternative means of simulating such injuries are extremely limited. Although animal models may enable simulation in the presence of a live circulation, both the very distinct anatomical differences between animals and humans and the complexities of animal welfare mean that such simulations are both limited in terms of realism and availability. Human cadaveric material, whether soft embalmed or unembalmed tissue is used, may satisfy some of the anatomical shortcomings of animal models, but is far from perfect. The frail elderly, from whom the material often originates, invariably have suffered major illnesses or undergone a range of medical procedures, which may then prevent trainees from executing the techniques required. The bone and soft tissues are often involuted and are of very limited value in key procedures such as vascular control manoeuvres, and especially fasciotomy and damage control external fixation. In both the animal model and cadaver options, it is also very difficult to simulate, control and repeatedly recreate different permutations of injuries.

This project aimed to create a prototype, realistic, complex trauma simulation system for injuries below the umbilicus reproducing massive lower limb/pelvic injury resulting from blast, gunshot wound/penetrating or blunt injury. It was proposed that the detail of the model and simulation system would be sufficient for complex wound/injury patterns to be reproduced involving major vessels, nerves, muscle groups/compartments and the skeleton. This would help to reduce the requirements for current animal models and human cadaver materials and create a repeatable and realistic option for surgical training. Specifically, the simulation system was designed to reproduce severe lower limb and pelvic trauma replicating the types of injuries encountered in current conflicts including those typified by improvised explosive devices and gunshot wounds.

5.33.2   Materials and methods

The model at the centre of the system was derived from existing computed tomography (CT) scans from a trauma patient who had consented to its use for research and development projects. This was crucial to obtain sufficiently detailed, anatomically correct information on which the simulation system could be based.

5.33.2.1   Step 1: data segmentation

The CT data (1-mm slice thickness, 0.5-mm increments and 0.424-mm pixel size on a Toshiba Aquilion Multislice scanner) was imported into Mimics® software (Mimics version 16, Materialise, Belgium). Mimics® was used to segment and create separate 'masks' of the bones, muscles, urinary tract, bowel and major blood vessels of the below trunk-level anatomy using a mix of automatic and manual techniques described in Chapter 3. Figure 5.238 shows 3D renders of the anatomical structures segmented. These structures were exported as high quality Standard Triangulation Language (STL) files for further modelling. Since the patient's right lower leg anatomy was affected by the surgical procedure, it was agreed that their left leg would be mirrored to create the necessary symmetry in subsequent stages.

5.33.2.2   Step 2: component and mould design

The STL data were imported into FreeForm® (Geomagic, 3D Systems, USA) for further modelling, smoothing and refinement. Each component of the system was individually tailored to fit as an assembly. Lifelike features, such as textures were added using 'emboss area' and 'emboss with image' tools to the muscle tissues, bladder and penis to improve the realism. To avoid duplication of work and avoid the areas affected by the patient's injuries, structures including the muscles and bones of the left leg were

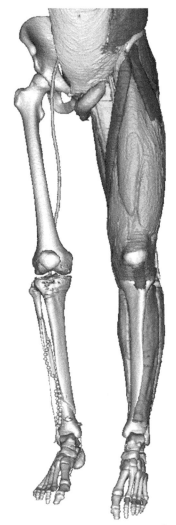

Figure 5.238 The segmented anatomical structures in Mimics®.

mirrored across the mid-sagittal plane. Remodelling of the skin was required to remove
any obvious indications of the mirroring procedure and a version replicating trauma
injuries to the right leg was also created. The completed model schematic is shown in
Figure 5.239(a) and (b).

FreeForm® was also used to undertake mould tool design of components
including the outer skin, bladder/penis and colon/rectum. The outer skin model
was segmented axially into sections representing crucial points at which artificial injuries would be reproduced and based on the maximum build volume of
an available Stereolithography machine (500 × 500 × 500 mm). The outer skin
mould was created by offsetting the model to the outside by 4 mm, identifying

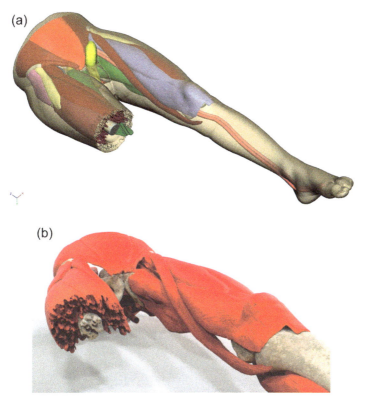

Figure 5.239 (a) Schematic of the model in FreeForm; (b) FreeForm rendering showing the modelled blast injury.

a split line that avoided significant undercuts and creating nonuniform rational B-spline surfaces that acted as splitting planes and the basis on which to design flange features. Inserts for the groin and buttock areas were also created to enable the creation of different simulated injuries. Flanges for the torso end and mid-thigh were designed to include locating features to hold the internal structures in place, and a separate a version of the right flange that simulated the effects of a blast injury was also created. The completed mould design for the pelvic region is shown in Figure 5.240.

5.33.2.3 Step 3: component fabrication

The bones and muscles components were shelled (to reduce the volume of material), split with a dovetail join line and fabricated using stereolithography (SLA 250-50, 3D Systems, USA). Once completed, the parts were assembled, filled with a pourable, two-part polyurethane, hand finished and used as master patterns to create flexible silicone tooling for vacuum casting production in tissue-mimicking materials. The circulatory system was split into sections before being fabricated

Figure 5.240 The outer skin pelvis region mould tool design in FreeForm®.

using a ProJet 3000 Plus HD (3D Systems, USA) machine to create patterns for dip moulding.

In parallel to the CAD process, materials that mimicked the physical handling and visual characteristics of human tissue were developed in silicone and polyurethane. This included mimics of superficial skin, fat, veins/arteries, muscle, facia and bone. The system was also designed to incorporate an anatomically correct, pulsatile circulatory system driven using a peristaltic pump.

Final-stage component fabrication required a mixture of vacuum casting, bench pouring, dip moulding and low-pressure injection processes depending on the tissue types. The silicone tools created from the bone and muscle master patterns were used to produce vacuum cast parts in the mimic materials. The sections of arteries and veins were used as tools for dip moulding in a silicone material. The sections were then assembled as a complete circulatory tree that could be attached to a peristaltic pump. The outer skin mould tool sections were fabricated using stereolithography (SLA 500, 3D Systems, USA) in a transparent (ClearVue, 3D Systems) resin to allow observation and evaluation of the moulding process. Additional flange strengthening components were fabricated using SLS® (EOS GmbH). Once completed, the mould sections were manually hand finished using grit paper to smooth split lines and improve the ability to assemble the multiple components and flanges.

5.33.2.4 Step 4: simulation model assembly and moulding

The moulding process involved numerous stages of tissue layering and material injection to ensure the internal structures remained in the correct anatomical position. Figure 5.241 shows part of the way through the tissue layering process with the full lower left leg internal structures in place ready for moulding the upper portions. The bones and muscles were supported by mounting features integrated into the outer mould sections. Figure 5.242 shows the closed mould with the internal structures visible through the superior flange.

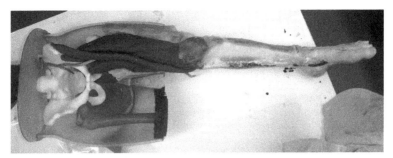

Figure 5.241 Layering the internal structures into the outer mould.

Figure 5.242 The closed mould with the internal structures visible.

5.33.3   Discussion and conclusions

This project demonstrated the potential of utilising patient CT data, advanced CAD techniques and AM to produce surgical training models with increased fidelity and realism over currently used alternatives. The prototype model developed included rendered and physical details such as specific wound/injury patterns encountered through combat situations. The model also included a dynamic circulation containing a finite volume of synthetic 'blood'. Active bleeding in the site of the simulated wound/injury allowed for the application of simulated emergency haemostatic techniques (novel dressings, tourniquet, etc.) with subsequent changes in the physiological measurements being fed back to those participating in the simulation.

The ability to more closely replicate what trauma surgeons encounter is an important step in improving the quality of training. The benefit to the defence and wider medical community from this innovation will be in terms of trauma training and readiness in the event of future conflicts and daily medical emergency scenarios. Free from the constraints of animal welfare, the human tissue act and the anatomy laws, the prototype systems provides the first platform for a more advanced simulation to be conducted in a range of environments, from simulated contact with the enemy for first responders, to specific surgical procedures and importantly, multiple-stage multidisciplinary resuscitation scenarios.

Given the prototype nature of the first model, further work is necessary to evaluate end user perceptions of the system and incorporate suggestions to improve the subsequent designs. It is also necessary to refine the fabrication and assembly process to ensure batch/mass production is more economically viable.

5.34   Research applications case study 6: three-dimensional bone surrogates for assessing cement injection behaviour in cancellous bone

Acknowledgements

The work described here was undertaken by Antony Bou Francis, Richard M Hall and Nikil Kapur, School of Mechanical Engineering, University of Leeds, UK in collaboration with PDR, Cardiff Metropolitan University, UK.

5.34.1   Introduction

Osteoporosis and other skeletal pathologies such as spinal metastasis and multiple myeloma compromise the structural integrity of the vertebra, thus increasing its fragility and susceptibility to fracture (Baroud, Falk, Crookshank, Sponagel, & Steffen, 2004a; Baroud, Vant, Giannitsios, Bohner, & Steffen, 2004b; Hughes, 2005; Ridler & Calvard, 1978). During vertebral augmentation procedures, bone cement is injected through a cannula into the cancellous bone of a fractured vertebra with the goal of relieving pain and restoring mechanical stability. However, prophylactic surgical stabilisation is often performed to reinforce a structurally compromised vertebra and decrease its susceptibility to fracture (Baroud, Crookshank, & Bohner, 2006; Widmer & Ferguson, 2012). The bone cements used are chemically complex, multicomponent and significantly non-Newtonian with their viscosity having differing degrees of time and shear rate dependency. These cements interact with the porous structures though which they flow and with other fluids present within the porous media. The most widely used cement, polymethyl-methacrylate, is generally assumed insoluble in any biofluid (bone marrow) it comes into contact with, thus the cement-marrow displacement is characterised as a two-phase immiscible flow in porous media (Bohner et al., 2003; Loeffel, Ferguson, Nolte, & Kowal, 2008). As vertebral cancellous bone has highly complex geometrical structures and displays architectural inhomogeneities over a range of length scales, the pore-scale cement viscosity varies due to its nonlinear dependency on shear rates, which are affected by variations in the local tissue morphology. Furthermore, the vertebral cancellous bone microarchitecture varies among the patients being treated, thus making the scientific understanding of the cement flow behaviour difficult in clinical or cadaveric studies (Chen, Shoumura, Emura, & Bunai, 2008; Mohamed et al., 2010). Previous experimental studies on cement flow (Gong et al., 2006; Hildebrand, Laib, Müller, Dequeker, & Rüegsegger, 1999; Hulme, Boyd, Ferguson, 2007; Lochmüller et al., 2008; Mohamed et al., 2010) have used open-porous aluminium foam to represent osteoporotic bone. Although the porosity was well controlled, the geometrical structure of the foams was inherently unique. We propose novel methodology using reproducible and pathologically representative 3D bone surrogates to help study biomaterial–biofluid interaction providing a clinical representation of cement flow distribution and a tool for validating computational simulations.

5.34.2   Methods

3D bone surrogates were developed to mimic the human vertebral body. Figure 5.243 shows the boundary of the 3D bone surrogates showing: (a) two identical and symmetrical elliptical openings 2 mm in height and 1 mm in width applied to mimic breaches due to anterior blood vessels, (b) one circular opening 3 mm in diameter applied to mimic posterior breaches because of the basivertebral veins and (c) the insertion channels that were incorporated to allow consistent needle placement during injection. The superior and inferior surfaces of the surrogates were kept open because of manufacturing restrictions. All dimensions are in millimetres.

The surrogates were designed in SolidWorks (Dassault Systèmes, Vélizy, France) then manufactured using RP (ProJet HD 3000 Plus, 3D Systems, USA). The structure of the surrogates was tailored to mimic three skeletal pathologies: osteoporosis (Osteo), spinal metastasis (Lesion) and multiple myeloma (MM). Figure 5.244 illustrates the developed bone surrogates and Table 5.28 describes the elements incorporated into each surrogate. Once the surrogates were manufactured, micro-CT (μCT 100, Scanco Medical, Switzerland) was used to assess the variability in their morphology. Eight of the Osteo surrogates were scanned at a spatial resolution of 24.6 μm (isotropic voxel size) with a 300-ms integration time, a 70-kV tube voltage, a 114-μA tube current and a 0.1-mm aluminium filter. Then, a cylindrical volume of interest 15 mm in diameter and 15 mm in length was consistently defined at the centre of each specimen. Within this volume of interest, 3D

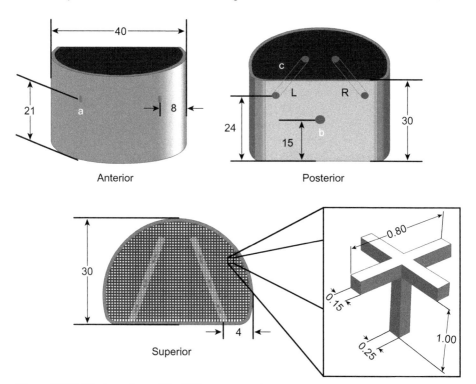

Figure 5.243 The boundary of the 3D bone surrogates.

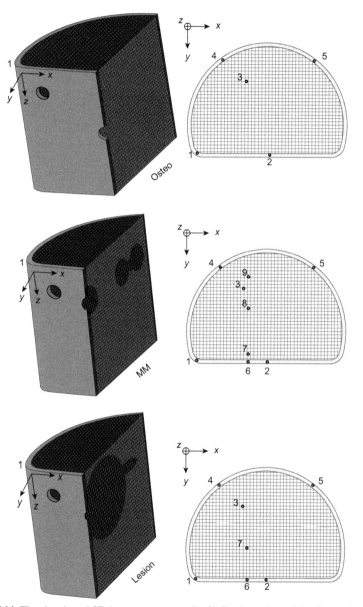

Figure 5.244 The developed 3D bone surrogates. (Left) Section-view of the Osteo, Lesion and Multiple Myeloma surrogates. (Right) Craniocaudal view showing the location of all elements incorporated into each surrogate (refer to Table 5.28).

morphometric indices were determined using the following settings: Sigma 1.2, Support 2.0, Threshold −120 HA mg/ccm (based on Ridler's method (Ridler & Calvard, 1978)), and proprietary software provided by the manufacturer. Only the bone volume fraction, BV/TV (%), trabecular thickness, Tb.Th (mm), and trabecular separation, Tb.Sp (mm), were compared. The porosity of the specimens was obtained from the μCT(100 − BV/

Table 5.28 **The location and size of all elements incorporated into the 3D bone surrogates. All coordinates are measured with respect to the geometrical centre of each element**

Surrogate	Element	Coordinate			Description
		x	y	z	
Osteo, lesion	1	0.0	0.0	0.0	Reference point
and MM	2	20.0	0.0	15.3	Outlet, circular Ø 3.0 mm
	3	13.5	−20.6	8.0	Inlet, circular Ø 2.1 mm
	4	8.2	−26.1	9.3	Outlet, elliptical width 1.0 and height 2.0 mm
	5	31.8	−26.1	9.3	Outlet, elliptical width 1.0 and height 2.0 mm
Lesion	6	14.3	0.0	10.0	Outlet, circular Ø 2.7 mm
	7	14.3	−8.9	10.0	Spherical void Ø 19.0 mm
MM	6	14.3	0.0	8.0	Outlet, circular Ø 2.6 mm
	7	14.3	−2.2	8.0	Spherical void Ø 6.0 mm
	8	14.3	−14.9	8.0	Spherical void Ø 6.0 mm
	9	14.3	−23.4	8.0	Spherical void Ø 6.0 mm

TV) and validated using Archimedes' suspension method of measuring volume (Hughes, 2005) which was performed using six cubes (2 cm³ volume) with the same structure as the Osteo surrogates. One of the six cubes was also used to measure the permeability of the Osteo structure using Darcy's law (Baroud et al, 2004a, 2004b; Widmer & Ferguson, 2012). Furthermore, static contact angle analysis (FTA 4000, First Ten Angstroms, VA, USA) was performed on the material to compare the surface wettability to that of cortical bone from a dry human femur.

5.34.3 Results

Figure 5.245 shows an example of the cylindrical volume of interest that was consistently defined at the centre of each 3D Osteo surrogates and used to obtain the bone volume fraction (BV/TV) in %, trabecular thickness (Tb.Th) in mm, and trabecular

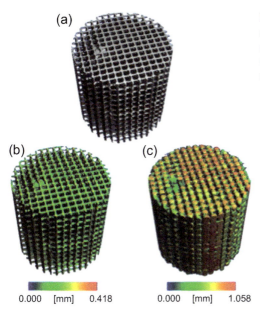

Figure 5.245 (a) Cylindrical volume of interest defined at the centre of the osteoporotic type 3D bone surrogates with its associated thickness map (b) and spacing map (c).

0.000   [mm]   0.418   0.000   [mm]   1.058

separation (Tb.Sp) in mm. Figure 5.246 shows the histograms associated with the respective thickness and spacing maps. The thickness map revealed three distinct peaks at approximately 0.18, 0.27 and 0.34 mm. The first two peaks were associated with the horizontal and vertical struts of the 3D bone surrogates, which had a nominal thickness of 0.15 and 0.25 mm, respectively. The third peak could be related to manufacturing artifacts associated with residual support material within the structure. On the other hand, the spacing map revealed one distinct peak at approximately 0.89 mm, which corresponded to the nominal spacings in the horizontal (i.e. between vertical struts) and vertical (i.e. between horizontal struts) planes of the 3D surrogates which were 0.8 and 1.0 mm, respectively. The variability in the morphology of the osteoporosis 3D bone surrogates was very low with an overall strut thickness (Tb.Th) of 0.25 ± 0.04 mm and an overall pore spacing (Tb.Sp) of 0.89 ± 0.03 mm. The average porosity which was obtained from the μCT data (100 − BV/TV) and validated using Archimedes' principle was $82.6 \pm 1.1\%$. The measured permeability of the Osteo structure was $57.1 \pm 6.1 \times 10^{-10}$ m². The surface wettability was comparable between all materials with contact angles ranging from 49° to 77° for the material used to manufacture the bone surrogates and 60°–75° for bone form a dry human femur.

5.34.4   Discussion

It is extremely important to control the surrogate morphology, as bone cement precursors are heterogeneous, especially their powder component which varies in composition, size and molecular weight of the pre-polymerized polymer beads as well as the morphology of the radiopacifier particles. All these factors have a significant

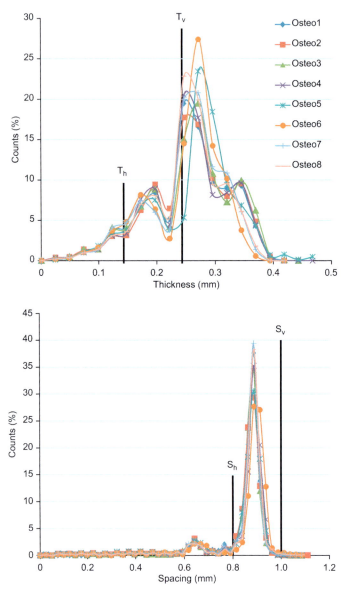

Figure 5.246 Histograms of the trabecular thickness (top) and trabecular separation (bottom) obtained from the CT data of eight osteoporotic type (Osteo) 3D bone surrogates. The black lines represent the nominal horizontal and vertical thickness (Th and Tv) and spacing (Sh and Sv), respectively.

effect on the interaction between the liquid and the powder components during mixing and injection, resulting in different flow behaviour for different cement formulations. The developed bone surrogates can be assumed constant in geometrical structure, as the variability in their morphology was very low. Achieving constant geometrical structure is crucial to reduce the variability, render the experiments reproducible and shift the focus onto understanding the influence of cement properties on the injection behaviour. The main disadvantage of the surrogates is that they have a uniform structure and do not simulate the highly complex geometrical structures and architectural inhomogeneities of vertebral cancellous bone. However, this was necessary to simplify the representation of the bone morphology, facilitate the manufacturing process and ensure the reliable reproduction of the surrogates. When complex structures are involved, more support material is required during the manufacturing process. Thus, it is more difficult to remove the support material without damaging the actual structure, especially when regions within the structure (i.e. vertebral cancellous bone) are thin and not well-connected.

Although the structure of the surrogates was uniform, the pore spacing of the Osteo surrogates (0.89 ± 0.03 mm) was comparable to that reported in the literature for human vertebral cancellous bone (Chen et al., 2008; Gong et al., 2006; Hildebrand et al., 1999; Hulme et al., 2007; Lochmüller et al., 2008). The surrogates also simulate the rheological environment within the vertebral body. The measured permeability of the 3D bone surrogates (57.1 ± 6.1 × 10^{-10} m²) was comparable to that reported by Nauman et al. (Nauman, Fong, & Keaveny, 1999) for human vertebral cancellous bone, which was 80.5 ± 47.5 × 10^{-10} and 35.9 ± 19.0 × 10^{-10} m² in the longitudinal and transverse directions, respectively. Furthermore, based on contact angle measurements, the surface wettability of the 3D surrogates matches that of bone. Previous experimental studies on cement flow (Baroud et al., 2006; Bohner et al., 2003; Loeffel et al., 2008; Mohamed et al., 2010) have used open-porous aluminum foam to represent osteoporotic bone. Although the porosity of the foam was well controlled, its geometric structure was inherently random. The proposed bone surrogates overcome the limitations of previous materials as their geometric structure is well controlled and can be tailored to mimic the morphology of specific bone conditions at different skeletal sites in the body. Another advantage is that the developed 3D bone surrogates have a boundary to simulate the vertebral shell which confines the flow and controls the intravertebral pressure, significantly affecting the filling pattern (Baroud et al, 2004a, 2004b). The boundary including the inlet and the flow exit points were kept constant for all the surrogates. The openings in the boundary simulate breaches through the cortex due to a fracture, a lesion and/or a blood vessel exchanging blood in and out of the vertebral body. This is important as such breaches create paths of least resistance providing means for leakage into the surrounding structures.

References

Baroud, G., Crookshank, M., & Bohner, M. (2006). High-viscosity cement significantly enhances uniformity of cement filling in vertebroplasty: an experimental model and study on cement leakage. *Spine (Phila Pa 1976)*, *31*(22), 2562–2568.

Baroud, G., Falk, R., Crookshank, M., Sponagel, S., & Steffen, T. (2004a). Experimental and theoretical investigation of directional permeability of human vertebral cancellous bone for cement infiltration. *Journal of Biomechanics*, *37*(2), 189–196.

Baroud, G., Vant, C., Giannitsios, D., Bohner, M., & Steffen, T. (2004b). Effect of vertebral shell on injection pressure and intravertebral pressure in vertebroplasty. *Spine (Phila Pa 1976)*, *30*(1), 68–74.

Bohner, M., Gasser, B., Baroud, G., & Heini, P. (2003). Theoretical and experimental model to describe the injection of a polymethylmethacrylate cement into a porous structure. *Biomaterials*, *24*(16), 2721–2730.

Chen, H., Shoumura, S., Emura, S., & Bunai, Y. (2008). Regional variations of vertebral trabecular bone microstructure with age and gender. *Osteoporosis International*, *19*(10), 1473–1483.

Gong, H., Zhang, M., Qin, L., Lee, K. K. H., Guo, X., & Shi, S. Q. (2006). Regional variations in microstructural properties of vertebral trabeculae with structural groups. *Spine (Phila Pa 1976)*, *31*(1), 24–32.

Hildebrand, T., Laib, A., Müller, R., Dequeker, J., & Rüegsegger, P. (1999). Direct three-dimensional morphometric analysis of human cancellous bone: microstructural data from spine, femur, iliac crest, and calcaneus. *Journal of Bone and Mineral Research*, *14*(7), 1167–1174.

Hughes, S. W. (2005). Archimedes revisited: a faster, better, cheaper method of accurately measuring the volume of small objects. *Physcis Education*, *40*(5), 468–474.

Hulme, P. A., Boyd, S. K., & Ferguson, S. J. (2007). Regional variation in vertebral bone morphology and its contribution to vertebral fracture strength. *Bone*, *41*(6), 946–957.

Lochmüller, E. M., Pöschl, K., Würstlin, L., Matsuura, M., Müller, R., & Link, T. M. (2008). Does thoracic or lumbar spine bone architecture predict vertebral failure strength more accurately than density? *Osteoporosis International*, *19*(4), 537–545.

Loeffel, M., Ferguson, S. J., Nolte, L. P., & Kowal, J. H. (2008). Vertebroplasty: experimental characterization of polymethylmethacrylate bone cement spreading as a function of viscosity, bone porosity, and flow rate. *Spine (Phila Pa 1976)*, *33*(12), 1352–1359.

Mohamed, R., Silbermann, C., Ahmari, A., Bohner, M., Becker, S., & Baroud, G. (2010). Cement filling control and bone marrow removal in vertebral body augmentation by unipedicular aspiration technique: an experimental study using leakage model. *Spine (Phila Pa 1976)*, *35*(3), 353–360.

Nauman, E. A., Fong, K. E., & Keaveny, T. M. (1999). Dependence of intertrabecular permeability on flow direction and anatomic site. *Annals of Biomedical Engineering*, *27*(4), 517–524.

Ridler, T. W., & Calvard, S. (1978). Picture thresholding using an iterative selection method. *Systems, Man Cybernetics, IEEE Transactions on*, *8*(8), 630–632.

Widmer, R. P., & Ferguson, S. J. (2012). On the interrelationship of permeability and structural parameters of vertebral trabecular bone: a parametric computational study. *Computer Methods in Biomechanics and Biomedical Engineering*, *16*(8), 908–922.

Future developments

6

6.1 Background

Several areas of development combine the advantages of computer-aided design (CAD), rapid prototyping (RP) and medical applications. Materials development is progressing in RP, which is opening up new opportunities to use objects made by RP techniques to be employed directly in medical treatment or to manufacture medical devices. For example, the direct manufacture of hearing aid shells is currently being developed and marketed by more than one medical device company.

The ability to process fully dense functional metals, particularly titanium alloys, in RP machines is also a particular interest in the manufacture of custom-designed implants. Several cases of custom-fitting implants have been demonstrated successfully in craniomaxillofacial reconstruction and complex pelvic restoration, and some series-produced parts such as acetabular cups have also been developed.

Tissue engineering is a fascinating area of development and RP/additive manufacturing (AM) approaches are being actively developed in many research centres to facilitate the direct manufacture of custom-fit tissue scaffolds and living tissue constructs. The potential to construct living tissue replacements for damaged or missing organs and bones is challenging but may ultimately reduce our dependence on organ donors and save many lives.

As can be seen, considering the technological development discussed below, the future has a great deal of potential for everyone working in this varied and challenging, but ultimately highly rewarding field.

6.2 Scanning techniques

Computed tomography (CT) scanning continues to develop, although this is incremental, as in detectors that are more sensitive, allowing X-ray doses to be lowered while capturing more data at smaller pixel sizes. The use of multiple arrays of detectors has drastically reduced scanning times and the improving ability to collimate the X-ray beams allows much thinner slices to be taken. In addition, increasing computing power and software that is more sophisticated is enabling ever-greater resolution and more control over how the images are handled and displayed.

Cone beam CT is being increasingly used, particularly in oral and orthodontic surgery, and is used for medical modelling. Cone beam CT is likely to improve in resolution and image quality over the coming years. The advantages of low-cost, low-dose X-ray and convenience make their increased use likely.

Medical Modelling. http://dx.doi.org/10.1016/B978-1-78242-300-3.00011-1

Magnetic resonance (MR) imaging has so far been used infrequently for three-dimensional (3D) virtual and physical modelling but the development of more tissue-specific protocols has enabled imaging of organs that was previously difficult to achieve using CT. It is likely the use of MR for many types of imaging requirement will increase as the speed and sophistication of the scanners improve. As with CT, the development of more sensitive detectors and the use of multiple arrays of detectors will improve speed and resolution. Software that is more sophisticated will be able to better filter out noise and produce sharper, clearer images.

The benefit of using non-ionising radiation will lead to a greater demand for MR compared with CT for some clinical situations that cannot justify exposure to X-rays. This is likely to lead to a growing application of modelling from MR. The growing interest in modelling soft tissues is also likely to encourage more work with MR data. Sophisticated techniques such as functional MR imaging scanning are also likely to increase in application, although the dynamic data they provide may not be suitable for physical modelling.

Different types of scanner are also likely to show potential in 3D modelling. Positron emission tomography, for example, shows the potential for scanning devices to be pathology targeted rather than anatomy targeted. Such modelling projects may enable better treatment planning or computer-aided, computer-guided or even robotic surgery.

Improvements in 3D ultrasound scanning may lead to the ability to model from the data, which has until now tended to show too much noise to be of practical use in modelling. However, some modelling work has been done and the ease of use, compact size and relative safety of ultrasound will encourage exploitation of the data it can provide in modelling.

6.3 Data fusion

One area currently of great interest is data fusion, which is taking data of the same patient from a number of different imaging modalities and combining them into a single, comprehensive 3D model. So far, work has been done to combine similar image types such as merging CT and MR images and combining different image types such as 3D CT models with surface models from optical scanners. However, to date much of this work has been research based and more work that is clinically based needs to be done to validate data fusion so that the effect of merging different data types is fully understood. The widespread use of common data formats such as DICOM will be a key enabling factor in allowing data fusion to be explored.

6.4 Rapid prototyping

The development of RP, or additive manufacturing (AM), may seem to have reached a plateau and current developments are often incremental. However, in general terms RP machines are becoming cheaper, faster and more reliable, which makes them ever more accessible to researchers and engineers working in a variety of applications. However, materials developments are also a major driver in medical applications of RP technologies. More biocompatible materials are still required, if much of the potential of RP is to be realised in mainstream medical treatment. Long-term stability of physical properties also remains a challenge in all areas of additive manufacturing, but it is of particular concern in medical applications. The shift in emphasis in research towards manufacture rather than modelling or prototyping proves to be a strong impetus in this respect, but much work remains to be done and this is the focus of research in the industrial sectors of aerospace and automotive. As it has in the past, this will lead to developments that require research and ingenuity on the part of engineers and clinicians to adopt adapt and implement these technologies in ways appropriate for medicine.

The increasing availability of multiple material and multicolor RP technologies will find applications in teaching and demonstration models and may facilitate the direct manufacture of medical and rehabilitation devices.

6.5   Tissue engineering

The ability to create implants that are entirely compatible with the tissue they are replacing is a major goal for reconstruction and rehabilitation, and many research groups are actively pursuing the area to such an extent that it has become a significant and exciting field in its own right. This is providing opportunities for more research into the ability to accurately capture anatomy, design for anatomy and specify multiple materials using CAD techniques. Many of the challenges of tissue engineering will occupy years of study from multidisciplinary teams, which will include not only clinicians and engineers but also biologists and biochemists. Many of the proposed developments also require a shift in scale to the micro (cellular) or nano (molecular) level, which is providing yet more challenges to researchers attempting to model human tissue.

Glossary and explanatory notes

Glossary of terms and abbreviations

Technical terms and abbreviations

3D Lightyear™	3D Systems SLA preparation software (3D Systems trademark)
3D printing	Commonly used term for RP/AM
3D Systems, Inc.	US-based company producing a wide range of RP machines
AM	Additive manufacturing
Arcam	Swedish RP metal-based RP machine manufacturer
Boolean	Simple mathematical addition or subtraction
CAD/CAE/CAM	Computer-aided design/engineering/manufacturing
CNC	Computer numerical control of machining centers, lathes and milling machines
Concept Laser	German manufacturer of metal-based laser melting machines
Connex	Objet multiple material RP machines
Digital materials	Objet terms for Connex technology-enabled materials
Digitizing	Capturing data points from an objects surface, 3D scanning
DLP™	Digital light processing (EnvisionTEC trademark)
EnvisionTEC	German manufacturer of resin-based RP machines
EOS GmbH	Electro Optical systems–German manufacturer of RP machines
FDM™	Fused deposition modelling–Stratasys term
FEA	Finite element analysis, computer testing of strength or stiffness
LOM	Laminated object manufacturing
LS	Laser sintering
Magics	STL manipulation and RP preparation software (Materialise NV)
Mimics	Software for processing medical scan data for RP (Materialise NV)
MJM	Multi-jet modelling
Objet	Stratasys AM/RP machines
Photopolymerizable	Liquid or resin that will solidify (cure) when exposed to certain wavelengths of light
Poly-jet modelling	Multiple jet printing process used on Objet machines
QuickCast™	SL build style for sacrificial investment casting patterns (3D Systems trademark)
Realizer	German manufacturer of metal-based laser melting machines
Renishaw	UK manufacturer of metal-based laser melting machines
Reverse engineering	To create a computer model of an object by digitizing or scanning it
RP	Rapid prototyping–general term for freeform build technologies
RP&M	Rapid prototyping and manufacturing
RP&T	Rapid prototyping and tooling

Sintering	Fusing a powder, usually by heat
SL	Stereolithography
SLA®	Stereolithography apparatus (3D Systems registered trademark)
SLC	Slice format for three-dimensional computer data
SLI	Slice format for three-dimensional computer data
SLM™	Selective laser melting (Realizer trademark)
SLM Solutions	German manufacturer of metal-based laser melting machines
SLS®	Selective laser sintering (3D Systems registered trademark)
Solidscape	Stratasys-owned company producing wax deposition-type RP machines
STL	Derived from stereolithography–computer file format used by most RP technologies
TCT	Time compression techniques (technologies)
ThermoJet™	Obsolete 3D printing type RP machine using wax material (3D Systems trademark)
UV	Ultraviolet light
Vacuum casting	Method of producing plastic parts from a silicone mold from a master pattern

Medical terms and abbreviations

Alloplastic	Artificial material
Artifact	Something artificial, a distortion that does not reflect normal anatomy or reflects pathology not usually found in the body, usually used in radiology
Arthroplasty	Surgical repair of a joint, usually joint replacement
Autologous	Derived from an organism's own tissues or DNA
Benign	Noncancerous; treatment or removal is curative
Collimation	Operation of controlling a beam of radiation
Craniofacial	Relating to both the face and head
Cranioplasty	Plastic surgery of the skull; surgical correction of a skull defect
Hounsfield number	Normalized value of the calculated X-ray absorption coefficient of a pixel in a computed tomogram (after the invention of CT)
In vitro	In glass, meaning in the laboratory
In vivo	In life, meaning within the body
Malignant	Said of cancerous tumors, tending to become progressively worse
Maxillofacial	Pertaining to the jaws and face, particularly with reference to surgery
Morphology	The configuration or structure of
Occlusion	Relationship between the upper and lower teeth; the bite
Orthosis	Wearable medical device to immobilize or support a limb, hand, or foot
Osteoporosis	Reduction in the amount of bone mass, leading to fractures after minimal trauma
Osteotomy	Surgical cutting of a bone
Pathology	Branch of medicine concerned with disease, especially its structure and functional effects on the body
Prosthesis	Artificial substitute for a missing body part, used for functional or cosmetic reasons, or both

Radiography	Making film records (radiographs) of internal body structures by passage of X-rays
Tibia	Bone in the lower leg; the shin bone
Valgus	Abnormal position in which part of a limb is twisted outward away from the midline
Varus	Abnormal position in which part of a limb is twisted inward towards the midline

Medical explanatory notes

Osseointegrated implants and implant-retained prostheses

Osseointegrated implants are titanium screws that are driven into patients' bones to form rigid and permanent fixation points for prostheses. The screw is driven into the bone in carefully selected positions during an operation. The screw is then typically left in place under the skin until the bone has healed and regrown around the screw. The screw is then exposed through the skin and an abutment is added. This abutment then forms the anchor point for the rigid and strong fixation of prostheses, i.e. implant-retained prostheses. These types of implants are extensively used in dental restoration and increasingly in facial prosthetics.

For more information see,

The Osseointegration Book: From Calvarium to Calcaneus.
Author(s)/Editor(s): Brånemark, Per-Ingvar
Quintessence Publishing, ISBN 0-86715-347-4.

Hypertrophic scar

A hypertrophic scar is a thick, raised scar resulting from skin injury such as burns. The formation of this kind of scar is not a part of normal wound healing and develops over time. They are more likely to be a problem in patients with a genetic tendency to scarring, and in deep wounds that require a long time to heal. In general, they are more likely to form in areas of the body that are subject to significant pressure or movement. Treatment is typically by applied pressure for long periods.

Distraction osteogenesis

Distraction osteogenesis is a method of lengthening bones by cutting through the bone and then controlling a steady gap between the two sections of bone. If the gap is kept small but constant, the bones attempt to grow to fill the gap as they would when healing from a fracture. The gap is typically maintained by mounting the two parts of the bone onto a precision screw mechanism. Adjusting the screw mechanism on a daily basis maintains the required gap. Typically the device is removed after the desired growth has occurred and been allowed to heal fully.

The technique can be used in three dimensions to correct any number of skeletal abnormalities and has been used successfully in orthopedic and maxillofacial surgery.

Benjamin double osteotomy

The Benjamin double osteotomy was a double transverse cut, one through the distal femur and one through the proximal tibia. The rationale was that it would relieve the interosseous pressure underneath the subchondral bone, which was thought to accelerate degeneration of the cartilage and sever the nerve in the subchondral bone plate, which was also thought to contribute to pain relief. A Charnley compression clamp was used to fix the osteotomy. The procedure developed a poor reputation because the condyles and tibial plateaux blood supplies were compromised and occasionally bone necrosis led to above-knee amputations. In addition, a long rehabilitation period was required for minimal benefit.

Bibliography

Further reading on anatomy

Dean, D., & Herbener, T. E. (2000). *Cross-sectional human anatomy*. LWW. ISBN-13: 978–0683303858.

Marieb, E. N., & Hoehn, K. (2012). *Human anatomy and physiology* (9th ed.). Pearson. ISBN-13: 978–0321743268.

Martini, F. H., Nath, J., & Bartholomew, E. F. (2014). *Fundamentals of anatomy and physiology* (10th ed.). Pearson. ISBN-13: 978–0321909077.

Moore, K. L., Agur, A. M. R., & Dalley, A. F. (2013). *Clinically orientated anatomy* (7th ed.). LWW. ISBN-13: 978–1451119459.

Sinnatamby, C. S. (2011). *Last's anatomy* (12th ed.). Churchill Livingstone. ISBN-13: 978–0702033957.

Tortora, G. J., & Derrickson, B. (2006). *Introduction to the human body: the essentials of anatomy and physiology* (7th ed.). Wiley. ISBN-13: 978–0471691235.

Weir, J., Abrahams, P. H., & Spratt, J. (2010). *An imaging atlas of human anatomy* (4th ed.). Mosby. ISBN-13: 978–0723434573.

Further reading on maxillofacial surgery and prosthetics

Bell, W. H., & Guerrero, C. A. (2007). *Distraction osteogenesis of the facial skeleton*. PMPH USA. ISBN-13: 978–1550093445.

Beumer, J., Esposito, S., & Marunick, M. (Eds.). (2011). *Maxillofacial rehabilitation: prosthodontic and surgical management of cancer-related, acquired and congenital defects of the head and neck* (3rd ed.). Quintessence. ISBN-13: 978–0867154986.

Branemark, P.-I., & de Oliveira, M. F. (1997). *Craniofacial prostheses: anaplastology & osseointegratio*. Quintessence Publishing. ISBN: 0867153210.

Branemark, P.-I., Higuchi, K. W., & de Oliveira, M. F. (1999). *Rehabilitation of complex craniomaxillofacial defects: the challenge of bauru*. Quintessence Publishing. ISBN: 0867153474.

Kau, C. H., & Richmond, S. (2010). *Three-dimensional imaging for orthodontics and maxillofacial surgery*. Wiley-Blackwell. ISBN-13: 978–1405162401.

McKinstry, R. L. (1995). *Fundamentals of facial prosthetics*. ABI Professional Publications. ISBN: 1886236003.

Seto, K.-ichi. (2004). *Atlas of oral and maxillofacial rehabilitation*. Quintessence Publishing. ISBN: 4874177964.

Taylor, T. D. (2000). *Clinical maxillofacial prosthetics*. Quintessence Publishing. ISBN: 0867153911.

Thomas, K. F. (1994). *Prosthetic rehabilitation*. Quintessence Publishing. ISBN: 1850970327.

Further reading on dental technology

Carr, A. B., & Brown, D. T. (2010). *McCracken's removable partial prosthodontics* (12th ed.). Mosby. ISBN-13: 978–0323069908.

Renner, R. P., & Boucher, L. J. (1987). *Removable partial dentures*. Quintessence Publishing. ISBN: 0867151897.

Stratton, R. J., & Wiebelt, F. J. (1988). *An atlas of removable partial denture design*. Quintessence Publishing. ISBN: 0867151900.

Further reading on computer-aided design and rapid prototyping

Gibson, I. (2002). *Software solutions for rapid prototyping*. Wiley-Blackwell. ISBN-13: 978–1860583605.

Hiller, J. D., & Lipson, H. (2009). *AMF/STL 2 format references*. STL 2.0: a proposal for a universal multi-material additive manufacturing file format. In D. Bourell (Ed.), *Proceedings of the twentieth annual international solid freeform fabrication symposium, August 3–5, 2009*. Austin, TX: University of Texas at Austin. Tangient LLC (2014). AMF – ASTM Additive Manufacturing File Format (AMF) [online]. Available at www.amf.wikispaces.com. Accessed 28.08.14.

Hopkinson, N., Hague, R., & Dickens, P. (2005). *Rapid manufacturing: an industrial revolution for the digital age*. Wiley. ISBN-13: 978–0470016138.

Kamrani, A. K., & Abouel Nasr, E. (2006). *Rapid prototyping: theory and practice*. Springer. ISBN-13: 978–0387232904.

Liou, F. W. (2007). *Rapid prototyping and engineering applications: a toolbox for prototype development*. CRC Press. ISBN-13: 978–0849334092.

McDonald, J. A., Ryall, C. J., & Wimpenny, D. I. (2001). *Rapid prototyping casebook*. Wiley. ISBN-13: 978–1860580765.

Pham, D. T., & Dimov, S. S. (2001). *Rapid manufacturing: the technologies and applications of rapid prototyping and rapid tooling*. Springer. ISBN-13: 978–1447111825.

Further reading on splinting

Coppard, B. M., & Lohman, H. (2001). *Introduction to splinting: a clinical-reasoning & problem-solving approach*. Mosby, Inc. ISBN-13: 978–0323009348.

Jacobs, M., & Austin, N. (2003). *Splinting the hand and upper extremity: principles and process*. LWW. ISBN-13: 978–0683306309.

Publications by the authors

Al Mortadi, N., Eggbeer, D., Lewis, J., & Williams, R. J. (2012). CAD/CAM/AM applications in the manufacture of dental appliances. *American Journal of Orthodontics and Dentofacial Orthopedics*, *142*(5), 727–733.

Al Mortadi, N., Eggbeer, D., Lewis, J., & Williams, R. J. (2013). Design and fabrication of a sleep apnea device using CAD/AM technologies. *Proceedings of the Institution of Mechanical Engineers, Part H: Journal of Engineering in Medicine*, *227*(4), 350–355.

Benham, M. P., Wright, D. K., & Bibb, R. J. (2001). Modelling soft tissue for kinematic analysis of multi-segment human body models. *Biomedical Sciences Instrumentation, 38*, 111–118. ISSN: 0067–8856.

Bibb, R. J. (2008). Chapter 3: modelling and rapid prototyping. In A. W. Sugar, & M. E. Ehrenfeld (Eds.), *Imaging and planning in surgery: a guide to research* (pp. 26–31). AO Publishing. ISBN: 3-905363-07-0.

Bibb, R. J. (2011). Technological developments in medical applications of rapid prototyping and manufacturing technology over the last decade. *Journal for New Generation Sciences, 8*(2), 1–12. ISSN: 1684–4998.

Bibb, R. J., Bocca, A., & Evans, P. (2002). An appropriate approach to computer aided design and manufacture of cranioplasty plates. *The Journal of Maxillofacial Prosthetics and Technology, 5*(1), 28–31. ISSN: 1366–4697.

Bibb, R. J., Bocca, A., Sugar, A., & Evans, P. (2003). Planning osseointegrated implant sites using computer aided design and rapid prototyping. *The Journal of Maxillofacial Prosthetics and Technology, 6*, 1–4. ISSN: 1366–4697.

Bibb, R. J., & Brown, R. (2000). The application of computer aided product development techniques in medical modelling. *Biomedical Sciences Instrumentation, 36*, 319–324. ISSN: 0067–8856.

Bibb, R. J., Eggbeer, D., Bocca, A., Evans, P., & Sugar, A. (2010). A custom-fitting surgical guide. In C. H. Kau, & S. Richmond (Eds.), *Three-dimensional imaging for orthodontics and maxillofacial surgery* (pp. 239–248). Wiley-Blackwell. ISBN: 978-1405162401.

Bibb, R. J., Eggbeer, D., & Evans, P. (2010). Rapid prototyping technologies in soft tissue facial prosthetics: current state of the art. *Rapid Prototyping Journal, 16*(2), 130–137. http://dx.doi.org/10.1108/13552541011025852.

Bibb, R. J., Eggbeer, D., & Williams, R. (2006). Rapid manufacture of removable partial denture frameworks. *Rapid Prototyping Journal, 12*(2), 95–99. http://dx.doi.org/10.1108/13552540610652438. ISSN: 1355–2546.

Bibb, R. J., Eggbeer, D., Williams, R. J., & Woodward, A. (2006). Trial fitting of a removable partial denture framework made using computer-aided design and rapid prototyping techniques. *Proceedings of the Institute of Mechanical Engineers, Part H: Journal of Engineering in Medicine, 220*(7), 793–797. http://dx.doi.org/10.1243/09544119JEIM62. ISSN: 0954–4119.

Bibb, R. J., Evans, P., Bocca, A., & Sugar, A. (2009). A rapid manufacture of custom-fitting surgical guides. *Rapid Prototyping Journal, 15*(5), 346–354. http://dx.doi.org/10.1108/13552540910993879.

Bibb, R. J., Freeman, P., Brown, R., Sugar, A., Evans, P., & Bocca, A. (2000). An investigation of three-dimensional scanning of human body parts and its use in the design and manufacture of prostheses. *Proceedings of the Institute of Mechanical Engineers, Part H: Journal of Engineering in Medicine, 214*(6), 589–594. http://dx.doi.org/10.1243/0954411001535615. ISSN: 0954–4119.

Bibb, R. J., Keating, A. P., Knox, J., & Zhurov, A. (2008). A comparison of plaster, digital and reconstructed study model accuracy. *Journal of Orthodontics, 35*, 191–201.

Bibb, R. J., & Sisias, G. (2002). Bone structure models using stereolithography: a technical note. *Rapid Prototyping Journal, 8*(1), 25–29. http://dx.doi.org/10.1108/13552540210413275. ISSN: 1355–2546.

Bibb, R. J., Thompson, D., & Winder, J. (2011). Computed tomography characterisation of additive manufacturing materials. *Medical Engineering and Physics, 33*(5), 590–596. http://dx.doi.org/10.1016/j.medengphy.2010.12.015.

Capel, A. J., Edmondson, S., Christie, S. D. R., Goodridge, R. D., Bibb, R., & Thurstans, M. (2013). Design and additive manufacture for flow chemistry. *Lab on a Chip: Microfluidic and Nanotechnologies for Chemistry, Biology, and Bioengineering*, *13*(23), 4583–4590. http://dx.doi.org/10.1039/C3LC50844G.

Eggbeer, D., Bibb, R. J., & Evans, P. (2006). Assessment of digital technologies in the design of a magnet retained auricular prosthesis. *The Journal of Maxillofacial Prosthetics and Technology*, *9*, 1–4. ISSN: 1366–4697.

Eggbeer, D., Bibb, R. J., & Evans, P. (2006). Toward identifying specification requirements for digital bone-anchored prosthesis design incorporating substructure fabrication: a pilot study. *International Journal of Prosthodontics*, *19*(3), 258–263. ISSN: 0893–2174.

Eggbeer, D., Bibb, R. J., & Evans, P. (2007). Digital technologies in extra-oral, soft tissue, facial prosthetics: current state of the art. *Journal of Maxillofacial Prosthetics and Technology*, *10*, 9–16.

Eggbeer, D., Bibb, R., & Ji, L. (2012). Evaluation of direct and indirect additive manufacture of maxillofacial prostheses. *Proceedings of the Institution of Mechanical Engineers, Part H: Journal of Engineering in Medicine*, *226*(9), 718–728. http://dx.doi.org/10.1177/0954411912451826. ISSN: 0954–4119.

Eggbeer, D., Bibb, R. J., & Williams, R. (2005). The computer aided design and rapid prototyping of removable partial denture frameworks. *Proceedings of the Institute of Mechanical Engineers, Part H: Journal of Engineering in Medicine*, *219*(3), 195–202. http://dx.doi.org/10.1243/095441105X9372. ISSN: 0954–4119.

Eggbeer, D., & Evans, P. (2011). Computer-aided methods in bespoke breast prosthesis design and fabrication. *Proceedings of the Institution of Mechanical Engineers, Part H: Journal of Engineering in Medicine*, *225*(1), 94–99.

Eggbeer, D., Evans, P., & Bibb, R. J. (2006). A pilot study in the application of texture relief for digitally designed facial prostheses. *Proceedings of the Institute of Mechanical Engineers, Part H: Journal of Engineering in Medicine*, *220*(6), 705–714. http://dx.doi.org/10.1243/09544119JEIM38. ISSN: 0954–4119.

Eggbeer, D., Williams, R. J., & Bibb, R. J. (2004). A digital method of design and manufacture of sacrificial patterns for removable partial denture metal frameworks. *Quintessence Journal of Dental Technology*, *2*, 490–499.

Evans, P., Eggbeer, D., & Bibb, R. J. (2004). Orbital prosthesis wax pattern production using computer aided design and rapid prototyping techniques. *The Journal of Maxillofacial Prosthetics and Technology*, *7*, 11–15. ISSN: 1366–4697.

Hughes, C. W., Page, K., Bibb, R. J., Taylor, J., & Revington, P. (2003). The custom-made titanium orbital floor prosthesis in reconstruction for orbital floor fractures. *British Journal of Oral and Maxillofacial Surgery*, *41*(1), 50–53. http://dx.doi.org/10.1016/S0266435602002498. ISSN: 0266–4356.

Kau, C. H., Zhurov, A., Richmond, S., Bibb, R. J., Sugar, A., Knox, J., et al. (2006). The 3-dimensional construction of the average 11-year-old child face: a clinical evaluation and application. *Journal of Oral and Maxillofacial Surgery*, *64*(7), 1086–1092. http://dx.doi.org/10.1016/j.joms.2006.03.013. ISSN: 0278–2391.

Minns, R. J., Bibb, R. J., Banks, R., & Sutton, R. A. (2003). The use of a reconstructed three-dimensional solid model from CT to aid the surgical management of a total knee arthroplasty: a case study. *Medical Engineering and Physics*, *25*, 523–526. http://dx.doi.org/10.1016/S1350-4533(03)00050-X. ISSN: 1350–4533.

Minns, R. J., Russell, S., Young, S., Bibb, R. J., & Moliter, P. (2006). Repair of articular cartilage defects using 3-dimensional tissue engineering textile architectures. In C. S. Anand, F. J. Kennedy, M. Miraftab, & S. Rajendran (Eds.), *Medical textiles and biomaterials for healthcare* (pp. 335–341). Woodhead Publishing Ltd. ISBN: 1-85573-683-7.

Paterson, A. M., Bibb, R. J., & Campbell, B. (2013). A comparison of additive manufacturing systems for the design and fabrication of Customised wrist splints. *Rapid Prototyping Journal*, *21*(3), in press.

Paterson, A. M. J., Bibb, R. J., & Campbell, R. I. (2010). A review of existing anatomical data capture methods to support the mass customisation of wrist splints. *Virtual and Physical Prototyping*, *5*(4), 201–207. http://dx.doi.org/10.1080/17452759.2010.528183. ISSN: 1745–2759.

Payne, T. M., Mitchell, S., & Bibb, R. (2013). Design of human surrogates for the study of biomechanical injury: a review. *Critical Reviews in Biomedical Engineering*, *41*(1), 51–89. http://dx.doi.org/10.1615/CritRevBiomedEng.2013006847.

Payne, T., Mitchell, S., Bibb, R., & Waters, M. (2014). Initial validation of a relaxed human soft tissue simulant for sports impact surrogates. *Procedia Engineering*, *72*, 533–538. http://dx.doi.org/10.1016/j.proeng.2014.06.092. ISSN: 1877–7058.

Sugar, A., Bibb, R. J., Morris, C., & Parkhouse, J. (2004). The development of a collaborative medical modelling service: organisational and technical considerations. *British Journal of Oral and Maxillofacial Surgery*, *42*, 323–330. http://dx.doi.org/10.1016/j.bjoms.2004.02.025. ISSN: 0266–4356.

Williams, R. J., Bibb, R. J., & Eggbeer, D. (2004). CAD/CAM in the fabrication of removable partial denture frameworks: a virtual method of surveying 3D scanned dental casts. *Quintessence Journal of Dental Technology*, *2*, 268–276.

Williams, R. J., Bibb, R. J., Eggbeer, D., & Collis, J. (2006). Use of CAD/CAM technology to fabricate a removable partial denture framework. *Journal of Prosthetic Dentistry*, *96*(2), 96–99. http://dx.doi.org/10.1016/j.prosdent.2006.05.029. ISSN: 0022–3913.

Williams, R. J., Bibb, R. J., Eggbeer, D., & Woodward, A. (2006). A patient-fitted removable partial denture framework fabricated from a CAD/CAM-produced sacrificial pattern. *Quintessence Journal of Dental Technology*, *4*(3), 200–204.

Williams, R. J., Bibb, R. J., & Rafik, T. (2004). A technique for fabricating patterns for removable partial denture frameworks using digitized casts and electronic surveying. *Journal of Prosthetic Dentistry*, *91*, 85–88. http://dx.doi.org/10.1016/j.prosdent.2003.10.002. ISSN: 0022–3913.

Williams, R. J., Eggbeer, D., & Bibb, R. J. (2008). CAD/CAM fabricated removable partial-denture alloy frameworks. *Practical Procedures and Aesthetic Dentistry*, *20*(6), 349–351.

Williams, R. J., Eggbeer, D., & Bibb, R. J. (2008). CAD/CAM rapid manufacturing techniques in the fabrication of removable partial denture frameworks. *Quintessence Journal of Dental Technology*, *6*(1), 42–50.

Winder, J., & Bibb, R. J. (2009). A review of the issues surrounding three-dimensional computed tomography for medical modelling using rapid prototyping techniques. *Radiography*, *16*, 78–83. http://dx.doi.org/10.1016/j.radi.2009.10.005.

Winder, R. J., & Bibb, R. J. (2005). Medical rapid prototyping technologies: state of the art and current limitations for application in oral and maxillofacial surgery. *Journal of Oral and Maxillofacial Surgery*, *63*(7), 1006–1015. http://dx.doi.org/10.1016/j.joms.2005.03.016. ISSN: 0278–2391.

Wolfaardt, J., King, B., Bibb, R. J., Verdonck, H., de Cubber, J., Sensen, C. W., et al. (2011). Digital technology in maxillofacial rehabilitation. Chapter 7. In J. Beumer, S. Esposito, & M. Marunick (Eds.), *Maxillofacial rehabilitation: prosthodontic and surgical management of cancer-related, acquired and congenital defects of the head and neck* (3rd ed.) (pp. P355–P373). Quintessence. ISBN: 978-0-86715-498-6.

For online journal papers, please go to http://dx.doi.org and use the digital object identifier (DOI).

Company contacts

3D Systems, 333 Three D Systems Circle Rock Hill, SC 29730, USA. www.3dsystems.com.

DICOM NEMA, Suite 1847, 1300 North 17th Street, Rosslyn, VA 22209, USA. http://medical.nema.org/.

DSM Somos, P.O. Box 6500 6401 JH Heerlen (NL). http://www.dsm.com/products/somos/en_US/home.html.

EnvisionTEC GmbH Elbestrasse 10, D-45768 Marl, Germany. www.envisiontec.de.

EOS GmbH Robert-Stirling-Ring 1, D-82152 Krailling, Munich, Germany. www.eos.info.

Materialise NV, Technologielaan 15, 3001 Leuven, Belgium. www.materialise.com.

PDR, National Centre for Product Design & Development Research (PDR), University of Wales Institute, Cardiff (UWIC), Western Avenue, Cardiff, CF5 2 YB, UK. www.pdronline.co.uk.

Renishaw plc, New Mills, Wotton-under-Edge, Gloucestershire, GL12 8JR, UK. www.renishaw.com.

Stratasys, 14950 Martin Drive, Eden Prairie, MN 55344–2020, USA. www.stratasys.com.

Index

Note: Page numbers followed by "f" and "t" indicate figures and tables respectively.

A

ABS (acrylonitrile butadiene styrene), 306, 442, 443t
  FullCure® 515/535, 301, 309f, 312t -M30i, 313
  Stratasys Dimension SST1200es models with, 301
Accura™
  Amethyst®, 359–362, 361f
  Xtreme® splint, 306f
Additive manufactured textiles (AMT), 295–297
  homogenous splint, 301–302, 302f, 304–305, 305f
Additive manufacturing (AM), 5–6, 173, 176–178
  applications in dental appliances manufacturing, 410–418
  fabrication, implant design for, 225–226, 225f
  format (AMF), 62, 63f
  of maxillofacial prostheses using 3D printing technologies, 256–272
  of removable partial denture framework design of, 372, 372f
  error analysis, 378
  error sources, 377–378, 378t
  finishing, 375
  rapid manufacture, 373–375, 373f–375f
  results, 375–376, 376f–377f
  3D scanning, 371–372
A-Footprint, 295–297
AM/RP technology, 95–96
American College of Radiology (ACR), 30–31
Amethyst®, 359–362, 361f
AnalyzeAVW Version 9.0, 441
Anatomical coverage
  computed tomography, 11
  magnetic resonance imaging, 22
  noncontact surface scanning, 25–26, 25f

Anatomical directional terms, 5f, 5t
Anatomical position, 4, 4f
Anatomical terminology, 3–4
Ancient Egyptian mummies CT imaging, producing physical models from, 450–457, 451f
  case studies, 452–456, 452f–453f, 455f–456f
  technology, 451
Andresen appliance, 411f, 413, 416f
Anthropometrics, 304f
  wrist splints, mass customisation of, 288–289
Arcam, 86
ArtCAM, 429–430
Artefacts, 125–132
  computed tomography, 12–14, 13f–16f
  cone beam, 19
  data import errors, 126
  gantry tilt distortion, 126, 127f
  image threshold artefact, 131–132, 132f
  irregular surface
    resulting from mathematical modelling, 128–130, 129f
    resulting from support structures, 128, 129f
  magnetic resonance imaging, 23–24
  metal artefact, 130, 130f
  model stair-step artefact, 126–127, 128f
  movement artefact, 131, 131f
Auricular prosthesis, 244–245, 246f
Autoclaving, 98
Autodesk® Maya® 2011, 300
Automatic import, of medical scan data, 31
Auto-surfacing, 56

B

Bespoke breast prosthesis design and fabrication, computer-aided methods in, 273–282, 275f–279f, 280t
Biomodeling, 1

Bone structure models, using
    stereolithography, 419–426
  cancellous bone, 420
  human sample data, 420
  multiple human samples, 423–425,
      424f–425f
  single human bone sample, 420–423,
      421f–422f
  software, 426
Boundary compensation, 68–69, 69f
Boundary representation (B-Rep) modelling,
    52, 57
Burns therapy conformers using noncontact
    scanning and rapid prototyping,
    producing, 208–214
  discussion, 214
  methods, 209–213, 210f–213f
    experimental moulds, 212–213
    mask manufacture, 213
  results, 213–214
    accuracy, 214
    treatment outcome, 213
    vacuum moulding, 213–214

C

CAD-Matrix™, 331
Cancellous bone, 420
  cement injection behavior assessment in,
      3D bone surrogates for, 465–472,
      466f–467f, 468t, 469f–470f
Carpal Skin, 298, 321
Casting, of removable partial denture
    frameworks, 361–362, 361f, 366–368
Cement injection behavior in cancellous
    bone, 3D bone surrogates for,
    465–472, 466f–467f, 468t, 469f–470f
Cleaning, 97–98
Collimation, 11
Colour model, 76–77, 78f, 80, 87, 88f,
    91–92
Comet 250, 365, 371–372
Communication, 30–31
Complex, mid-face osteotomies, 194–200
  3D CT scanning and virtual mode
      creation, 195, 195f
  computer-aided surgical planning,
      195–197, 196f
  discussion, 199–200
  guide additive manufacture, 197–198, 198f
  results, 162, 165, 198f–199f

surgical positioning guide, design of, 197,
    197f
Compression, of medical scan data, 31
Computed tomography (CT), 8–16, 9f, 35
  additive manufacturing materials and
      human tissues, comparison of,
      439–449, 440t, 441f, 444f–446f
  anatomical coverage, 11
  artefacts, 12–14, 13f–16f
    data import errors, 126
    gantry tilt distortion, 126, 127f
    image threshold artefact, 131–132
    irregular surface resulting from
        mathematical modelling, 130
    metal artefact, 130
    movement artefact, 131
    stair-step artefact, 126–127
  background, 8–9
  complex, mid-face osteotomies, 195, 195f
  cone beam. See Cone beam CT (CBCT)
  custom-fit surgical guides, rapid
      manufacture of, 145–154
  data, using, 39–44, 40f–43f
  future developments in, 473
  gantry tilt, 12, 12f
  kernels, 14–16, 16f–17f
  medical modelling using rapid prototyping
      techniques, guidelines for, 101–107,
      102f
    anatomical coverage, 105f
    data transfer, 107
    gantry tilt, 105–106
    image reconstruction kernels, 106–107
    noise, 106, 106f
    parameters, 105
    patient arrangement, positioning and
        support, 103–104, 104f–105f
    slice thickness, 105
    X-ray scatter, 106
  microcomputed tomography, 419
  orientation, 12
  partial pixel effect, 9–10, 10f
  reconstructed 3D solid model from, for
      total knee arthroplasty, 155–159
    discussion, 158
    materials and methods, 155–157,
        156f–157f
    postoperative management and
        follow-up, 158
  sequential, 284

slice thickness, 11–12
spiral, 284
wrist splints, mass customisation of,
    284–285
Computer-aided design (CAD), 1, 5–6, 45,
    65, 69–70, 101
  of ancient Egyptian mummies, producing
      physical models from, 450–457, 451f
    case studies, 452–456, 452f–453f,
        455f–456f
    technology, 451
  applications in dental appliances
      manufacturing, 410–418
  bespoke breast prosthesis design and
      fabrication using, 273–282,
      275f–279f, 280t
  complex, mid-face osteotomies, 195–197,
      196f
  cranioplasty plates. See Cranioplasty
      plates; computer-aided design and
      manufacture of
  custom-fit surgical guides, rapid
      manufacture of, 147
  data generation, 47, 47f
  for digitised splint design and fabrication,
      335–343, 336f
    current techniques, 336–337, 337f
    experimental procedures, 338–340,
        338f–340f
    results, 340–341
  osseointegrated implants using, planning,
      137–144
    approach, 138
    benefits and future development, 143
    case study, 140–142, 140f–142f
    results, 142–143, 143f
    scanning problems, 138, 139f
    software problems, 139–140, 139f
    update, 144
  removable partial denture frameworks,
      fabrication of, 353–363
  skin texture relief using, recreating, 427–438
    assessment, 431
    computer representation and
        manipulation, 429–430
    experimental series, 432–436,
        433f–436f
    limitations of, 437
    pilot study, 431–432, 432f–433f
    rapid prototyping reproduction, 430–431

results, 436–437
specification requirements, 429
technologies, identification of, 431
sleep apnea device, design and fabrication
    of, 401–408, 402f–406f
STL modeling, 57
voxel modeling, 56–57
Computer-aided manufacturing (CAM), 65.
    See also Computer-aided design
    (CAD)
  applications in
    dental appliances manufacturing,
        410–418
    sleep apnea device, design and
        fabrication of, 401–408, 402f–406f
Computer numerical controlled (CNC)
    machining, 93–95, 96–97, 96f,
    123–124, 124f, 173–174, 212–214
  advantages and disadvantages of, 96t
  skin textures, reproduction of, 430
Cone beam CT (CBCT), 17–20. See also
    Computed tomography (CT)
  advantages of, 17–19
  applications of, 19–20
  background, 17
  future developments in, 473
  limitations of, 19
Continuous functional grading (CFG),
    324–325, 324f
Cranioplasty plates, computer-aided design
    and manufacture of, 208–227
  AM fabrication, implant design for,
      225–226, 225f
  future development and benefits,
      226–227, 226t
  initial case, 217–220, 218f–221f
  press tool design, 223, 223f–224f
  second case, 222–223, 222f
Creaform, 402–403
Custom-fit surgical guides, rapid manufacture
    of, 145–154
  case study, 148–150, 149f–150f
  discussion, 151
  methods, 146–148
    computer-aided design, 147
    finishing, 148
    rapid manufacture, 147–148
    3D CT scanning, 146–147
  results, 150, 151f
  update, 152

Custom-made prostheses
  breast prostheses. *See* Bespoke breast
    prosthesis design and fabrication;
    computer-aided methods in
  titanium orbital floor prosthesis, in orbital
    floor fractures reconstruction,
    160–166
    case report, 157f, 162, 163f–165f
    imaging, 161
    model construction, 161–162
    prosthesis construction, 162

D

Data clean up, 45–47, 46f
Data fusion, 474
Decimation, 45–46, 46f
Digital All Wales Network (DAWN), 116
Digital imaging, 30–31
  and communications in medicine
    (DICOM), 12, 30–33, 105–107, 115,
    126, 146–147, 155–156, 167–168,
    186, 441, 474
Digital light processing (DLP), 79–81,
    79–81
  advantages and disadvantages of, 81t
  principle, 79
Digital Materials™ (DM), 324
Digital micro-mirror device (DMD), 79
Digitised splinting approach
  with multiple-material functionality,
    319–334
    aesthetically appealing and personalised
      lattice integration, 324
    aim and objectives, 322–323
    future work, 331–332
    key features/characteristics in virtual
      environment, replication of, 323
    multiple-material integration/
      heterogeneous structures, 324–326,
      324f, 326f
    prototype, 326–329, 327f–329f
    results and discussion, 330–331
    splinting process, 320–321
    upper extremity splint fabrication,
      additive manufacturing for, 321–322,
      322f
  for splint design and fabrication, CAD
    strategy for, 335–343, 336f
    current techniques, 336–337, 337f

    experimental procedures, 338–340,
      338f–340f
    results, 340–341
Digitizing, 7–8. *See also* Digitised splinting
    approach
Direct Dimensions, 288
Direct Metal Laser Sintering (DMLS), 85–86
Dremel handheld power tool, 375
D-rings, 302–303
DSM Somos®, 359–360
  9120 Epoxy, 297
  10110-epoxy resin, 235
DTM Corp., 66

E

Easy3DScan Tower Graphics, 383
Electron beam melting, 226
EnvisionTEC GmbH, 66–67, 79
EOS P100 LS system, 297–298, 321
EOS P730 LS machine, 321
Ethylene oxide, 98
European Medical Device Directive, 313

F

Facial disproportion
  multidisciplinary management of, 3D
    technology in, 167–172
    discussion, 171
    materials and methods, 167–169,
      168f–170f
    results, 169–170, 170f
Facial scanning, preliminary trial of,
    202–203
File transfer protocol (FTP), 115–116
Filtering, 28, 29f
Finishing, in removable partial denture
    frameworks, 362, 362f
Finite element analysis (FEA), 44, 51, 323,
    332, 419–420
Finite element meshes, 51
FreeForm® Plus software, 57, 147, 167–169,
    175–176, 179, 182–184, 187,
    195–196, 210–211, 216–217,
    222–223, 230–231, 233–234,
    236f, 243–245, 257, 259, 274, 300,
    338–339, 338f, 459–460
  removable partial denture frameworks,
    fabrication of, 354–355, 365–366,
    372, 372f

skin textures, reproduction of, 429–432, 436–437

FullCure® 515/FullCure® 535, 301, 309f, 312t

Fused deposition modeling (FDM)™, 71–83, 81–84, 82f–83f, 83t, 123–125, 128, 130, 206, 301, 420
  advantages and disadvantages of, 83t
  principle, 81
  splint, 306f

Future developments, 473–476
  data fusion, 474
  rapid prototyping, 474–475
  scanning techniques, 473–474
  tissue engineering, 475

G

G604 Primer, 264

Gamma sterilisation, 98

Gantry tilt, 31
  computed tomography, 12, 12f, 105–106, 126, 127f

Gel-discs/pads, 325

GeoDigm, 386–392

Geomagic Studio, 26, 57, 287–288, 300, 323, 338, 338f

Glass transition temperature (GTT), 336

Grasshopper plug-in, 300–301

Guiding gigs, 412f, 413–415, 414f–415f

H

Helisys machines, 93

Heterogeneous splint using Object Connecx technologies, 303, 304f, 308–311, 310f–311f, 312t

High Consistency HC20 silicone, 264

Homogenous AM textile splint, 301–302, 302f, 304–305, 305f

Homogenous circumferential build designs, 302–303, 303f, 306–308, 306f–309f

Hounsfield scale, 8

Human body surfaces, 3D scanning of, 201–207
  methods, 202–206
    facial scanning, preliminary trial of, 202–203
    prosthesis manufacture, 205–206
    surgical subject, scanning of, 203–205, 204f–205f

results, 206–207
  accuracy, 206
  outcome analysis, 206–207
  update, 207

Human form, 2–3

I

Image threshold artefact, 131–132, 132f

Implant retained facial prosthesis, 228–240
  advanced technologies in, 229–231
    data capture, 230
    design, 230–231
    manufacture, 231
  case study, 231–235
  discussion, 237–238
    data capture, 237
    design, 238
    manufacture, 238
  existing techniques, 229, 233–235
  results, 237

Initial graphics exchange specification (IGES), 54
  contours, 50
  format and exchange, 62

Irregular surface, resulting from mathematical modelling, 128–130, 129f
  support structures, 128, 129f

ISDN line, 116

ISEL-RF1, 383

Isometric parameters (isoparms), 54, 54f

J

JANET (Joint Academic Network), 116

Joining, 74, 74f

K

Kernels, 161–162
  computed tomography, 14–16, 16f–17f
  image construction, 106–107

Konica-Minolta Vivid 900 laser scanners, 231–232

L

Laminated object manufacture (LOM), 93, 94f–95f, 124, 205–206, 205f, 211, 212f, 214
  advantages and disadvantages of, 95t

Laser Cusing, 85–86

Laser melting, 85–86, 86t
  advantages and disadvantages of, 86t
  Selective Laser Melting, 85–86, 147–148,
    150–152, 152f, 189–190, 234–235,
    236f, 237–238, 245, 246f–247f,
    247–248
Laser sintering (LS), 84–86, 84–85, 85f, 146
  Selective Laser Sintering, 124, 206, 420
  wrist splint design and fabrication using,
    295–297, 300–302
Layer-additive manufacturing, 67, 68f,
  241–242
LayerWise (technology), 181, 189–190
Light cure acrylic material, 257
Likert scale, 264, 266
Lloyds LR50KPlus, 266
Low-temperature thermoplastic (LTT),
  320–321

M

Magics software, 147–148, 210, 233–234,
  278, 373, 423
Magnetic keepers, 233–234
Magnetic resonance imaging (MRI), 20–24,
  21f, 35, 102–103
  anatomical coverage, 22
  artefacts, 23–24
  background, 20–22
  future developments in, 474
  image quality and protocol, 23
  missing data, 22
  orientation, 23
  scan distance, 22–23
  wrist splints, mass customisation of,
    285–286
Manual import, of medical scan data,
  31–32, 32t
Mask manufacture, 213
Mass customization (MC), of wrist splints,
  283–293
  contact data acquisition methods,
    288–289, 290t
    anthropometrics, 288–289
  future work, 289–291
  non-contact data acquisition methods,
    284–288, 290t
    computed tomography, 284–285
    magnetic resonance imaging, 285–286
    optical-based systems, 286–288, 286f,
      288f

Massive lower limb/pelvic injury, trauma
  simulation of, 458–464
  discussion, 463–464
  materials and methods, 459–462
    component and mould design,
      459–461, 461f–462f
    component fabrication, 461–462
    data segmentation, 459, 460f
    simulation model assembly and
      moulding, 462, 463f
Materialise Magics (software), 57
Material jetting technology, 88–92, 88–93,
  88f–92f
  advantages and disadvantages of, 92t
Mathematical curve-based surfaces, 52–56,
  52f–55f
Mathematical modelling, irregular surface
  resulting from, 128–130, 129f
MATLAB®, 301
Maxillofacial prostheses using 3D
  printing technologies, direct and
  indirect additive manufacture of,
  256–272
  discussion, 269–270, 270f
  methods, 257–266, 258f
    additive manufacture, 262, 263f
    aesthetic outcome, qualitative rating
      of, 264
    direct approach, 263–264, 264f
    material testing, 264–266, 265f–266f,
      265t
    mould design, 260–262, 261f–262f
    patient scanning and design, 257–260,
      259f–260f
  results, 266–268
    aesthetic outcomes, 266, 267t
    mechanical properties, 266–268, 268f,
      268t–269t
Maxillofacial rehabilitation, 201–282
  bespoke breast prosthesis design and
    fabrication. *See* Bespoke breast
    prosthesis design and fabrication;
    computer-aided methods in
  burns therapy conformers using noncon-
    tact scanning and rapid prototyping,
    producing, 208–214
  cranioplasty plates, computer-aided
    design and manufacture of, 208–227
  human body surfaces, 3D scanning of,
    201–207

implant retained facial prosthesis,
    228–240
maxillofacial prostheses using 3D printing
    technologies, direct and indirect
    additive manufacture of, 256–272
soft-tissue facial prosthetics, rapid
    prototyping technologies in, 241–255
Maya® 2011, 338
McNeel Rhinoceros, 300, 303, 323
MED610™ transparent, 313
Media, 32–33
Medical Device Directive (MDD), 95–97
Medical imaging, 7–34
    computed tomography. *See* Computed
        tomography (CT)
    cone beam CT, 17–20
    magnetic resonance imaging. *See* Magnetic
        resonance imaging (MRI)
    media, 32–33
    medical scan data. *See* Medical scan
        data
    noncontact surface scanning, 24–30
    point cloud data, 32
Medical modeling, 1
    collaboration, 101–119
        aims of, 111
        disconnected procedure, 112f
        implementation of, 112–115
        integrated procedure, 113f–114f
        organisational issues, 117
        technical issues, 115–116
        update, 119
    using rapid prototyping, CT guidelines
        for, 101–107, 102f
        anatomical coverage, 105f
        data transfer, 107
        gantry tilt, 105–106
        image reconstruction kernels, 106–107
        noise, 106, 106f
        parameters, 105
        patient arrangement, positioning and
            support, 103–104, 104f–105f
        slice thickness, 105
        X-ray scatter, 106
    stages of, 2t
Medical rapid prototyping (MRP), 120–135
    applications of, 124–125
    computer controlled milling, 123–124,
        124f
    defined, 120

fused deposition modeling, 123
laminated object manufacturing, 124
model artefacts, 125–132
    CT data import errors, 126
    CT gantry tilt distortion, 126, 127f
    image threshold artefact, 131–132,
        132f
    irregular surface resulting from
        mathematical modelling, 128–130,
        129f
    irregular surface resulting from support
        structures, 128, 129f
    metal artefact, 130, 130f
    model stair-step artefact, 126–127, 128f
    movement artefact, 131, 131f
selective laser sintering, 124
stereolithography, 123
3D image acquisition and processing for,
    121–122
update, 133
Medical scan data, 30–32, 35–64
    automatic import, 31
    communication, 30–31
    compression, 31
    using CT data, 39–44, 40f–43f
    digital imaging, 30–31
    file management and exchange, 58–62
        additive manufacturing format/STL2.0,
            62, 63f
        initial graphics exchange specification, 62
        object, 61
        STEP format, 62
        stereolithography, 58–61, 58f–61f
        virtual reality modelling language/
            X3D, 61
    manual import, 31–32, 32t
    pixel data operations, 35–39
        region growing, 36
        thresholding, 35, 36f–38f
    point cloud data operations, 44–47,
        44f–45f
        CAD data generation, 47, 47f
        data clean up, 45–47, 46f
    pseudo 3D formats, 48–51, 48f–49f
        initial graphics exchange specification
            contours, 50
        slice file formats, 50–51
    two-dimensional formats, 48
    true 3D formats, 51–57
        finite element meshes, 51

Medical scan data (*Continued*)
  mathematical curve-based surfaces, 52–56, 52f–55f
  mesh optimization, 51
  polygon faceted surfaces, 51
  STL modeling, 57
  T-splines, 56, 56f
  voxel modeling, 56–57
Medpor®, 160
Mesh(es/ing)
  finite element, 51
  optimization, 51
  remeshing, 46–47, 46f
  smoothing of, 45
  tessellation of, 46–47
Metal artefact, 130, 130f
Metal implants, shadowing by
  computed tomography, 13, 13f–14f
  magnetic resonance imaging, 24
Microcomputed tomography (μCT), 419
Microsoft Access 2010, 327–329
Microsoft Xbox Kinect™, 30
Mimics, 39, 112–113, 115, 175, 179, 182, 440, 460f
Minns™ meniscal-bearing total knee, 157
Minolta VIVID 900, 383, 394
Morita 3D Accuitomo 170 Cone Beam CT Scanner, 18f
Moulage, 24–25
Movement artefact, 131, 131f
Multipart reconstruction, 182
  data segmentation, 182, 183f
  defect reconstruction, 182–183, 183f
  implant design, 183–185, 185f
  results, 185–186, 185f
MultiTool 1.0.1 Maya, 340

N

Nasal prosthesis, 245–247, 246f
9850 Shore 50 elastomer, 308–311
Noise
  in computed tomography, 14, 15f–16f, 106, 106f
  in magnetic resonance imaging, 24
  in noncontact surface scanning, 28–30, 29f
Noncontact surface scanning, 7–8, 24–30
  anatomical coverage, 25–26, 25f
  background, 24–25

burns therapy conformers using, producing, 208–214
  low-cost and open-source methods, 30
  missing data, 26, 27f
  movement, 26–27, 28f
  noise, 28–30, 29f
Nonuniform rational B-spline (NURBS)
  surfaces, 47, 52–53, 52f–54f, 55–56, 338–340, 342
NVivo, 329

O

Objet, 66–67, 88–89
  MED610™ transparent, 313
  PolyJet Matrix Technology, 324
Objet Connex
  heterogeneous splint using, 303, 304f, 308–311, 310f–311f, 312t
  Objet Connex3, 330
  Objet Connex 500, 264–265
    AM splints fabrication using, 321–322, 329f
Objet Eden 330, 440
Objet Poly-jet modelling, 430
Object Studio™, 308
Objet 3D printing, 248
Optical-based systems, mass customisation of wrist splints, 286–288, 286f, 288f
Orbital floor fractures reconstruction, custom-made titanium orbital floor prosthesis in, 160–166
  case report, 157f, 162, 163f–165f
  imaging, 161
  model construction, 161–162
  prosthesis construction, 162
Orbital floor implant incorporating placement guide, 178
  data segmentation, 179, 179f
  defect reconstruction, 179
  fitting and surgery, 181
  implant and
    guide fabrication, 181, 181f
    positioning guide design, 179–180, 180f
  results, 181–182
Orbital prosthesis, 243–244, 243f
Orbital rim augmentation implant, 173–178, 174f
  additive manufacture, 176

fitting and surgery, 176–177, 177f–178f
implant design, 175–176, 176f
results, 177–178
3D data acquisition and transfer, 175
Orientation, 70–73, 71f, 73f
build failure, risk of, 72, 73f
computed tomography, 12
data quality, 72–73
magnetic resonance imaging, 23
model quality, 71–72
stair-step effect of, 71–72, 71f
support, 72
surface finishing, 71–72
time and cost, building, 71
Osseointegrated implants using CAD and
        rapid prototyping, planning, 137–144
approach, 138
benefits and future development, 143
case study, 140–142, 140f–142f
results, 142–143, 143f
scanning problems, 138, 139f
software problems, 139–140, 139f
update, 144

P

Partial pixel effect, in CT, 9–10, 10f
Partial removable denture framework
        (RPD), 85–86, 86f
Patches, 52
Perfactory®, 79, 80f, 248
removable partial denture frameworks,
        fabrication of, 359–363, 373
skin textures, reproduction of, 430
Phantom®Desktop™ haptic interface, 147,
        339–340, 342–343, 354–355, 355f,
        403, 404f
Photogrammetry, 258, 274, 286–287
Photoshop, 225
Physical reproduction, 65–98
computer numerical controlled
        machining, 93–95, 96f, 96t
digital light processing, 79–81, 79–81,
        80f, 81t
fused deposition modelling, 81–84,
        82f–83f, 83t
laminated object manufacture, 93,
        94f–95f, 95t
laser melting, 85–86, 86t
laser sintering, 84–86, 85f

material jetting technology, 88–93,
        88f–92f, 92t
powder bed 3D printing, 86–88, 87f,
        88t
rapid prototyping, 65–75
boundary compensation, 68–69, 69f
data input, 69–70
history of, 65–67
joining, 74, 74f
layer-additive manufacturing, 67, 68f
orientation, 70–73, 71f, 73f
sectioning, 73, 74f
separating, 73–74
trapped volumes, 74–75
stereolithography, 75–79, 77f–78f, 78t
Picture Archiving and Communication
        System, 32–33
Pixel data operations, 35–39
region growing, 36
thresholding, 35, 36f–38f
Plaster, digital and reconstructed study
        model accuracy, comparison of,
        380–384, 381t, 386–395
discussion, 386–393, 394t
future work, 395
manual measurements, 382–383
null hypothesis, 382
reconstructed models, measurement of,
        384, 385f
results, 386, 386t–391t, 387f–392f
statistical analysis, 384
virtual measurements, 383–384, 384f
Point cloud data, 32
operations, 44–47, 44f–45f
CAD data generation, 47, 47f
data clean up, 45–47, 46f
Polyethyleneterephthalate glycol (PETG),
        208, 213–214
Polygon faceted surfaces, 51
Polygonisation, 44–45, 44f, 209
Polygon meshes, 61
PolyJet Matrix Technology, 263, 301, 324
Polymerisation, 76
Polyworks, 233–234, 354, 365, 371–372
Pop rivets, 337
Post traumatic zygomatic osteotomy and
        orbital floor reconstruction, 186–192
data segmentation, 186–187, 187f
device fabrication, 189–190, 190f

Post traumatic zygomatic osteotomy
    and orbital floor reconstruction
    (*Continued*)
  results, 191–192
  surgery, 190, 191f
  surgical planning and device design,
    187–189, 188f–190f
Powder bed 3D printing, 86–88, 86–88, 87f
  advantages and disadvantages of, 88t
Press tool design, 223, 223f–224f
PROCESS acronym, 345
Pro Engineer®
  Interactive Surface Design Extension
    (ISDX), 339–340, 339f
  Wildfire 4.0, 338
ProJet systems, 248
  CP range
    of machines, 89–90
    of printers, 238
  models, 88–89, 89f, 91
  3000 Plus 3D printing system, 406, 461–462
Proprietary, 197–198
Pseudo 3D formats, 48–51, 48f–49f
  initial graphics exchange specification
    contours, 50
  slice file formats, 50–51
PTC Pro/Engineer Wildfire 5.0, 300

Q

Q10 (electron beam melting), 226
Qualitative rating of aesthetic outcome, 264
Quality, 248
QuickCast™, 477

R

Rapidform software, 231–232
Rapid manufacturing, 5–6
Rapid prototyping (RP), 1, 5–6, 39–44, 65–75
  boundary compensation, 68–69, 69f
  burns therapy conformers using,
    producing, 208–214
  data input, 69–70
  future developments in, 474–475
  history of, 65–67
  joining, 74, 74f
  layer-additive manufacturing, 67, 68f
  medical. *See* Medical rapid prototyping
    (MRP)

medical modeling for, 2t
medical modelling using, CT guidelines
    for, 101–107, 102f
  anatomical coverage, 105f
  data transfer, 107
  gantry tilt, 105–106
  image reconstruction kernels, 106–107
  noise, 106, 106f
  parameters, 105
  patient arrangement, positioning and
    support, 103–104, 104f–105f
  slice thickness, 105
  X-ray scatter, 106
orientation, 70–73, 71f, 73f
osseointegrated implants using, planning,
    137–144
  approach, 138
  benefits and future development, 143
  case study, 140–142, 140f–142f
  results, 142–143, 143f
  scanning problems, 138, 139f
  software problems, 139–140, 139f
  update, 144
removable partial denture frameworks,
    fabrication of, 353–363
sectioning, 73
separating, 73–74
skin texture relief using, recreating,
    427–438
  assessment, 431
  computer representation and
    manipulation, 429–430
  experimental series, 432–436,
    433f–436f
  limitations of, 437
  pilot study, 431–432, 432f–433f
  rapid prototyping reproduction, 430–431
  results, 436–437
  specification requirements, 429
  technologies, identification of, 431
trapped volumes, 74–75
RapMan 3D print extrusion system, 297
Reconstructive implants, computer-aided
    design and manufacture of, 173–193
multipart reconstruction, 182
  data segmentation, 182, 183f
  defect reconstruction, 182–183, 183f
  implant design, 183–185, 185f
  results, 185–186, 185f

orbital floor implant incorporating
   placement guide, 178
   data segmentation, 179, 179f
   defect reconstruction, 179
   fitting and surgery, 181
   implant and guide fabrication, 181,
      181f
   implant and positioning guide design,
      179–180, 180f
   results, 181–182
orbital rim augmentation implant,
   173–178, 174f
   3D data acquisition and transfer, 175
   additive manufacture, 176
   fitting and surgery, 176–177, 177f–178f
   implant design, 175–176, 176f
   results, 177–178
post traumatic zygomatic osteotomy,
   186–192
   data segmentation, 186–187, 187f
   device fabrication, 189–190, 190f
   results, 191–192
   surgery, 190, 191f
   surgical planning and device design,
      187–189, 188f–190f
Reference planes, 4, 6f
Region growing, 36
Relief creation, 357, 358f
Remeshing, 46–47, 46f
Removable partial denture (RPD)
   frameworks
   CAD and RP fabrication of, 353–363
      3D scanning, 354
      casting, 361–362, 361f
      finishing, 362, 362f
      framework design, 358–359, 358f–360f
      FreeForm® modelling, 354–355
      pattern comparison, 360–361
      pattern manufacture, 359–360, 361f
      relief creation, 357, 358f
      surveying, 355, 356f
      unwanted undercuts, removing, 356,
         357f
      useful undercuts, identifying, 356–357
   direct additive manufacture of
      3D scanning, 371–372
      design of, 372, 372f
      error analysis, 378
      error sources, 377–378, 378t

finishing, 375
   rapid manufacture, 373–375, 373f–375f
   results, 375–376, 376f–377f
   trial fitting of
      case, 364–365
      data capture and digital design,
         365–366, 365f–367f
      discussion, 369
      investment casting, 366–368, 368f
      results, 368–369, 369f
Retention bar experiment, direct
   manufacture of, 247, 247f
Reverse engineering, 7–8
Rhinoceros® 4.0, 338, 338f–339f
RP&M technologies, 242–245, 248–251,
   253–254, 274
   specification, 251t, 252, 253t

S

Sectioning, 73, 74f
Segmentation, 39, 121–122
Selective Laser Melting (SLM™), 85–86,
   147–148, 150–152, 152f, 189–190,
   234–235, 236f, 237–238, 245,
   246f–247f, 247–248. See also Laser
   melting.
Selective Laser Sintering (SLS®), 124, 206,
   420. See also Laser sintering (LS)
Separating, 73–74
Siemens Somatom Plus4 Volume Zoom
   scanner, 161
Silastic®, 160
Skin texture, definition, 428
Skin texture relief using CAD and RP,
   recreating, 427–438
   case studies, 431–436
      experimental series, 432–436,
         433f–436f
      pilot study, 431–432, 432f–433f
   limitations of, 437
   methods, 431
      assessment, 431
   results, 436–437
   technologies, identification of, 429–431
      computer representation and
         manipulation, 429–430
      rapid prototyping reproduction,
         430–431
      specification requirements, 429

Sleep apnea device, 416–417
  using CAD/AM technologies, design and
      fabrication of, 401–408, 402f–406f
Slice file format (SLC), 49–51, 72–73, 157,
      423–424
Slice thickness, computed tomography,
      11–12, 105
SLM Realizer
  machine, removable partial denture
      frameworks fabrication of, 373
  software, 374
Soft-tissue facial prosthetics, rapid
      prototyping technologies in, 241–255
  accuracy, 248–249
  auricular prosthesis, 244–245, 246f
  economics, 250–252, 251t–252t
  fit, 248
  methodology, 242
  nasal prosthesis, 245–247, 246f
  orbital prosthesis, 243–244, 243f
  physical properties, 249–250
  resolution texture, 249
  retention bar experiment, direct
      manufacture of, 247, 247f
  RP&M specification, 252, 253t
  texture experiment, 244, 244f
Solidscale skin textures, reproduction of,
      430–431
Solidscape, 88–91, 248
  removable partial denture frameworks,
      fabrication of, 359–361, 373
SolidWorks (in surrogate designing),
      466–468
Spider, 233–234, 354, 365, 371–372
Stair-step artifact, 126–127, 128f
Standard for the Exchange of Product Model
      Data format, 62, 63f
Standard Triangulation Language (STL) file
      format, 420–421, 423, 440
Static wrist immobilization (SWI) splints,
      319, 320f, 321, 344, 346
  volar-based, 336, 337f
Stepped functional grading (SFG), 324–325,
      324f
Stereolithography (SLT), 28–30, 32, 44–45,
      50–51, 57, 69–70, 75–79, 77f–78f,
      102–103, 123, 128, 138–140, 146–147,
      203–205, 210–212, 223, 245

  advantages and disadvantages of, 78t
  bone structure models using, 419–426
    cancellous bone, 420
    human sample data, 420
    multiple human samples, 423–425,
      424f–425f
    single human bone sample, 420–423,
      421f–422f
    software, 426
  format and exchange, 58–61, 58f–61f
  removable partial denture frameworks,
      fabrication of, 359–360, 373
  principle, 75–79
  skin textures, reproduction of, 430
  wrist splint design and fabrication using,
      297, 301, 306–307, 307f–308f
Stereolithography apparatus (SLA®), 151,
      156, 161–162, 163f, 186–187,
      197–198, 200, 245, 246f
Sterilisation, 98
Sterility Assurance Level (SAL), 98
STL2.0, 62, 63f
Stratasys, Inc., 66, 81
  ABS-M30i claim ISO 10993, 313
Support structures, irregular surface
      resulting from, 128, 129f
Surface modeling, 52–56
Synthes Cortex screws, 181, 184, 188–189
Synthetic anisotropy, 298

T

TangoBlackPlus, 301
TangoPlus, 263–265, 268t–269t, 324
Technical terminology, 5–6
Texture experiment, 244, 244f
ThermoJet®, 88–89, 212, 234, 237–238,
      241–244, 246f, 248, 253t
  removable partial denture frameworks,
      fabrication of, 359–361, 373
  skin textures, reproduction of, 430–432,
      436–437
3D-Coat, 57
3D formats
  pseudo 3D formats, 48–51, 48f–49f
    initial graphics exchange specification
      contours, 50
    slice file formats, 50–51

true 3D formats, 51–57
  finite element meshes, 51
  finite element meshes, 51
  mathematical curve-based surfaces, 52–56, 52f–55f
  mesh optimization, 51
  polygon faceted surfaces, 51
  STL modeling, 57
  T-splines, 56, 56f
  voxel modeling, 56–57
3D Lightyear™, 360, 423–424
3D printers, 81, 238
Three-dimensional (3D) printing, 5–6
3D Studio Max, 57, 429
3D surface capture, 241–242
3D Systems, Inc., 66, 75–76, 88–89, 91–92
3D Systems SLA 250, 329f
3D topographic scanningThreshold(ing), 35, 36f–38f, 274
  artefact, 131–132, 132f
3-Matic (CAD package), 57
Tissue engineering, 475
Toshiba Aquilion Multislice scanner, 175
Total knee arthroplasty, reconstructed 3D solid model from CT for, 155–159
  discussion, 158
  materials and methods, 155–157, 156f–157f
  postoperative management and follow-up, 158
Touch probe digitizers, 7–8
Trapped volumes, 74–75
True 3D formats. See 3D formats
T-splines, 56, 56f
Two-dimensional formats, 48

U

Ultrasound, 102–103
Undercuts
  unwanted, removing, 356, 357f
  useful, identifying, 356–357
Upper extremity splint design, refined 3D CAD workflow for, 344–351, 348f

V

Vacuum moulding, 213–214
Velcro, 302–303, 337
VeroBlue, 314
VeroWhitePlus, 301

Virtual reality modelling language (VRML), 61
Volumetric CT, 102–103
Voxel modeling, 56–57
VX elements, 403

W

Waterclear™, 360, 362
WaterShed XC resin, 278
Wrist splint design and fabrication, additive manufacturing systems comparison for, 294–318, 295f
  aim and objectives, 299
  future work, 313–316, 315f
  method, 299–303
    heterogeneous splint using Object Connecx technologies, 303, 304f
    homogenous AM textile splint, 301–302, 302f
    homogenous circumferential build designs, 302–303, 303f
  results, 304–311
    heterogeneous splint using Object Connecx technologies, 308–311, 310f–311f, 312t
    homogenous AM textile splint, 304–305, 305f
    homogenous circumferential splints, 306–308, 306f–309f
  traditional process, 296f
Wrist splints, mass customisation of, 283–293
  contact data acquisition methods, 288–289, 290t
    anthropometrics, 288–289
  future work, 289–291
  non-contact data acquisition methods, 284–288, 290t
    computed tomography, 284–285
    magnetic resonance imaging, 285–286
    optical-based systems, 286–288, 286f, 288f

X

X3D, 61
Xbox Kinect, 287
X-ray scatter, computed tomography, 106

Z

ZBrush, 57, 429–430
Z-Corp, 66–67, 86–87, 188, 206, 442
  3D Printing system, 444–445
  AM system, 445

ZScanner 800 3D laser scanner, 299–300,
  346–347
ZPrinter 310 Plus, 440
ZPrinter 450, 314

CPI Antony Rowe
Chippenham, UK
2020-10-05 19:14